Mathematical Epidemiology of Infectious Diseases

Part two

With best wishes,

Hans

D1388546

WILEY SERIES IN MATHEMATICAL AND COMPUTATIONAL BIOLOGY

CHAPLAIN/SINGH/MCLACHLAN—On Growth and Form: Spatio-temporal Pattern
Formation in Biology

CHRISTIANSEN—Population Genetics of Multiple Loci

DIEKMANN/HEESTERBEEK—Mathematical Epidemiology of Infectious Diseases:
Model Building, Analysis and Interpretation

Reflecting the rapidly growing interest and research in the field of mathematical biology, this outstanding new book series examines the integration of mathematical and computational methods into biological work. It also encourages the advancement of theoretical and quantitative approaches to biology, and the development of biological organisation and function.

The scope of the series is broad, ranging from molecular structure and processes to the dynamics of ecosystems and the biosphere, but unified through evolutionary and physical principles, and the interplay of processes across scales of biological organisation.

Topics to be covered in the series include:

- Cell and molecular biology
- Functional morphology and physiology
- Neurobiology and higher function
- Immunology
- Epidemiology
- Ecological and evolutionary dynamics of interacting populations

A fundamental research tool, the *Wiley Series in Mathematical and Computational Biology* provides essential and invaluable reading for biomathematicians and development biologists, as well as graduate students and researchers in mathematical biology and epidemiology.

Mathematical Epidemiology of Infectious Diseases

Model Building, Analysis and Interpretation

O. Diekmann

University of Utrecht, The Netherlands

J. A. P. Heesterbeek

Centre for Biometry Wageningen, The Netherlands

JOHN WILEY & SON, LTD

Chichester • New York • Weinheim • Brisbane • Singapore •Toronto

Other Wiley Editorial Offices

John Wiley & Sons, Inc., 605 Third Avenue,
New York, NY 10158-0012, USA

Weinheim • Brisbane • Singapore • Toronto

Library of Congress Cataloging-in-Publication Data

Diekmann, O.
  Mathematical epidemiology of infectious diseases : model building, analysis, and
interpretation / O. Diekmann, J.A.P. Heesterbeek.
      p. cm. — (Wiley series in mathematical and computational biology)
  Includes bibliographical references and index.
  ISBN 0-471-98682-8 (cased : alk. paper) — ISBN 0-471-49241-8 (pbk. : alk. paper)
    1.   Communicable diseases—Epidemiology—Mathematical models. I. Heesterbeek,
J.A.P. II. Title. III. Mathematical and computational biology
  RA643.D54 2000   99-052964
  614.5'01'51—dc21

British Library Cataloguing in Publication Data

A catalogue record for this book is available from the British Library

ISBN 0 471 98682 9 (cased); 0 471 9241 8 (pbk)

Produced from PostScript files supplied by the author
Printed and bound in Great Britain by Antony Rowe Ltd, Chippenham
This book is printed on acid-free paper responsibly manufactured from sustainable forestry, in which at least two trees are planted for each one used for paper production.

Contents

Series Preface

Theoretical biology is an old subject, tracing back centuries. At times, theoretical developments have represented little more than mathematical exercises, making scant contact with reality. At the other extreme have been those works, such as the writings of Charles Darwin, or the models of Watson and Crick, in which theory and fact are intertwined, mutually nourishing one another in inseparable symbiosis. Indeed, one of the most exciting developments in biology within the last quarter-century has been the integration of mathematical and theoretical reasoning into all branches of biology, from the molecule to the ecosystem. It is such a unified theoretical biology, blending theory and empiricism seamlessly, that has inspired the development of this series.

This series seeks to encourage the advancement of theoretical and quantitative approaches to biology, and to the development of unifying principles of biological organisation and function, through the publication of significant monographs, textbooks and synthetic compendia in mathematical and computational biology. The scope of the series is broad, ranging from molecular structure and processes to the dynamics of ecosystems and the biosphere, but unified through evolutionary and physical principles, and the interplay of processes across scales of biological organisation.

The principal criteria for publication beyond the intrinsic quality of the work are substantive biological content and import, and innovative development or application of mathematical or computational methods. Topics will include but not be limited to cell and molecular biology, functional morphology and physiology, neurobiology and higher function, immunology and epidemiology, and the ecological and evolutionary dynamics of interacting populations. The most successful contributions, however, will not be so easily categorised, crossing boundaries and providing integrative perspectives that unify diverse approaches; the study of infectious diseases, for example, ranges from the molecule to the ecosystem, involving mechanistic investigations at the level of the cell and the immune system, evolutionary perspectives as viewed through sequence analysis and population genetics, and demographic and epidemiological aspects at the level of the ecological community.

The objective of the series is the integration of mathematical and computational methods into biological work; the volumes published hence should be of interest both to fundamental biologists and to computational and mathematical scientists, as well as to the broad spectrum of interdisciplinary researchers who comprise the continuum connecting these diverse disciplines.

Simon Levin

Preface, Creed and Apology

What it is all about and what not

Epidemiology is the study of the spread of diseases, in space and time, with the objective to trace *factors* that are responsible for, or contribute to, their occurrence.

In 1854, John Snow demonstrated that cholera could be transmitted via drinking water. He combined incidence data from an area surrounding Broad Street in London with a sketch of the location of water pumps, and noticed that the cases were clustered around a particular pump. This is a classic example of how *description* and *data analysis* may lead to an *explanation*, which then can be used for *prevention* (or *prediction*). The general usefulness of maps in this context is brilliantly illustrated in the *Altas of Disease Distributions*, by Cliff & Haggett (1988).

To achieve an understanding of complex phenomena, one needs many tools. This book is not about maps and not about statistics (for the latter we refer to Becker, 1989). It is about caricatural mathematical descriptions of the mechanisms of transmission of infectious agents and about the analysis of the models that result from such descriptions. It is about *translating* assumptions concerning biological (behavioural, immunological, demographical, medical) aspects into mathematics, about mathematical *analysis* of certain classes of equations aided by interpretation, and about the *drawing of conclusions* where results from the mathematical analysis are translated back into biology.

Another way of phrasing this is to say that this book is about *thought experiments* that help to create conceptual clarity, to expose hidden working hypotheses and to find mechanistic links between different observable quantities. It tries to unravel the relationship between assumed mechanisms at the individual level and the resulting phenomena at the population level.

The top down approach

It is the mathematicians' inclination to strive for an abstract and general theory, the hope being that, once such a theory exists, one can make it operational by mere specification and elaboration. We are aware that this view is too rosy. To apply general qualitative theory to the concrete and the specific in order to generate quantitative conclusions is a highly non-trivial affair requiring its own type of ingenuity. Top down does not make bottom up redundant. It is an art to discern pattern in facts and figures concerning a specific disease agent affecting a specific host population.

We believe that collaboration between specialists in different fields is the designated way to make top down and bottom up meet. This believe was triggered by our own experience in working, sometimes indirectly, together with experts from many areas of human, animal (both wild and farm) and plant epidemiology. We realise that this book only offers the top down part of the story. It provides *tools* (concepts, methods, techniques, results) and *qualitative insights* (e.g. the Threshold Theorem of Section 1.2.1 and the asymptotic speed of propagation of Chapter 8) but does not embark upon, for example, a quantitative study of the advantages and disadvantages of various vaccination strategies against rubella in the Netherlands, or the question of how many hospital beds are likely to be needed for AIDS patients in the next twenty years.

However, we do claim that, for a mathematically oriented book, we are exceedingly conscientious about the meaning and interpretation of the assumptions that are made. Even though most of the models discussed are deterministic, we think in terms of individuals and their behaviour, and hence 'deterministic' just means that there are assumed to be very many such individuals and that, as a consequence, chance fluctuations are of relatively minor importance. We strongly advocate the formulation of model assumptions in terms of behaviour of individuals, and in this respect the book is written with missionary zeal (even top experts have been known to sometimes misinterpret quadratic/mass-action terms by jumping too recklessly from the individual to the population level and back). Our hope is that through formulation of population models in terms of behaviour of individuals one can facilitate the linking up of the top down with the bottom up approach.

A workbook

Some books offer wisdom. They can be read at leisure in an armchair near a fireplace, provided one pauses every now and then for contemplation. This is not such a book.

This book has a zillion exercises and begs to be read with pencil and paper at hand (or, perhaps, in a more modern way, using a computer with some program for symbolic manipulation). Some of the exercises you may just want to read to see what statements they concern. This reading is essential, since usually the exercises are an integrated part of the exposition. For many exercises, however, mere reading is not enough: one actually needs to do them. Learning to translate, model, analyse and interpret involves training. Some exercises are ridiculously simple since we have tried not to omit arguments or to tire the reader with details; where other writers would state 'one easily sees' or 'as a simple argument shows', etc., we have inserted an exercise. Other exercises, however, are very hard and elaborate. We anticipate that our readers will feel at times frustrated or even irritated. We therefore provide complete elaborations of all exercises, even of the 'ridiculously simple ones', as an integrated part of the book. When you cannot do a specific exercise, we advise that you initially only briefly glance at the elaboration as a kind of hint and then try again. Only if all glances fail to inspire should you study the elaboration in detail.

We are not sadists who like to pester their readers with exercises. We are convinced that the reward is enormous. In literally working through this book the reader acquires modelling skills that are also valuable outside of epidemiology, certainly within population dynamics, but even beyond that. The reader receives training in mathematical argumentation, modelling and analysis.

The book is primarily aimed at self-study. To study the book in a small group that meets for discussion sessions, however, can be more stimulating and rewarding (if only because by dividing the labour it is easier to keep the spirits high). Finally, we trust that a lecturer can use the book as a basis for a course on epidemic modelling. We invite readers to send us their comments and constructive criticism.

Portrait of the reader as a young person

What is the audience that we have in mind? The answer depends on our temper. In optimistic moods we expect that anybody with an interest in epidemic modelling should be able to digest most of the text and benefit from it. When realism strikes, we appreciate that a certain background in mathematics is required, and we narrow down the description to applied mathematicians (to-be) with an interest in population biology and epidemiology and to theoretical biologists and epidemiologists with a strong inclination to persevere when mathematical language at first seems to complicate, rather than simplify, the modelling.

Our hope is that the applied mathematicians learn to see i) the subtleties of model assumptions; ii) that continuous-time models not necessarily take the form of a system of ordinary differential equations ('ODE') and iii) that often biological interpretation suggests how to proceed with the mathematical analysis.

Our hope is that the theoretical biologists and epidemiologists i) enlarge their tool kit considerably and ii) conclude that sometimes abstraction may actually make things simpler and more transparent.

Our ideal reader feels attracted by these educational aims.

A brief outline of the book

This book is divided into three parts and twelve chapters. In Part I, we shall introduce the key questions, basic ideas, fundamental concepts and mathematical arguments in as simple a context as possible. This entails in particular that we treat all host individuals as identical with respect to behaviour and physiology and that we deal with such concepts as thresholds, final sizes for epidemics, repeated outbreaks, the endemic state and population regulation.

When the host population is heterogeneous, we need more advanced mathematics. To describe the initial phase of epidemic spread, we can restrict attention to linear mathematics and a systematic approach is possible. The theory, with many examples, is presented in Part II. In addition we pay some (but not much) attention to nonlinear aspects in a general setting. We shall pay some more attention to age structure and spatial structure in separate chapters, since these are particularly relevant for understanding of the population dynamics of many infective agents. To analyse nonlinear structured models one is often forced to make debatable simplifying assumptions. Even then, one needs to resort to tricks, for lack of a powerful general theory. We therefore do not forage deeply into nonlinear theory. For most of the examples in the book we have those infective agents in mind that are usually collectively called 'microparasites', but in Chapter 9 we briefly touch upon some aspects where 'macroparasites' differ from 'microparasites' (and where they do not), and concentrate on the consequences that these differences and agreements have for the mathematical treatment of invasion. In the final chapter of Part II we pay attention to one of the fundamental and conceptually most difficult aspects of epidemic theory: the myriad ways in which one can model contacts between individuals.

Part III consists of two chapters. As a consequence of our educational aim, we provide, in Chapter 11 and 12, complete elaborations of all exercises. These elaborations are detailed and sometimes lengthy, and in this way often serve as an extension and deepening of the main text. This makes the elaborations into an integrated part of the book.

A final remark concerns our way of referring to the literature. The literature of epidemic theory is extensive and growing steadily. It would be very difficult, bordering on the impossible, to do justice to all valuable contributions to the literature. We have deliberately chosen to write a textbook and not a review of the state-of-the-art in epidemic theory. As a consequence we have two types of references: local specialist literature (mostly papers) and global general texts for further reading (mostly books). The local literature is included in places where it is necessary for the exposition at that point and is given in footnotes. The global references are given in a short list near the end of the book. They are ordered thematically and include background mathematics books. We do not usually refer to our own papers, since much of the relevant material is represented somehow in the text. The research was performed over a ten-year period with many collaborators. Apart from the people mentioned explicitly in the next section, the main collaborators whose efforts are presented here are Mick Roberts and Klaus Dietz.

Acknowledgements

Originally it was intended that the focus of this book would be broader with four additional authors: Mart de Jong, Mirjam Kretzschmar, Hans Metz and Denis Mollison. In the hectic modern academic world, time budgets and other priorities are frequently decisive and good plans all too often come to grief. Yet the catalytic influence of these four people still permeates much of the book and we are grateful to them for jointly launching the project. Much of the insight we may have derives from collaboration and extensive discussions with them.

For many many years the CWI (Centre for Mathematics and Computer Science) in Amsterdam supported activities in mathematical biology, among them were two courses/seminars on epidemic modelling. Both of us started our scientific careers at this institute and both of us were involved in these activities where the seeds for this book were sown. Our current employers, the University of Utrecht and DLO (Agricultural Research Organisation) supported the process of germination, growth and maturation of the project. Financial support was also provided by the veterinary institute (ID-DLO, Lelystad), which triggered us to pay more attention to the operationalisation of theory. We thank all institutions for providing stimulating boundary conditions. In addition, we thank fellow lecturers during the courses and the many participants. In particular we want to mention Hisashi Inaba, Aline de Koeijer and Marieke Goree for their detailed comments on (very early) versions of our text.

Odo Diekmann gave two courses based on the material at the University of Utrecht and the students Jennifer Baker, Jos van Schalkwijk, Annemarie Pielaat, Peter van der Wal, Li Chi Wang, Don Klinkenberg, Bob Rink, Martin Bootsma and Sybren Botma provided substantial impetus. He also enjoyed teaching mini-courses on epidemic modelling in the context of two special years devoted to mathematical biology: one at the University of Utah (1995/1996) and one in Cuernavaca (Mexico) in 1998. Such concentrated efforts helped to shape the text of this book. So, for providing stimuli of a most pleasant kind, he thanks Fred Adler, Mark Lewis, Hans Othmer, Carlos Castillo-Chavez, Lourdes Esteva and Cristobal Vargas.

Comments on the text were also provided by Herb Hethcote, Fraser Lewis and Klaus Dietz. These certainly also led to improvements. We are also very grateful to Rampal Etienne for preparing all the figures for us.

Hopefully, the long prehistory of the book and the help of the above-mentioned persons served to eliminate mistakes and obscurities. As is traditional, we take full responsibility for all that remain.

And what about reality?

So far our aim has been to position the book in a scientific landscape of which we provided some topography. Here are a few final remarks in this direction.

Infectious agents have had decisive influences on the history of mankind.[1] The future asks for predictions and rational control decisions.[2] Such is the grand context for this modest book that concentrates on the language of mathematical models and the tools for their analysis. We hope it will help our readers to probe the intricacies of the real world.

It is a healthy attitude to compare models with data. But, we claim, insight also very often derives from comparing models to models. We see the danger though. In 1965, D.G. Kendall[3] wrote, referring in particular to A.G. McKendrick and R. Ross, who shaped epidemic theory in the early twentieth century using their medical background as a starting point:

> 'Mathematicians may be blamed for subsequently carrying the game too far, but its highly respectable medical origin should not be overlooked.'

We wonder whether this cap of carrying the game too far fits us. We do not think so and certainly do not hope so, but it is you, reader, who decides.

Biological reality is complex and mathematical models are only caricatures of it. Apparently,[4] Picasso once said:

> 'Art is a lie that helps us to discover the truth'

In our context, 'art' stands for simple models that describe relations between key components of an essentially much more complex reality.

---

[1] W.H. McNeill: *Plagues and Peoples.* Penguin Books, London, 1979; C.E.A. Winslow: *The Conquest of Epidemic Disease. A Chapter in the History of Ideas.* University of Wisconsin Press, Madison, 1980; J. Diamond: *On Guns, Germs and Steel.* Vintage, Random House, London, 1997.

[2] L. Garrett: *The Coming Plague.* Penguin Books, London, 1994; P.W. Ewald: *The Evolution of Infectious Disease.* Oxford University Press, Oxford, 1994.

[3] D.G. Kendall: Mathematical models of the spread of infection. In: *Mathematics and Computer Science in Biology and Medicine.* HMSO, London, 1965, pp. 213-225.

[4] L.A. Segel: *Modeling Dynamic Phenomena in Molecular and Cellular Biology.* Cambridge University Press, Cambridge, 1984.

Finally, we give the floor to the medical doctor who arguably is the founding father of modern epidemic theory, Sir Ronald Ross, who wrote (Ross, 1911, p. 651):

'As a matter of fact all epidemiology, concerned as it is with variation of disease from time to time or from place to place, *must* (sic) be considered mathematically (...), if it is to be considered scientifically at all. (...) And the mathematical method of treatment is really nothing but the application of careful reasoning to the problems at hand.'

Part I

The bare bones: basic issues explained in the simplest context

1

The epidemic in a closed population

1.1 The questions (and the underlying assumptions)

In general, populations of hosts show demographic turnover: old individuals disappear by death and new individuals appear by birth. Such a demographic process has its characteristic time scale (for humans on the order of 1-10 years). The time scale at which an infectious disease sweeps through a population is often much shorter (e.g. for influenza it is on the order of weeks). In such a case we choose to ignore the demographic turnover and consider the population as 'closed' (which also means that we do not pay any attention to emigration and immigration).

Consider such a closed population and assume that it is 'virgin' (or, 'naive'), in the sense that it is completely free from a certain disease-causing organism in which we are interested. Assume that, in one way or another, the disease-causing organism is introduced in at least one host. We may ask the following questions:

- Does this cause an epidemic?
- If so, with what rate does the number of infected hosts increase during the rise of the epidemic?
- What proportion of the population will ultimately have experienced infection?

Here we assume that we deal with *microparasites*, which are characterised by the fact that a single infection triggers an autonomous process in the host. We assume in addition that this process finally results in either death or immunity, so that no individual can be infected twice (this assumption is somewhat implicitly contained in the formulation of the third question).

In order to answer these questions, we first have to formulate assumptions about transmission. For many diseases transmission can take place when two hosts 'contact' each other, where the meaning of 'contact' depends on the context (think of 'mosquito biting man' for malaria, sexual contact for gonorrhea, sitting in the same streetcar for influenza, tuberculosis,...) and may, in fact, sometimes be a little bit vague (for fungal plant diseases transmitted through air transport of spores it is even far-fetched to think in terms of 'contact'). It is then helpful to follow a three-step procedure:

(1) model the contact process;

(2) model the mixing of susceptibles and infectives; that is, specify what fraction of the

contacts of an infective are with a susceptible, given the population composition in terms of susceptibles and infectives;

(3) specify the probability that a contact between an infective and a susceptible actually leads to transmission.

An easy phenomenological approach to (1) is to assume that individuals have a certain expected number of contacts per unit of time, say c, with other individuals. (We postpone a discussion of how c may relate to population size and/or density.)

1.2   Initial growth

1.2.1   Initial growth on a generation basis

During the initial phase of a potential epidemic, there are only few infected individuals amidst a sea of susceptibles. So if we focus on an infected individual we may simply assume that all its contacts are with susceptibles. This settles the second step in the procedure sketched above.

For many diseases the probability that a contact between a susceptible and an infective actually leads to transmission depends on the time elapsed since the infective was itself infected. To be specific, let us assume that this probability equals

$$\begin{cases} 0 & \text{if} \quad \tau < T_1, \\ p & \text{if} \quad T_1 \leq \tau \leq T_2, \\ 0 & \text{if} \quad T_2 < \tau, \end{cases}$$

where τ denotes the *infection-age* (i.e. the time since infection took place), $0 < p \leq 1$, and where we have assumed that there is a latency period (i.e. the period of time between becoming infected and becoming infectious) of length T_1 followed by an infectious period of length $T_2 - T_1$. (What happens at the end of the infectious period is unspecified at this point; it may be that the host dies or it may be that its immune system managed to conquer the agent, with a now-immune host happily living on; we shall come back to this point later on.)

In order to distinguish between an avalanche-like growth and an almost-immediate extinction, we introduce the *basic reproduction ratio*:

$$R_0 \quad := \quad \text{expected number of secondary cases per primary case}$$
$$\text{in a 'virgin' population.}$$

In other words, R_0 is the initial *growth rate* (more accurately: multiplication factor; note that R_0 is dimensionless) when we consider the population on a *generation basis* (with 'infecting another host' likened to 'begetting a child'). Consequently, R_0 has threshold value 1, in the sense that an epidemic will result from the introduction of the infective agent when $R_0 > 1$, while the number of infecteds is expected to decline (on a generation basis) right after the introduction when $R_0 < 1$. The advantage of measuring growth on a generation basis is that for many models one has an explicit expression for R_0 in terms of the parameters. Indeed, from the assumptions above, we find

$$R_0 = pc(T_2 - T_1). \tag{1.1}$$

<u>We conclude</u> that whether or not the introduction of an infectious agent leads to an epidemic explosion is determined by the value of the generation multiplication factor R_0 relative to the threshold value one. At least for simple submodels for the contact process and infectivity, one can determine R_0 explicitly in terms of parameters of these submodels.

Exercise 1.1 Female mosquitoes strive for a fixed number of blood meals per unit of time, in order to be able to lay eggs (the impression on hot summer nights that they are sadistic creatures, that try to bite as many people as often as they can, is wrong). Show that consequently the mean number of bites that *one* human receives per unit of time is proportional to $D_{\text{mosquito}}/D_{\text{human}}$, i.e. the ratio of the two densities D.

Exercise 1.2 Consider one infected mosquito. Assume it stays infected for an expected period of time T_1 during which it bites (different) people at a rate c. Assume that each bite results in successful transmission with probability p_1. How many people is this mosquito expected to infect?

Exercise 1.3 Consider one infected human. Assume it stays infected for an expected period of time T_2 during which it is bitten by (different) mosquitoes at a rate κ. Let each bite result in successful transmission with probability p_2. How many mosquitoes is this human expected to infect?

Exercise 1.4 Argue that for the above crude description of malaria transmission the quantity

$$c^2 T_1 T_2 p_1 p_2 \frac{D_{\text{mosquito}}}{D_{\text{human}}}$$

is a threshold parameter with threshold value 1. Spell out the meaning of 'threshold parameter' in some detail.

1.2.2 *The influence of demographic stochasticity*

Within the idealised deterministic description of disease transmission, we found in the preceding subsection that a newly introduced infective agent starts to spread exponentially when $R_0 > 1$, while going extinct more or less immediately when $R_0 < 1$. But for the deterministic description to be warranted, we need not only a large number of susceptibles but also a large number of infectives. Yet the very essence of the *introduction* of the agent is that it is present in only a few hosts. So we need to take account of demographic stochasticity, i.e. the chance fluctuations associated with the fact that individual hosts are discrete units, counted with integers and either infected or not (rather than fractionally). Only when the infectives form a small *fraction* of the large population, and not just a small number, does the deterministic description apply.

To refine the analysis, we need branching processes, as introduced below, but we keep counting on a generation basis. As infecting another host is very similar to producing offspring, we shall freely use the standard terminology of true reproduction, even though it does not apply literally to the context of disease spread that we consider here. We repeat that the number of susceptibles is assumed to be very large so that we can, in the initial phase of an epidemic, neglect the depletion of susceptibles by their conversion into infectives. So the finite population we are going to consider in the following is the subpopulation of infected individuals.

Consider this finite population from a generation perspective and assume that individuals reproduce independently from each other, the number of offspring for each being taken from the same probability distribution $\{q_k\}_{k=0}^{\infty}$. This means that any individual begets k offspring with probability q_k and that $\sum_{k=0}^{\infty} q_k = 1$. We note right away that the expected number of offspring R_0 can be found from $\{q_k\}$ as

$$R_0 = \sum_{k=1}^{\infty} k q_k. \tag{1.2}$$

As an auxiliary tool we introduce the *generating function* g defined by

$$g(z) = \sum_{k=0}^{\infty} q_k z^k, 0 \leq z \leq 1. \tag{1.3}$$

(Recall the convention $z^0 = 1$.)

Exercise 1.5 Check that

i) $g(0) = q_0$,

ii) $g(1) = 1$,

iii) $g'(1) = R_0$,

iv) $g'(z) > 0$,

v) $g''(z) > 0$.

Hint for iv) and v): make the sign condition on q_k, which is implicitly contained in the interpretation, explicit.

Now assume that $q_0 > 0$, which means that there is a positive probability that an individual will beget no offspring at all. Let us start the process with just one individual. Then clearly q_0 is also the probability that the population will be extinct after one step along the generation ladder.

Let z_n denote the probability that the population will be extinct after n steps along the generation ladder. Then our last observation translates into $z_1 = q_0$. We claim that the z_n can be computed recursively from the difference equation

$$z_n = g(z_{n-1}). \tag{1.4}$$

To substantiate this claim, we argue as follows. If in the first generation there are k offspring then the lines descending from each of these should go extinct in $n-1$ generation-steps in order for the population to go extinct in or before the nth generation. For each separate line the probability is, by definition, z_{n-1}. By independence, the probability that all k-lines go extinct in $n-1$ steps is then simply $(z_{n-1})^k$. It remains to sum over all possible values of k with the appropriate weight q_k. Thus we find

$$z_n = q_0 + \sum_{k=1}^{\infty} q_k (z_{n-1})^k = g(z_{n-1}),$$

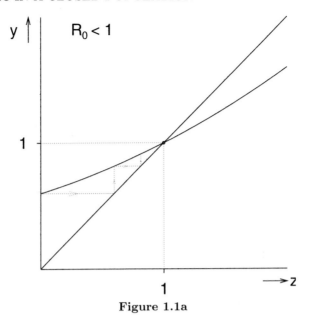

Figure 1.1a

as claimed.

Since g is increasing, the sequence z_n must be increasing and so it has a limit $z_\infty = \lim_{n\to\infty} z_n$. By definition z_∞ is the probability that the population started by the first individual will go extinct. When $z_\infty = 1$, the population goes extinct with certainty. When $0 < z_\infty < 1$, there exists a complementary probability $1 - z_\infty$ that, as further arguments show (see books on branching processes, such as Jagers (1975), Mode (1971) and Harris (1963)), exponential growth sets in and the deterministic description applies. So we expect that $R_0 < 1$ implies $z_\infty = 1$, while for $R_0 > 1$ the inequality $0 < z_\infty < 1$ holds. That our expectation is correct follows most easily from a graphical consideration, see Figures 1.1a and 1.1b.

Exercise 1.6 Argue that z_∞ is the *smallest* root in $[0, 1]$ of the equation

$$z = g(z) \tag{1.5}$$

(and interpret this equation as a consistency condition that the probability to go extinct should satisfy; hint: recall the derivation of the difference equation (1.4)).

Exercise 1.7 Use Exercise 1.5-iii to show analytically that $z_\infty = 1$ for $R_0 < 1$ while $0 < z_\infty < 1$ for $R_0 > 1$. Hint: Also use Exercise 1.5-v.

Exercise 1.8 Determine z_∞ for the critical case $R_0 = 1$.

We conclude that even in the situation where the infective agent has the potential of exponential growth, i.e. $R_0 > 1$, it still may go extinct due to an unlucky (for the agent) combination of events while numbers are low. The probability that such an extinction happens, when we start out with exactly one primary case, can be computed as a specific root of the equation $z = g(z)$.

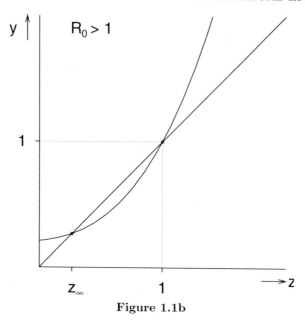

Figure 1.1b

But how do we derive the function g (or, equivalently, the probabilities q_k) from the kind of specification of a transmission model that we employed so far? (Recall Sections 1.1 and 1.2.1.)

When during a time interval of length $\triangle T$ contacts are made according to a Poisson process with intensity c, the probability that k contacts are made equals

$$\frac{(c\triangle T)^k}{k!}\, e^{-c\triangle T}.$$

In other words, the number of contacts follows the Poisson distribution with parameter $c\triangle T$. When contacts lead to successful transmission with probability p, the number of 'successful' contacts is again Poisson-distributed with the parameter modified to $pc\triangle T$.

Exercise 1.9 Prove the last statement.

If we recall that the generating function for the Poisson distribution with parameter λ equals

$$g(z) = e^{\lambda(z-1)} \tag{1.6}$$

and that $R_0 = pc\triangle T = pc(T_2 - T_1)$, we arrive at the conclusion that for this particular submodel z_∞ is, for $R_0 > 1$, the unique root in $(0, 1)$ of the equation

$$z = e^{R_0(z-1)} \tag{1.7}$$

(recall Figure 1.1). In order to avoid the wrong impression that this is a general result, we add that

$$z_\infty = \frac{1}{R_0} \tag{1.8}$$

for another submodel for infectivity, viz. the one where $\triangle T$ above is not a fixed quantity but a random variable following the exponential distribution. The following exercises provide the details underlying this assertion.

We conclude that the probability z_∞ that the introduction of an infected host from outside does not lead to an epidemic can, for some simple submodels, be either determined by a graphical construction or be expressed explicitly in terms of the parameters. Different submodels that yield the same value for R_0 may lead to different values of z_∞.

Exercise 1.10 Convince yourself that when $\triangle T$ is exponentially distributed with parameter α,

$$
\begin{aligned}
g(z) &= \alpha \int_0^\infty e^{pc\triangle T(z-1)} \, e^{-\alpha\triangle T} \, d(\triangle T) \\
&= \frac{\alpha}{\alpha - pc(z-1)}.
\end{aligned}
$$

Hint: interchange summation with respect to k and integration with respect to $\triangle T$ in the definition of g.

Exercise 1.11 Compute that for this submodel $R_0 = pc/\alpha$. Do this in two different ways:

i) using Exercise 1.5-iii;

ii) in a way involving the interpretation; more precisely, compute the expected length of the infectious period and multiply with the expected number per unit of time of 'successful' contacts.

Exercise 1.12 Check equation (1.8).

1.2.3 Initial growth in real time

By looking at initial growth on a generation basis, we now understand that a threshold phenomenon occurs governed by a parameter R_0, allowing for a clear biological interpretation, and that even for R_0 above threshold, there exists a positive probability z_∞ that introduction of one primary case does not lead to explosive exponential growth. For simple submodels of contact and infectivity, we are able to derive explicit expressions (or simple equations) for R_0 and z_∞ in terms of the parameters of the submodels.

The disadvantage of measuring growth on a generation basis is that, due to the fact that generations usually overlap each other in real time, it does not correspond to what we actually observe. Indeed, when we speak about exponential increase during the initial phase of an epidemic, we mean that

$$
I(t) \approx Ce^{rt}
$$

for some $r > 0$ (and some constant $C > 0$), where $I(t)$, the *prevalence*, is the number of cases notified up to time t (and time is measured relative to some convenient, but otherwise arbitrary, starting point). The *incidence* $i(t)$ (i.e. the number of new cases per unit of time) will be proportional to dI/dt, hence to $\exp(rt)$. (As a side-remark we note that the constant of proportionality may involve the probability that cases are actually reported—a quantity that is often less than one in the real world.)

Now let us return to the world of models. Is it possible to compute the exponential growth rate r in terms of the parameters of our submodels for contact and infectivity?

New cases at time t result from contacts with individuals that were infected themselves before t and that are infectious at time t. In a deterministic description we pretend that the actual number of new cases equals the expected number of new cases. These arguments, our assumptions on contact and infectivity above, and elementary bookkeeping together lead to the equation

$$i(t) = pc \int_{T_1}^{T_2} i(t - \tau)\, d\tau \tag{1.9}$$

for the incidence in the initial phase of an epidemic.

Exercise 1.13 Give a detailed interpretation of the equality in (1.9), to check both the equation and your understanding of it.

Substituting the Ansatz $i(t) = ke^{rt}$, we find that r should satisfy the so-called characteristic equation

$$1 = pc \int_{T_1}^{T_2} e^{-r\tau}\, d\tau. \tag{1.10}$$

(An *Ansatz* is an assumed relation that will be motivated, justified, or refuted by the consequences that one is going to derive from it. A *characteristic* equation is an equation in terms of a scalar quantity—usually complex and often denoted by λ— which constitutes a solvability condition (both necessary and sufficient) for a more complicated problem. For example, the equation $\det(M - \lambda I) = 0$ is a characteristic equation to the eigenvalue problem $Mv = \lambda v$, where M is a matrix and v a vector.)

A usual procedure in showing the existence of solutions for relations such as (1.10) is to let $f(r)$, say, be defined by the right-hand side of (1.10) and to use monotonicity arguments as follows. For $r = 0$ we obtain $f(0) = R_0$. Moreover, f is a decreasing function of r, $\lim_{r \to \infty} f(r) = 0$, and $f(r) \to +\infty$ for $r \to -\infty$. See also Figures 1.2a and 1.2b for a qualitative picture.

Exercise 1.14 Check these statements. Hint: do *not* evaluate the integral, as this will make the proof much more cumbersome.

We conclude:

- there exists a unique real root r; in other words, equation (1.10) tells us unambiguously what the exponential growth rate is for this model;
- we do not have an explicit formula for r (whereas we do for R_0);
- $r > 0$ if and only if $R_0 > 1$ and $r < 0$ if and only if $R_0 < 1$; this means that we have growth in real time if and only if we have growth on a generation basis.

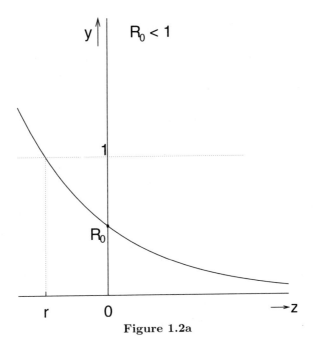

Figure 1.2a

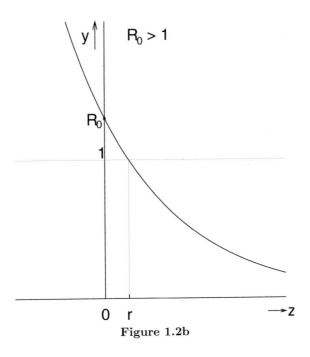

Figure 1.2b

In general, a high value of R_0 does not necessarily imply a high value of r. To use demography as a metaphor: if people have many children, but only when they themselves are already quite old, population growth may still be slow. And in more mathematical language: the formula for R_0 depends only on the difference $T_2 - T_1$, whereas in the equation for r the magnitude of T_1 and T_2 matters a lot.

Exercise 1.15 The aim of this exercise is to elaborate on the last observation. The task is to choose T_1^*, T_2^* and T_1^{**}, T_2^{**} such that $R_0^* > R_0^{**}$ but $r^* < r^{**}$, meaning that for the $*$ model growth on a generation basis is faster while for the $**$ model growth in real time is faster. By 'to choose' we do not mean that you have to provide numerical values, but rather that you have to describe some procedure by which such T values could be obtained. (Incidentally, does this exercise shed any light on the one-child policy of the Chinese government?)

Exercise 1.16 If an epidemic has growth rate r in the initial phase, what is the doubling time?

The reason to consider the generation perspective is that the formula for R_0 is explicit, whereas for r we only have an equation. In Chapter 5 we will find that for more complicated situations, R_0 is also only characterised by some equation. That characterisation, however, is still much more explicit than the corresponding characterisation of r. The underlying reason is that for r the time course of infectivity matters, whereas for R_0 it does not.

Remark 1.17. We refrain from a formulation and analysis of the real-time branching process that corresponds to our description of contact and infectivity. Conditional on the infective agent not going extinct, it would predict exponential growth with the rate r derived from the deterministic description in this section. But even when exponential growth occurs, there is an initial phase in which numbers are still low and stochastic effects manifest themselves. In particular one should think about the *duration* of this initial phase as a random variable. In other words, the continuous-time branching process predicts that either the agent goes extinct or exponential growth with rate r sets in after a random delay.[1]

At this point we have more or less answered the first two of our three questions and it is time to turn to the third question concerning the final 'size' of the epidemic. It will turn out that this question necessitates that we make additional model specifications.

1.3 The final size

1.3.1 The standard final-size equation

In a closed population and with infection leading to either immunity or death, the number of susceptibles can only decrease and so it must have a limit for time tending to infinity. Will this limit be zero? Or will some fraction of the population escape from

---

[1] J.A.J. Metz: The epidemic in a closed population with all susceptibles equally vulnerable; some results for large susceptible populations and small initial infections. *Acta Biotheoretica*, **27** (1978), 75-123.

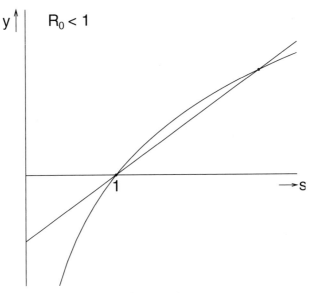

Figure 1.3a

ever getting infected? If so, what fraction (i.e. how does the fraction depend on the parameters)?

These questions were posed—and answered!—by Kermack & McKendrick in 1927.[2] We shall now first formulate and analyse the answers and only thereafter present the arguments to derive them, while scrutinising the underlying assumptions.

Let $s = S/N$ denote the proportion s of susceptibles S in a total population of size N. Let $s(\infty)$ denote this proportion at the end of the outbreak. The complementary quantity $1 - s(\infty)$ we shall call the *final size* of the epidemic, since it gives the fraction of the population that became infected sooner or later. The key result is that $s(\infty)$ is a root of the *final-size equation*

$$\ln s(\infty) = R_0(s(\infty) - 1).   \tag{1.11}$$

When $R_0 < 1$ the relevant root is $s(\infty) = 1$, meaning that the introduction of the agent does not lead to a major outbreak. When $R_0 > 1$ there exists a unique root between 0 and 1, which is the relevant one (the root $s(\infty) = 1$ persists but becomes irrelevant).

Exercise 1.18 Verify our statements about roots of equation (1.11). Hint: derive inspiration from Figures 1.3a and 1.3b.

Exercise 1.19 Show that

i) $s(\infty) \sim e^{-R_0}$ for $R_0 \to \infty$;

ii) $s(\infty) \sim 1 - 2(R_0 - 1)$ for $R_0 \downarrow 1$.

---

[2] W.O. Kermack & A.G. McKendrick (1927): Contributions to the mathematical theory of epidemics, part I. *Proc. Roy. Soc. Lond. A*, **115** (1927), 700-721. Reprinted (with parts II and III) as *Bull. Math. Biol.*, **53** (1991), 33-55.

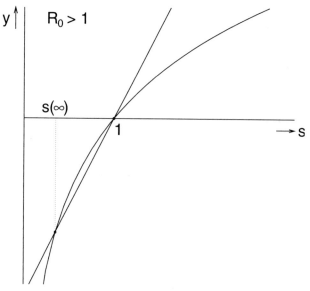

Figure 1.3b

We conclude that, within the framework of our assumptions, a certain fraction $s(\infty)$ escapes from ever getting the disease and that this fraction is completely determined by R_0 through equation (1.11). The larger R_0 is, the smaller is the fraction that escapes. In fact, the fraction is negligibly small for large values of R_0.

Below we shall derive the final-size equation by manipulating (integro)differential equations that correspond to a deterministic description. First, however, we present a quite different (less formal but much more direct) derivation in terms of a probabilistic consistency consideration. We advocate this method, since it generalises much more easily to far more complicated situations (see Section 6.3).

An individual that is susceptible at time t_0 and experiences a force of infection $\lambda(t)$ for $t > t_0$ will escape from being infected with probability $\mathcal{F}(t)$ defined by

$$\left.\begin{array}{r}\frac{d\mathcal{F}}{dt} = -\lambda\mathcal{F} \\ \mathcal{F}(t_0) = 1\end{array}\right\} \Rightarrow \mathcal{F}(t) = e^{-\int_{t_0}^{t}\lambda(\tau)\,d\tau}.$$

For large populations we shall have $s(\infty) = \mathcal{F}(\infty)$, i.e. the fraction that remains susceptible equals the probability to remain susceptible. Let us call $\int_{t_0}^{\infty}\lambda(\tau)\,d\tau$ the total cumulative force of infection. The fraction $z = 1 - s(\infty)$ that falls victim to the infection generates a total cumulative force of infection equal to

$$\frac{1}{N}pc\Delta TzN = R_0 z$$

(we have to divide by N since contacts are with probability $1/N$ with the susceptible individual that we consider). Hence

$$s(\infty) = \mathcal{F}(\infty) = e^{-R_0(1-s(\infty))},$$

and by taking logarithms we arrive at (1.11). The fact that the equation, in this form, is identical to equation (1.7) for the probability of a minor outbreak is not a coincidence:

Remark 1.20 In Section 10.4.2, it will be shown, by way of example, how to relate $s(\infty)$ to the probability of extinction in a backward branching process.

1.3.2   Derivation of the standard final-size equation (and reflection upon the underlying assumptions)

We start this section with a rather long exercise that introduces the Kermack-McKendrick ODE model that is described by a system of ODE (ordinary differential equations). This model is a special case of the Kermack-McKendrick model (Kermack & McKendrick loc. cit), which is formulated in terms of integral equations (see e.g. equation (1.16)). The aim of the exercise is to derive (1.11) from this system.

Exercise 1.21 Let S denote the size of the subpopulation of susceptibles, I the size of the subpopulation of infectives and R the size of the subpopulation of removed individuals (meaning immune or dead or in quarantine, but neither susceptible nor infective). We postpone a discussion of the precise meaning of 'size', in particular whether it refers to numbers or to spatial density. The *force of infection* is by definition the probability per unit of time for a susceptible to become infected. As an analogue of the law of mass action from chemical kinetics, we assume that the force of infection is proportional to I (see also Section 10.2). The constant of proportionality we shall call β (the transmission rate constant).

We assume that an infected individual becomes immediately infectious. In other words, upon infection, an individual labelled S turns into an individual labelled I.

We assume that infectives have a constant probability per unit of time α to become removed. Stated differently: the infectious period has an exponential distribution with parameter α, i.e., the probability to still be infectious τ units of time after infection is $e^{-\alpha\tau}$.

The system of differential equations

$$\begin{aligned}
\frac{dS}{dt} &= -\beta SI \\
\frac{dI}{dt} &= \beta SI - \alpha I \\
\frac{dR}{dt} &= \alpha I
\end{aligned} \tag{1.12}$$

concisely summarises our assumptions. One speaks about an SIR model or the SIR system. (See Exercises 2.2 and 3.9 for an SEIR model, and Exercise 3.17 for an SIS model. These are also called *compartmental* models, with each letter referring to a 'compartment' in which an individual can reside, often (but not in the case of S) with an exponentially distributed sojourn time; for an exposition of theory and applications of this type of models see, e.g. Jacquez (1998).) Note, however, that the last equation is redundant, since the first two form a closed system. Moreover, $S + I + R = \text{constant} = N$, where N denotes the total population size.

i) Convince yourself that for this model $R_0 = \beta N/\alpha$ and $r = \beta N - \alpha$. Use these results to reformulate the threshold condition as one involving population size N.

ii) Show that $\frac{\alpha}{\beta} \ln S - S - I$ is a conserved quantity. Hint: Consider $\frac{dI}{dS}$ and separate variables.

iii) Assume that $R_0 > 1$. If time runs from $-\infty$ (before the epidemic) to $+\infty$ (after the epidemic), argue that

$$\frac{\alpha}{\beta} \ln S(+\infty) - S(+\infty) = \frac{\alpha}{\beta} \ln S(-\infty) - S(-\infty)$$

and that $S(-\infty) = N$. Derive (1.11) from these identities.

iv) Now let time run from 0 (the start of the epidemic) to $+\infty$ (after the epidemic). Put $N = S(0) + I(0)$ (so $R(0) = 0$). Derive the following variant of (1.11) (recall that $s(t) = S(t)/N$):

$$\ln s(\infty) = R_0(s(\infty) - 1) + \ln s(0). \tag{1.13}$$

Analyse this equation graphically in the spirit of Exercise 1.18, paying attention to both cases $R_0 < 1$ and $R_0 > 1$. Deduce in particular what happens for $I(0) \downarrow 0$ or, equivalently, $s(0) \uparrow 1$.

v) Draw the phase portrait describing the reduced (S, I) system

$$\frac{dS}{dt} = -\beta SI,$$
$$\frac{dI}{dt} = \beta SI - \alpha I.$$

Hint: use the conserved quantity (first integral) and restrict attention to the positive quadrant.

vi) For what value of S does the epidemic reach its peak, in the sense that I is maximal? Can you understand this? Hint: Recall the definition of R_0 and pretend that we can 'freeze' the dynamic variable S at any value we wish to consider.

vii) Verify that the root $s(\infty)$ of equation (1.11) becomes smaller when we increase N. (Observe first that R_0 increases as N increases.) This is essentially an overshoot phenomenon. Try to figure out what we mean by 'overshoot phenomenon' by combining the results of v) and vi).

viii) Reformulate (1.12) in terms of *fractions* of individuals. What happens to the parameter β? Reflect upon the dimension of β and its relation to observable quantities, such as the c we introduced in Section 1.1.

After this lengthy exercise we take up our general discussion and set out to derive (1.11) while forming a clear idea of the underlying assumptions. So far we asserted that individual hosts have an expected number c of contacts per unit of time. Can we sustain this assumption when the disease makes victims? The answer presumably depends on what it really means to fall 'victim' to the disease.

When at the end of the infectious period hosts become immune and take part in the contact process as before, the epidemic does not interfere with the contact process and we can consider c as a fixed constant during the epidemic. This seems the easiest situation, so let us deal with it first.

If we focus on the contacts of an infected individual and contacts are 'at random', the probability that the partner in the contact is susceptible equals $s = S/N$, the proportion s of susceptibles S in the total population of size N. The equation for the incidence then reads

$$i(t) = s(t)pc \int_{T_1}^{T_2} i(t - \tau) \, d\tau. \tag{1.14}$$

(Note that N does not depend on time and that we recover equation (1.9) for the incidence in the initial phase by putting $s(t)$ equal to 1.) Our assumption that the population is closed implies

$$i(t) = -\dot{S}(t) \tag{1.15}$$

i.e. the incidence equals the change in the susceptibles. (A side-remark on notation: the time derivative of $S(t)$ is sometimes denoted by dS/dt, sometimes by $\dot{S}$, and sometimes by S'. Through this warning we attempt to avoid any possible confusion on this point.) After dividing both sides of (1.14) by N we can rewrite the equation as

$$\dot{s}(t) = s(t)c \int_0^\infty A(\tau) \, \dot{s}(t - \tau) \, d\tau, \tag{1.16}$$

where we have introduced the function A defined by

$$A(\tau) = \begin{cases} p & \text{if } T_1 \leq \tau \leq T_2, \\ 0 & \text{otherwise.} \end{cases} \tag{1.17}$$

Exercise 1.22 Derive (1.11) from (1.16). Hint: Divide by $s(t)$ and integrate from $t = -\infty$ to $t = +\infty$; use that $R_0 = c \int_0^\infty A(\tau) \, d\tau$ (see Exercise 1.28).

(A side-remark on a fundamental point: In the following, we will more and more work with a general non-negative infectivity function $A = A(\tau)$ as a basic model ingredient. Such a function incorporates information about

- the degree to which an individual, that was infected τ units of time ago is expected to take part in the contact process (e.g. an individual that has died does not take part at all);
- the probability of transmission, given a contact between such an individual and a susceptible individual.

At present, we still keep a multiplicative constant c, denoting the rate at which individuals are engaged in contacts, in the formulation, to facilitate the comparison of various alternative assumptions concerning the contact process. Later on, when dealing with other issues, we shall absorb the c in the A, to simplify the notation. Of course, this has an impact on the interpretation of A and we encourage our readers to be aware of the difference.)

For many situations it is reasonable to assume that c, and hence R_0, is proportional to population *density*, i.e. proportional to the number of individuals per unit area (but of course there are exceptions as well, e.g. when sexual contacts are concerned). The threshold for R_0 may be reformulated as a threshold for the density: below a critical density, introduction of the disease is harmless; above the critical density, an outbreak will result (for instance, infectious diseases of humans became much more prominent when cities were formed; dense agricultural crops are very vulnerable for pests). Moreover, if densities are larger, the epidemic will affect a larger proportion of the population (this is an overshoot phenomenon: the expected number of successful transmissions will gradually decrease during the epidemic, but even when this number drops below one, there will be many more new cases due to the substantial part of the population that is infectious at that moment, cf. Exercise 1.21-v,-vii).

Next let us look at another extreme, where each individual is supposed to die at the end of the infectious period. Then total population size N depends on t. It is therefore quite conceivable that the expected number of contacts per unit of time per individual depends on t as well. Let us assume it is proportional to N, with constant of proportionality θ. (The difference between θ here and β in Exercise 1.21 is that the latter incorporates a factor corresponding to the probability of transmission and, moreover, that in Exercise 1.21 we consider an exponentially distributed infectious period. Here we have incorporated information about the probability of transmission in the function A.) Then we may write the equation for the incidence as

$$\dot{S}(t) = \frac{S(t)}{N(t)}\theta N(t) \int_0^\infty A(\tau)\dot{S}(t - \tau)\, d\tau. \tag{1.18}$$

When $N(-\infty)$ denotes the original population size and $c = \theta N(-\infty)$ we obtain, after dividing by $N(-\infty)$, again equation (1.16) but now with

$$s(t) = \frac{S(t)}{N(-\infty)} \tag{1.19}$$

(i.e. $s(t)$ measures susceptibles as a fraction of the original population size). So all our previous conclusions still apply, the only difference being that now $1 - s(\infty)$ is the fraction of the population that died, whereas before it was the fraction that became immune.

Exercise 1.23 Suppose that at the end of the infectious period an individual dies with probability $1 - f$, and otherwise becomes immune, while we still assume that $c = \theta N(t)$. Does the final size depend on f?

We conclude that the final-size equation (1.11) holds when the disease is harmless (in the sense that it interferes in no way with the contact process) or, if it is not harmless, (1.11) holds when contact intensity is proportional to population density.

1.3.3   The final size of epidemics within herds

The foregoing does not exhaust the catalogue of assumptions that may be appropriate for certain specific real-life situations. For instance, phocine distemper virus is transmitted from seal to seal when they come to a sandbank to rest and sunbathe.

When group size diminishes (by death due to the disease), they occupy less area, while maintaining roughly the same contact intensity (in other words, the effective density remains constant).

More generally we claim that when animals live in herds, the number of contacts per unit of time per individual could very well be (almost) independent of herd size. If herd size diminishes due to a fatal infectious disease, the force of infection does not go down as quickly as it does in the model of the foregoing section, which assumed that the density, and hence the per capita number of contacts per unit of time, is proportional to population size. This has an effect on the final size. Exactly how much effect it has depends on the probability f to survive an infection. The final-size equation will now involve two parameters: R_0 and f. And it will involve two fractions: the fraction $s(\infty)$ that escapes from ever getting the infection and the fraction $n(\infty)$ that is still alive at the end of the epidemic.

Exercise 1.24 Argue that the interpretation of $n(\infty)$, $s(\infty)$ and f implies the relation

$$n(\infty) - s(\infty) = f(1 - s(\infty)). \qquad (1.20)$$

A second relation between these quantities and R_0 reads

$$n(\infty) = s(\infty)^{\frac{1-f}{R_0}} \qquad (1.21)$$

which will be derived below in Exercise 1.27 and beyond.

There are various ways in which we can combine and/or rewrite these equations and each of these may be helpful for a certain purpose.

Exercise 1.25 i) Derive the equation

$$\ln s(\infty) = \frac{R_0}{1-f} \ln(f + (1-f)s(\infty)). \qquad (1.22)$$

ii) Analyse this equation graphically.

iii) Show that in the limit $f \uparrow 1$ we find (1.11).

iv) Show that $s(\infty)$ is a monotonically increasing function of f.

v) Assume $R_0 > 1$. Show that $n(\infty) \downarrow 0$ when $f \downarrow 0$.

Exercise 1.26 When f and R_0 are unknown, but data about the situation after the epidemic allow us to make an estimate of $s(\infty)$ and $n(\infty)$, we can try to determine the parameters from the final-size equation. Therefore show that

$$\begin{cases} f = \frac{n(\infty) - s(\infty)}{1 - s(\infty)} \\ R_0 = \frac{1 - n(\infty)}{1 - s(\infty)} \frac{\ln s(\infty)}{\ln n(\infty)}. \end{cases} \qquad (1.23)$$

Let us now discuss the difference between the situations described in this and the previous section. Death has two effects: a direct one and an indirect one. The direct one is simply that a fraction $1 - f$ of $1 - s(\infty)$ dies. The indirect one is that $s(\infty)$ itself decreases since, while immunes hinder contacts between infectives and susceptibles, dead individuals do not. The indirect effect makes the difference between the two models. When R_0 is big, $s(\infty)$ is very small anyhow $(\sim e^{-R_0})$, and the indirect effect is negligible. If, on the other hand, R_0 is only little above the threshold value one, the indirect effect can be substantial.[3]

In the case of seals affected by phocine distemper virus the survival probability f may very well depend on physiological and immunological conditions, which, in turn, are determined by environmental aspects such as food availability and pollution.

We conclude that the precise form of the final-size equation depends on our assumptions concerning the contact process (and hence the force of infection), as affected by population size, and that this really matters when R_0 is only a little above threshold and the survival probability is low.

Exercise 1.27 i) Argue that the ODE system

$$
\begin{aligned}
\frac{dS}{dt} &= -\gamma \frac{SI}{N}, \\
\frac{dI}{dt} &= \gamma \frac{SI}{N} - \alpha I, \\
\frac{dN}{dt} &= -(1-f)\alpha I
\end{aligned}
$$

describes the spread of an infectious disease when the per capita number of contacts per unit of time is independent of the population size N. The parameter γ can be considered as the product of two more basic parameters. Which are these? The equations reflect a specific assumption about mortality as a result of infection. What assumption is this?

ii) Show that $R_0 = \gamma/\alpha$ (independent of population size!) and $r = \gamma - \alpha$.

iii) Show that, for any $-\infty \le \sigma \le t \le \infty$,

$$
\frac{S(t)}{S(\sigma)} = \left(\frac{N(t)}{N(\sigma)} \right)^{\frac{R_0}{1-f}}, \tag{1.24}
$$

and conclude that (1.21) is correct. Hint: consider dS/dN.

iv) Derive that $N(\infty) - N(-\infty) = (1-f)(S(\infty) - S(-\infty))$ and next rewrite this identity in the form (1.20).

Hint: Integrate both $\frac{dN}{dt} = -(1-f)\alpha I$ and $\frac{dI}{dt} = -\frac{dS}{dt} - \alpha I$ from $-\infty$ to ∞.

v) Note from the ODE system that the force of infection equals $\gamma I/N$, whereas the force of mortality $-\frac{1}{N}\frac{dN}{dt}$ equals $\alpha(1-f)I/N$. Hence the two are proportional, with constant of proportionality $(1-f)/R_0$. Derive the identity (1.24) from this. (Note: mathematically this amounts, in the end, to the same separation-of-variables method as in iii) above; this part of the exercise, however, is meant to prepare the reader for the assumption (1.28) below.)

We now set out to derive (1.20) and (1.21) for a more general class of models. This derivation may be skipped by readers who feel unhappy when abstract-looking formulae are manipulated.

---

[3] A. de Koeijer, O. Diekmann & P. Reijnders: Modelling the spread of phocine distemper virus among harbour seals. *Bull. Math. Biol.*, **60** (1998), 585-596.

Recalling the identity $i(t) = -\dot{S}(t)$ (see (1.15)), which expresses the incidence as the decrease in susceptibles, we start out from

$$\dot{S}(t) = c\frac{S(t)}{N(t)} \int_0^\infty \dot{S}(t - \tau)A(\tau)\, d\tau, \tag{1.25}$$

where c is the number of contacts per unit of time, and where $A(\tau)$ describes the infectivity at infection-age τ. The total population $N(t)$ at time t is composed of susceptibles and individuals who were infected τ units of time ago (at time $t - \tau$), and are still alive at t. Symbolically,

$$N(t) = S(t) - \int_0^\infty \dot{S}(t - \tau)\mathcal{F}(\tau)\, d\tau, \tag{1.26}$$

where $\mathcal{F}$ denotes the survival probability as a function of infection-age τ. It is then reasonable to write

$$A(\tau) = a(\tau)\mathcal{F}(\tau), \tag{1.27}$$

where $a(\tau)$ measures the output of infectious material at infection-age τ, given that the individual is still alive.

It is unclear whether a final-size equation can be derived without further assumptions. What we shall do is add a relation between a and $\mathcal{F}$, which allows for a clear biological interpretation and which enables us to derive (1.20) and (1.21).

The hazard rate of death, $\mu(\tau)$, is by definition the probability per unit of time of dying at infection-age τ, given that one survived until τ. Mathematically this translates into

$$\mathcal{F}'(\tau) = -\mu(\tau)\mathcal{F}(\tau). \tag{1.28}$$

We now assume that μ is proportional to a, or, in other words, that the probability per unit of time of dying is proportional to the output rate of infectious material.

Exercise 1.28 Call the constant of proportionality q, i.e. assume that

$$\mu(\tau) = qa(\tau). \tag{1.29}$$

The goal of this exercise is to express q in terms of c, f and R_0. The result reads

$$q = c\frac{1 - f}{R_0}. \tag{1.30}$$

i) Argue that

$$R_0 = c \int_0^\infty A(\tau)\, d\tau. \tag{1.31}$$

ii) Derive that $f - 1 = -q \int_0^\infty A(\tau)\, d\tau$. Hint: Integrate, using (1.29) and (1.27), the identity (1.28) from 0 to $+\infty$.

Exercise 1.29 Our motivation for (1.29) did not derive from the interpretation, but rather from formula manipulation. We want to be able to integrate the identity obtained from (1.25) by dividing by $S(t)$. So we want that

$$\frac{1}{N(t)} \int_0^\infty \dot{S}(t-\tau)A(\tau) \, d\tau \propto \frac{d}{dt} \ln N(t)$$

(where $\propto$ means 'is proportional to'). Show that this is indeed true if we assume (1.29). Hint: Write the integral term in (1.26) as

$$\int_{-\infty}^t \dot{S}(\tau)\mathcal{F}(t-\tau) \, d\tau.$$

A combination of the above two exercises leads to the conclusion that

$$\frac{N(t)}{N(\sigma)} = \left(\frac{S(t)}{S(\sigma)} \right)^{\frac{1-f}{R_0}} \qquad \text{for all } t, \sigma, \tag{1.32}$$

also for this more general class of models. And, by letting $\sigma \to -\infty$, $t \to +\infty$, we thus arrive at (1.21).

Exercise 1.30 Derive, as a check on our model formulation, (1.20) by letting $t \to \infty$ in (1.26). Hint: See the hint in the foregoing exercise.

Exercise 1.31 Recapitulate the differences in the assumptions underlying the models of this and the foregoing section. Check that the difference does not really matter in the initial phase and that, in particular, R_0, r and the probability to go extinct are determined in exactly the same manner. Next imagine two widely separated seal colonies, one being twice as large as the other. Assume that the same virus is introduced in both. Compare the predictions made by the two models concerning R_0 and the final size. (If you wish to include r and the probability to go extinct in the comparison, go ahead.)

Exercise 1.32 With a sexually transmitted disease in mind, contemplate how the average number of sexual contacts per individual per unit of time depends on population size (see also Chapter 10).

1.3.4   The final size in a finite population

In a deterministic description of an epidemic in a closed population, as given above, we found a characterisation of the *fraction* $1 - s(\infty)$ that is ultimately affected. In this section we will briefly indicate how that relates to the limit, for population size going to infinity, of the *distribution* of final size when we consider a finite population. So imagine a population consisting of N individuals, one of whom is infected from an outside source. Once we specify a submodel for contact and transmission, we can calculate the probability distribution for the number of its 'offspring' (cf. Section 1.2.2; but since we now consider a finite population of susceptible hosts, we will, typically, encounter binomial distributions instead of Poisson distributions). With a lot more effort, we can calculate the probability distribution for the sum of the first and second generation. At least in principle, we can extend this to any number of generations and so, by taking the limit, compute the distribution of the final size of the epidemic (note that there necessarily is a well-defined limit, because of monotonicity). There exist various sophisticated ways to compute the final-size distribution much more effectively than sketched above (see Ball (1995), Lefèvre & Picard (1995)[4] and the references given there), but we emphasise that it is not at all an easy task. Here we shall first summarise some qualitative aspects of the result and then, by way of exercises, delineate a trick that leads to a convenient computational scheme for the simplest submodel for infectivity.

When $R_0 < 1$, the final-size distribution is concentrated near 1. When we rescale by using k/N rather than k as the variable, and let $N \to \infty$, the distribution becomes more and more concentrated at zero, meaning that a negligible *fraction* is affected.

When $R_0 > 1$ the final size distribution has a double peak. The first peak corresponds to so-called *minor outbreaks,* in which the infective agent goes extinct before affecting a substantial *fraction* of the population. The second peak corresponds to *major outbreaks,* in which approximately a fraction $1 - s(\infty)$ is infected. This distinction between minor and major becomes more and more prominent if we rescale to k/N and let $N \to \infty$. Then a fraction z_∞ of the total probability one becomes concentrated at zero, while a fraction $1 - z_\infty$ becomes concentrated at $1 - s(\infty)$ (with appropriate scaling the distribution around $1 - s(\infty)$ is accurately described by a normal distribution with standard deviation of order $\sqrt{N}$; see Martin-Löf (1986)[5]).

We conclude that the deterministic description has to be complemented by two observations deriving from the stochastic analysis:

- even for $R_0 > 1$, introduction may lead to a minor outbreak only (cf. Section 1.2.2);
- $1 - s(\infty)$ is the *mean* size of major outbreaks.

Exercise 1.33 The setting for this exercise is as described in Exercise 1.21 (the Kermack-McKendrick ODE model). Note that this is identical to the setting of Exercise 1.27 under the parameter specification $f = 1$ (infection always leads to immunity, never to death) and $\gamma = \beta N$, with $N =$ population size. Recall that $R_0 = \beta N / \alpha$. Let $P_{(n,m)}(t)$ denote the probability that at time t there are n susceptibles and m infectives. Clearly, the indices are constrained by the inequalities $n, m \geq 0$ and $n + m \leq N$. It is helpful to introduce the convention that $P = 0$ when the indices violate these constraints.

When the population is currently in state (n, m), two transitions are feasible: a susceptible is infected, in which case the new state is $(n - 1, m + 1)$, or an infective may lose its infectivity, in which case the new state is $(n, m - 1)$. The first event has probability per unit of time $\beta n m$ and the second αm.

i) Argue that P should satisfy the ODE system (for a more precise derivation

---

[4] F.G. Ball: Coupling methods in epidemic theory. In: Mollison (1995), pp. 34-52; C. Lefèvre & P. Picard: Collective epidemic processes: a general modelling approach to the final outcome of SIR epidemics. In: Mollison (1995), pp. 53-70.

[5] A. Martin-Löf : Symmetric sampling procedures, general epidemic processes and their threshold limit theorems. *J. Appl. Prob.,* **23** (1986), 265-282.

see the Appendix)

$$\frac{dP_{(n,m)}(t)}{dt} = -\beta nm P_{(n,m)}(t) + \beta(n+1)(m-1)P_{(n+1,m-1)}(t)$$
$$-\alpha m P_{(n,m)}(t) + \alpha(m+1)P_{(n,m+1)}(t).$$

(Some jargon: we are dealing with a continuous-time Markov chain (see e.g. Taylor & Karlin (1984)). Quite naturally, we have indexed the states by the combination (n,m). One can translate the current formulation to the standard formulation in terms of vectors and matrices by employing an appropriately defined (e.g. via lexicographic ordering) map $(n,m) \longmapsto i$. This helps to see the connection with the general theory, but otherwise it just complicates the bookkeeping. So we stick to the index (n,m). For another example of this type of mapping see Section 7.3.)

ii) Describe in words the situation embodied in the initial condition

$$P_{(n,m)}(0) = \begin{cases} 1 & \text{for } n = N-1, \ m = 1, \\ 0 & \text{otherwise.} \end{cases}$$

iii) The states $(n,0)$ are *absorbing states*. First formulate the precise meaning of this statement. Next decide whether or not you agree. Are there any other absorbing states?

iv) Make plausible that $P_{(n,m)}(\infty) := \lim_{t\to\infty} P_{(n,m)}(t)$ exists and that this vector is concentrated on the set of absorbing states, i.e. $P_{(n,m)}(\infty) = 0$ for $m \neq 0$. Hint: Represent the set of states as a triangle of lattice points in the (S,I) plane. Indicate possible transitions by arrows.

v) Make the connection between $P_{(n,m)}(\infty)$ and the final-size distribution explicit.

vi) We proceed to derive an efficient computational procedure. First derive that the expected sojourn time in state (n,m) equals $(\beta nm + \alpha m)^{-1}$ provided $m > 0$. Next argue that the probability that the transition out of state (n,m) will lead to state $(n-1, m+1)$ equals

$$\frac{\beta nm}{\beta nm + \alpha m},$$

and, with the complimentary probability

$$\frac{\alpha m}{\beta nm + \alpha m},$$

the next state will be $(n, m-1)$.

vii) Inspired by the generation point of view, we now base our bookkeeping on events (or, equivalently, jumps from one state to another). Let $Q_{(n,m)}(l)$ denote the probability that the lth event brought the population in state (n,m). (So now we are dealing with a discrete-time Markov chain; see e.g. Taylor & Karlin (1984).) Show that, provided $m > 1$,

$$Q_{(n,m)}(l+1) = \frac{\beta(n+1)}{\beta(n+1) + \alpha}Q_{(n+1,m-1)}(l) + \frac{\alpha}{\beta n + \alpha}Q_{(n,m+1)}(l),$$

while, for $m = 0, 1$,

$$Q_{(n,m)}(l+1) = \frac{\alpha}{\beta n + \alpha} Q_{(n,m+1)}(l)$$

and

$$Q_{(n,m)}(0) = \begin{cases} 1 & \text{for } n = N - 1, m = 1, \\ 0 & \text{otherwise.} \end{cases}$$

viii) Show how it is reflected in the recurrence relations for Q that $(n, 0)$ is absorbing.

ix) Show that the number of events is bounded by $2N - 1$.

x) How can one compute the final-size distribution by using the recurrence relations for Q?

xi) To prepare for the next exercise, rewrite the recurrence relation of vii) in the form

$$Q_{(n,m)}(l+1) = \frac{R_0(n+1)}{R_0(n+1) + N} Q_{(n+1,m-1)}(l) + \frac{N}{R_0 n + N} Q_{(n,m+1)}(l)$$

for $m > 1$ and

$$Q_{(n,m)}(l+1) = \frac{N}{R_0 n + N} Q_{(n,m+1)}(l)$$

for $m = 0, 1$.

Exercise 1.34 Next consider the setting of Exercise 1.27 with $f < 1$. Denote the population state with n susceptibles, m infectives and k immune individuals by (n, m, k), with the constraints $n, m, k \geq 0$ and $n + m + k \leq N$. In the spirit of the last exercise, derive the recurrence relation

$$Q_{(n,m,k)}(l+1) = \frac{R_0(n+1)}{R_0(n+1) + n + m + k} Q_{(n+1,m-1,k)}(l)$$
$$+ (1 - f) \frac{n + m + k + 1}{R_0 n + n + m + k + 1} Q_{(n,m+1,k)}(l)$$
$$+ f \frac{n + m + k}{R_0 n + n + m + k} Q_{(n,m+1,k-1)}(l)$$

for $m > 1$, and

$$Q_{(n,m,k)}(l+1) = (1 - f) \frac{n + m + k + 1}{R_0 n + n + m + k + 1} Q_{(n,m+1,k)}(l)$$
$$+ f \frac{n + m + k}{R_0 n + n + m + k} Q_{(n,m+1,k-1)}(l)$$

for $m = 0, 1$. What are the absorbing states? What is the maximum number of events? What is now the final size distribution and how can we compute it?

1.4   The epidemic in a closed population: summary

We have introduced several numbers to classify the 'infectiousness' of a disease, and we have indicated, by way of examples, how these might be calculated from submodels for the contact process and the probability that transmission occurs, given a contact between a susceptible and an infective which was itself infected τ units of time earlier. These numbers give consistent information in the sense that

$$R_0 > 1 \Leftrightarrow r > 0 \Leftrightarrow s(\infty) < 1 \Leftrightarrow z_\infty < 1.$$

With other aspects of ordering we should be careful: a disease with a long latent period may have a big R_0 and yet a small r (Exercise 1.15). But the ordering of R_0 and $s(\infty)$ of two different non-lethal diseases is always identical.

Exercise 1.35 Check that this is true. Do you understand why we added 'non-lethal'?

     The tacit underlying assumption of deterministic models is that the number of individuals involved in the changes is large. The reward is the clear picture of a sharp threshold: $R_0 < 1 \Rightarrow$ no epidemic; $R_0 > 1 \Rightarrow$ epidemic affecting ultimately a fraction $1 - s(\infty)$ of the population.

     In reality, populations are finite and thresholds manifest themselves at the individual level: a contact does or does not take place, and when it does, it either does or does not result in transmission. This blurs the picture: even when $R_0 > 1$, introduction of the agent may only lead to a minor outbreak, whereas for R_0 slightly less than 1 it may result in a substantial fraction of the population being affected. Rather than one precisely determined final size, we have to consider a probability distribution for final size.

     The relationship between the intensity of contacts between individuals on the one hand and population size on the other, is a complicated issue that deserves a great deal of attention (much more than it usually gets). The question 'what actually is population density?' is much harder than the seemingly obvious answer 'number divided by unit area' suggests. Behavioural patterns, as well as heterogeneity of the area (some parts may be attractive habitat, while other parts are only inhabited or visited for want of something better), create non-homogeneous distributions. Consequently, typical 'distance to' and 'contact with' nearest neighbours are the relevant quantities. (Recalling that Part I concentrates on the 'simplest context', we stop here and refer to Chapter 8 for further remarks.) To compute the final size in the case of a potentially lethal infection, and to compare the final size of the same infection in different populations, one needs to address these issues. We have presented two consistent sets of assumptions leading to, respectively, (1.11) with R_0 proportional to population density, and (1.22) with R_0 independent of population size. Other meaningful variants may yet be uncovered. We do not claim completeness.

Exercise 1.36 The objective of this exercise is to illustrate how knowledge about R_0 comes in helpful when evaluating the chances that a particular control strategy will be effective. Suppose a perfect vaccine is available and we can keep a fraction q of the population vaccinated. Show that $q > 1 - 1/R_0$ leads to eradication of the infective agent.

Remark: By 'perfect' we mean that a vaccinated individual is completely protected. Usually this is interpreted as 'showing no clinical symptoms after exposure'. In the present context the key feature is 'being not infectious at all after exposure'.

Exercise 1.37 If $R_0 < 2$, the probability z_∞, that introduction leads to a minor outbreak only, satisfies $z_\infty > \frac{1}{2}$. Reflect upon the possibility to draw conclusions from one or two field observations or experiments.

Exercise 1.38 Argue that the probability that a major outbreak occurs is

$$1 - (z_\infty)^k$$

when we are certain that in some early 'generation' there are exactly k cases.

Exercise 1.39 Suppose $R_0 < 1$. Argue that the expected size (in terms of number of victims') of the epidemic equals $(1 - R_0)^{-1}$.

Exercise 1.40 We have repeatedly switched between a generation perspective and a real-time perspective. But so far we did not derive the deterministic final size in the generation framework. So let us do that now.

We start one generation 'step' with S_0 susceptibles and I_0 infectives, all of infection-age zero. We put

$$\frac{dS}{dt}(t) = -cA(t)\frac{S(t)}{N}I_0, \quad S(0) = S_0,$$

where t is now both infection-age (which we usually denote by τ) and time in between two generations and where $c, A(t)$ and N have the same meaning as before. We assume that the disease is non-lethal and that, consequently, population size N is constant. By integration, we find

$$S(t) = S_0 e^{-\frac{c}{N}\int_0^t A(\tau)\,d\tau\, I_0},$$

and in the limit $t \to \infty$

$$S(\infty) = S_0 e^{-R_0\frac{I_0}{N}},$$

where we have used that

$$R_0 = c\int_0^\infty A(\tau)\,d\tau.$$

Clearly then there are

$$S_0(1 - e^{-R_0\frac{I_0}{N}})$$

new cases created in this generation. Therefore the generation process is described by the recurrence relations

$$S_{k+1} = S_k e^{-R_0\frac{I_k}{N}},$$
$$I_{k+1} = S_k(1 - e^{-R_0\frac{I_k}{N}}).$$

i) Show that $I_k + S_k - \frac{N}{R_0}\ln S_k$ is a conserved quantity.

ii) Derive the final size equation.

iii) When you think about it, this is really strange. Since generations are not separated but overlap, each infective of a particular generation experiences a different 'environment', in the sense of the time course of the susceptible fraction of the population that can potentially be infected. So the recurrence relations above, for non-overlapping generations, describe an altogether different process. Yet the final size coincides exactly with the final size of the real-time integral equations. Why?

(The authors admit that this exercise is beyond their reach. But they know one argument that makes the result less miraculous than it seems at first: for the final size it does not matter by whom you are infected, but only whether you are infected. To elaborate the argument, one has to think of a process in which infectives 'mark' other individuals, the mark indicating that the individual is now infected. For this process, the bookkeeping in generations and the bookkeeping in real time differ only in the order in which marks are made. So when at the end we count what fraction of the population carries a mark, we find that the outcome does not depend on our way of bookkeeping.)

iv) In the literature one sometimes finds discrete-time epidemic models formulated as

$$S_{k+1} = S_k - \beta I_k S_k,$$
$$I_{k+1} = \beta I_k S_k.$$

What is wrong with this formulation? How would that possibly show up in the behaviour of the solution?

Exercise 1.41 Next consider an epidemic within a herd, as in Section 1.3.3. Show that the generation process is described by the recurrence relations

$$S_{k+1} = S_k \left(1 - (1-f)\frac{I_k}{N_k}\right)^{R_0/(1-f)},$$
$$I_{k+1} = S_k - S_{k+1},$$
$$N_{k+1} = N_k - (1-f)I_k.$$

Derive the final-size equations (1.20) and (1.21) from these.

Hint: Consider the ODE system

$$\frac{dS}{dt}(t) = -ca(t)\frac{S(t)}{N(t)}I(t),$$
$$\frac{dI}{dt}(t) = -\mu(t)I(t),$$
$$\frac{dN}{dt}(t) = -\mu(t)I(t),$$

and assume that $\mu(t) = qa(t)$ with $q = c\frac{1-f}{R_0}$ and

$$f = e^{-\int_0^\infty \mu(\tau)\, d\tau}$$

as before. Show that $I(\infty) = fI_0$, $N(\infty) - N_0 = (f-1)I_0$ and $S(t)/S_0 = (N(t)/N_0)^{R_0/(1-f)}$. Deduce the recurrence relations from these identities. To derive (1.21), first show that

$$S_{k+1}/S_k = (N_{k+1}/N_k)^{R_0/(1-f)}$$

and then iterate. To derive (1.20), introduce as an auxiliary variable the number of immunes R_k, defined by $R_0 = 0$ (beware: here R_0 is the value of R in the zeroth generation and not the basic reproduction ratio) and $R_{k+1} = R_k + fI_k$. Show that $N_k = S_k + I_k + R_k$ and $R_{k+1} = f(N_0 - S_k)$. Take the limit $k \to \infty$.

Exercise 1.42 Check that the recurrence relations of Exercise 1.40 are recovered from those of Exercise 1.41 in the limit $f \uparrow 1$.

2

Heterogeneity: the art of averaging

2.1 Differences in infectivity

We start with an example. Assume that the latency period and the infectious period of all individuals are the same, but that their infectivities during the infectious period may differ. With reference to the situation and notation introduced in Section 1.2.1, we say that T_1 and T_2 are fixed while p may differ from one individual to another. To describe the whole population, rather than each individual separately, we can specify the *distribution* of p-values (so we imagine that all individuals carry a label specifying the p-value they will have, should they happen to become infected).

During the very first period of the epidemic, the p-values of the few infected individuals, which are determined by chance, matter a lot. Once exponential growth takes off, all p-values are represented among the many infectives. Under the assumption that all individuals are equally susceptible (i.e. the p-label has no influence whatsoever on susceptibility), the occurrence of p-values among those actually infected is accurately described by the a priori given distribution. The *force of infection* (i.e. the per susceptible capita probability per unit of time of becoming infected) is obtained by summing all contributions of the infectives. So effectively the force of infection is determined by the *mean value* of p. In other words, in our deterministic description (which ignores the demographic stochasticity of the very early stages) we can work with the *expected* infectivity, while ignoring the variance and other characteristics of the distribution. (We have already used this implicitly when working with c, the *expected* number of contacts per unit of time.)

This argument is not restricted to the special example. Given a stochastic submodel that determines infectivity at the individual level, we may invoke the law of large numbers to work with expected values at the population level when describing the dynamics of exponential growth and convergence to a final size (in the very late stages demographic stochasticity comes in again, but as this by definition concerns only few individuals, it hardly influences the final size in a large population).

So let us define

$$A(\tau) := \text{expected infectivity at time } \tau \text{ after infection took place}, \qquad (2.1)$$

having in mind that often the 'shape' of A will be as depicted in Figure 2.1.

The easiest interpretation of 'infectivity' is again the probability of transmission given a contact between a susceptible and an infective of disease age τ. But note that whenever death is a possibility, A necessarily incorporates the probability to survive

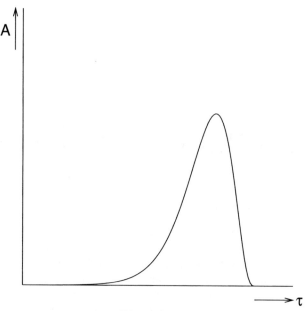

Figure 2.1

to age τ (as a factor when the risk of dying is independent of the output of infectious material but in a more complicated implicit manner otherwise). Moreover, there are contexts that ask for a slightly different interpretation, such as the rate of spore production in the case of fungal plant diseases. As attentive readers have undoubtedly noted, our notation in Chapter 1 anticipated this definition of $A(\tau)$. So it should come as no surprise that all results concerning R_0, r and $s(\infty)$ apply in this generality:

$$R_0 = c \int_0^\infty A(\tau) \, d\tau, \qquad (2.2)$$

$$1 = c \int_0^\infty e^{-r\tau} A(\tau) \, d\tau,$$

and either (1.22) or $\ln s(\infty) = R_0(s(\infty) - 1)$.

Exercise 2.1 Show that for the Kermack-McKendrick ODE model of Exercise 1.21,

$$cA(\tau) = \beta N e^{-\alpha\tau}.$$

Exercise 2.2 To include a latency period and yet work with a system of ODE, the following, so-called SEIR, system is used (here E means 'exposed but not yet infectious'):

$$\frac{dS}{dt} = -\beta SI,$$

$$\frac{dE}{dt} = \beta SI - \theta E,$$

$$\frac{dI}{dt} = \theta E - \alpha I,$$

$$\frac{dR}{dt} = \alpha I.$$

i) Show that in this case

$$cA(\tau) = N\beta \frac{\theta}{\alpha - \theta}(e^{-\theta\tau} - e^{-\alpha\tau}).$$

Hint: In order to be in the state 'infectious' at time τ, one should have entered this state at some time $\tau_0 \in (0, \tau)$ and remained in it in the interval (τ_0, τ). Compute the probability for arbitrary τ and then integrate with respect to τ_0.

ii) Deduce that $R_0 = \frac{\beta N}{\alpha}$, independently of θ. Are you surprised?

iii) Demonstrate that in general, i.e. for arbitrary patterns of infectivity, the length of a latency period has no influence on R_0 whenever individuals do not die. What about r?

Exercise 2.3 So far we have pretended that infected individuals take part in the contact process in just the same way as healthy individuals do. Now imagine that the infectious period is from $\tau = T_1$ to $\tau = T_2$, but that individuals have, due to illness, a reduced contact rate in the second half of this period. Can you incorporate this in $A(\tau)$? How does this change the interpretation of $A(\tau)$?

Exercise 2.4 i) Explain why a death rate μ (from 'natural', i.e. disease-unrelated causes) is expressed as a factor $\exp(-\mu\tau)$ in $A(\tau)$.

ii) Deduce from this that a latency period of fixed length T_L is reflected in a factor $\exp(-\mu T_L)$ in R_0.

iii) Derive from i), or directly, that for the Kermack-McKendrick ODE model with death rate μ we have $R_0 = \beta N/(\alpha + \mu)$.

iv) To calculate R_0 for the SEIR system of Exercise 2.2 with death rate μ, we use a trick that we have encountered already in Exercise 1.33-vi. During the latency period there are competing 'risks', viz. to die or to become infectious. If the latter occurs, we are again in the same situation as in iii) of the present exercise. Check that these arguments lead to the expression $R_0 = \frac{\theta}{\theta + \mu} \frac{\beta N}{\alpha + \mu}$.

v) If μ is small, death can be ignored. But what does 'small' mean? What are the quantities that one should compare μ with? Hint: Pay attention to the dimension.

Exercise 2.5 Consider a herpes infection, which, after a first infective period, may become latent, resurge with a certain probability per unit of time etc. Let 1 denote the infective state and 2 the latent state. Let σ denote the rate of going from 1 to 2 and ν the rate of going from 2 to 1, while μ denotes the death rate. Assume that a newly infected individual is in state 1.

i) Give a priori arguments why $R_0 = \infty$ when $\mu = 0$ but $\nu > 0$.

ii) Show that $\int_0^\infty A(\tau)d\tau = h_1 \frac{\nu + \mu}{\mu(\sigma + \nu + \mu)}$, where h_1 measures the infectivity in state 1. Hint: First read on a bit, then return to the exercise.

Let us reflect upon some implicit underlying assumptions that correspond to the image we have of what is going on in a host after an infectious agent has invaded. We imagine that infection triggers an autonomous process within the infected individual. In particular, the invading organism reproduces within the host at such a rate that further infections with the same agent are irrelevant. Examples include measles, influenza, rabies and HIV, and fall under the general heading of *microparasites* (to distinguish them from another class, the *macroparasites*, such as worms, for which infection is not a unique event but rather a repeated process; we come back to these in Section 4.1.1 and Chapter 9). When taking the function $A(\tau)$ as our chief modelling ingredient, we give up on a precise biochemical-physical description of what is going on in the host and take instead the pragmatic view that all that matters for spread at the population level is the output of infectious material by the host. One increasingly important area not treated in this book is the interaction between the agent and the immune system of the host, and specifically the way in which this interaction shapes the infectious output A (see e.g. Dushoff (1996)[1]). In principle one could try to determine $A(\tau)$ experimentally, but in practice this is almost always impossible. (A heartening counterexample: spore production by a fungal plant pathogen can often be measured.) A first conclusion is that we should only have confidence in those inferences that are *robust*, i.e. independent of the details of A. The threshold $R_0 > 1$ is a clear example. A second is that we might try to work with parameter scarce, coarse and quasi-mechanistic, submodels of the autonomous process within the host, compute as many characteristics of A (such as R_0) from such a description as we need, and try to get insight and understanding by analysing how the result depends on the parameters. (For this to work, the parameters should allow a quasi-mechanistic interpretation, of course.) If we are ambitious, we might try to fit the parameters on the basis of whatever field and experimental data are available.

An attractive class of submodels, take the form of continuous-time Markov chains (see e.g. Taylor & Karlin (1984) for background). Here we assume that an infected individual can be in a finite number of states (which we shall call *d-states*, with 'd' denoting disease). The traditional SIR, SEIR, SIS, ..., systems of ordinary differential equations are all based on such a description. Here we shall present a straightforward systematic procedure to compute R_0 for such models.

Label the d-state by $i = 1, 2, ..., n$. Do not include 'dead' or 'removed' in this list, which correspond to the definite loss of infectivity. Let $\Theta = (\theta_1, ..., \theta_n)^\top$ (where $\top$ indicates taking the transpose) with $\sum_{i=1}^{n} \theta_i = 1$ denote the probability distribution for d-state at the moment immediately following infection. (Often one will arrange things such that $\theta_1 = 1, \theta_i = 0$, for $i \neq 1$, and let infected individuals progress through states $1, ..., n$ in consecutive order.) Let Σ denote the $n \times n$-matrix of transition probability per unit of time. By this we mean that σ_{ij}, $i \neq j$, is the probability per unit of time for an individual's state to change from j to i. The diagonal elements are defined by $\sigma_{jj} = -\sum_{i \neq j} \sigma_{ij}$ (columns therefore sum to zero). If states are passed through in their natural ordering $1, 2, ..., n$, we will have that $\sigma_{ij} = 0$ for all $i \notin \{j, j+1\}$ and $\sigma_{jj} = -\sigma_{j+1,j}$. Here and everywhere else in the text we adopt the convention for the elements of transition matrices that the second index gives the state of departure and that the first index gives the state of arrival (in contrast to common practice in

[1] J. Dushoff: Incorporating immunological ideas in epidemiological models. *J. theor. Biol.*, **180** (1996), 181-187.

probability theory). Let D denote the diagonal matrix of death/removal rates. Then

$$\frac{dx}{d\tau} = \Sigma x - Dx,$$
$$x(0) = \Theta,$$

where the vector $x(\tau)$ describes the probability to be in the various states at time τ (note that $x(\tau)$ may be defective, i.e. $\sum_{i=1}^{n} x_i(\tau) < 1$, due to the possibility of death/removal). We will refer to the matrix $\Sigma - D$ as the transition matrix.

Let h denote the vector of infectivities associated with the various d-states. Then

$$A(\tau) = h \cdot x(\tau) = \sum_{i=1}^{n} h_i x_i(\tau) \tag{2.3}$$

where '$\cdot$' represents the inner product. (We are deliberately a bit vague about the interpretation of the 'infectivity' vector h. When infectivity is interpreted as 'probability of transmission, given a contact', one should multiply $A(\tau)$ by a contact intensity parameter c and integrate with respect to τ from 0 to ∞ to obtain R_0. The alternative is to incorporate the contact intensity in h, in which case the integral of A itself equals R_0.) A key point is now to observe that in order to determine the integral of A, we do not need to determine $x(\tau)$. For if we integrate the differential equation from 0 to ∞, while noting that $x(+\infty) = 0$ by death/removal, we find

$$-\Theta = (\Sigma - D) \int_0^\infty x(\tau) \, d\tau \Rightarrow \int_0^\infty x(\tau) \, d\tau = -(\Sigma - D)^{-1}\Theta$$
$$\Rightarrow \int_0^\infty A(\tau) \, d\tau = -h \cdot (\Sigma - D)^{-1}\Theta. \tag{2.4}$$

So all we have to do is to invert the matrix $\Sigma - D$ and apply the result to the vector Θ.

Exercise 2.6 Interpret $(-(\Sigma - D)^{-1})_{ij}$ as the expected total sojourn time in state i given that the system currently has state j.

Exercise 2.7 i) Do Exercise 2.5.

ii) Analyse how R_0 depends on the recrudescence parameter ν.

Exercise 2.8 Take $n = 3$. Let Σ and D be described by the scheme in Figure 2.2. Let $\theta_1 = 1$, $\theta_2 = \theta_3 = 0$, $h_1 = 0$, $h_2 = 1$, $h_3 = 2$. Calculate $\int_0^\infty A(\tau) \, d\tau$.

Exercise 2.9. In Section 1.3.3 we introduced the condition $\mu(\tau) = qa(\tau)$, see (1.29). Show that this translates into the condition $\mu = qh$ in the present context, where μ now denotes the vector of death rates on the diagonal of D. Go over the interpretation of this condition once more.

In concluding this section, we analyse the influence of variability in infectivity on the probability of a minor outbreak. An abstract but very convenient approach is the following. Distinguish individuals from one another according to 'type', and label types with a variable ξ taking values in a set Ω. Let the measure m on Ω describe

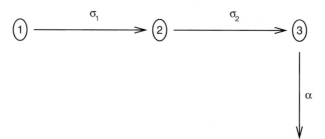

Figure 2.2

the distribution of types (i.e. for any measurable subset ω of Ω, the number $m(\omega)$ equals the fraction of the population with type $\xi \in \omega$; in particular, $m(\Omega) = 1$). (An illuminating example, for those unfamiliar with the idea of a measure, may be when ξ denotes p, as discussed at the beginning of this section; then, with $\omega = [p_1, p_2)$, we have that $m(\omega)$ is the probability $\Pr\{p_1 \leq p < p_2\}$ that an arbitrary individual has p-value such that $p_1 \leq p < p_2$. For a very brief introduction to measures see Section 4.2.) Let $a(\tau; \xi)$ be the infectivity at infection-age τ of individuals of type ξ. Then

$$A(\tau) = \int_\Omega a(\tau; \xi) m(d\xi), \tag{2.5}$$

while, reasoning as in Section 1.2, the generating function for the associated branching process is found to be

$$g(z) = \int_\Omega \exp\left(c(z-1) \int_0^\infty a(\sigma; \xi)\, d\sigma\right) m(d\xi). \tag{2.6}$$

(Recall the assumptions: contacts are made according to a Poisson process with intensity c and transmission will occur with probability $a(\tau; \xi)$; so the probability that an infective of type ξ will have k offspring equals

$$\frac{(c \int_0^\infty a(\sigma; \xi)\, d\sigma)^k}{k!} e^{-c \int_0^\infty a(\sigma; \xi)\, d\sigma}$$

and this now has to be averaged over ξ, multiplied by z^k and summed over k to obtain g; by interchanging the integration with respect to ξ with the summation over k we arrive at (2.6).)

Exercise 2.10 Verify that $g(z) = \exp(R_0(z-1))$ whenever infectivity is deterministic (i.e. there is but one type; formally Ω then consists of one point only). Conclude that the probability of a minor outbreak equals the (deterministic) compliment $s(\infty)$ of the final size whenever infectivity is deterministic.

Exercise 2.11 i) Prove that in general $g(z) \geq \exp(R_0(z-1))$. Hint: apply Jensen's inequality.

ii) Deduce from this that in general $z_\infty \geq s(\infty)$.

iii) Variability in infectivity has the tendency to increase the probability that introduction of the agent leads to a minor outbreak only. Do you agree? What exactly do we mean by such a statement? Do you find it intuitively plausible?

2.2   Differences in infectivity and susceptibility

Consider a population divided into two subpopulations, which we shall indicate by the labels 1 and 2. Although we are going to consider hypothetical numbers to illustrate a general phenomenon, it may be helpful to think of a sexually transmitted disease with sexual intercourse as a 'contact'. In such a context one then has to imagine a promiscuous population without any form of partnership. Suppose there are $N_1 = 10^5$ individuals of type 1, which have $c_1 = 100$ contacts/year, and $N_2 = 10^7$ individuals of type 2, which have $c_2 = 10$ contacts/year. Assume that a fraction $c_2 N_2/(c_1 N_1 + c_2 N_2)$ of the contacts of a 1-individual are with 2-individuals and a fraction $c_1 N_1/(c_1 N_1 + c_2 N_2)$ of the contacts of a 2-individual are with 1-individuals (note that these rules satisfy the consistency condition that there are as many contacts between 1-individuals and 2-individuals as there are between 2-individuals and 1-individuals). Consider an infective agent that yields an infectious period of exactly one year, during which transmission occurs with probability 1 for each contact with a susceptible. What is R_0?

The average value of c equals

$$\frac{c_1 N_1 + c_2 N_2}{N_1 + N_2} \approx 11,$$

and where new cases to be distributed over the types according to the relative subpopulation sizes (i.e. as $N_1/(N_1 + N_2) : N_2/(N_1 + N_2)$) then this would, given our assumptions about infectivity, be equal to R_0. Actually, however, type i individuals account for a fraction $c_i N_i/(c_1 N_1 + c_2 N_2)$ of all $c_1 N_1 + c_2 N_2$ contacts and since transmission occurs during contact, the new cases are distributed over the types according to these relative numbers of contacts (i.e. as $c_1 N_1/(c_1 N_1 + c_2 N_2) : c_2 N_2/(c_1 N_1 + c_2 N_2)$), and therefore

$$R_0 = c_1 \frac{c_1 N_1}{c_1 N_1 + c_2 N_2} + c_2 \frac{c_2 N_2}{c_1 N_1 + c_2 N_2} = \frac{c_1^2 N_1 + c_2^2 N_2}{c_1 N_1 + c_2 N_2} \approx 18.$$

We conclude that the distinction of 'types' can have a profound influence on R_0 when it involves a correlation between infectivity and susceptibility (see also Section 10.4.1).

Exercise 2.12 Intuitively, one would think that whenever a group decreases its contact intensity (i.e. the value of the parameter c), the basic reproduction ratio would decrease. Yet this is not necessarily the case. What may happen is that when a less active group reduces its activity even further, a more active group is 'forced' to increase the fraction of 'within-group' contacts, in order to sustain its contact intensity, which then leads to an increase of R_0.

Substantiate this verbal argument by first computing the derivative of

$$R_0 = \frac{c_1^2 N_1 + c_2^2 N_2}{c_1 N_1 + c_2 N_2}$$

with respect to c_1, and then to find a condition for the result to be negative.

Exercise 2.13 The ratio of type 1 to type 2 individuals is 1:100. What is this ratio among new cases in the early phase of the epidemic? And what is this ratio among those responsible for transmission in the early phase of the epidemic?

In the present example there is correlation between the susceptibility and the infectivity of an individual, yet we can figure out directly how new cases will be distributed with respect to type. This allows us to compute the appropriate average directly. The key point is that the properties of the *two* individuals involved in the transmission event have *independent* influence. A much more complicated situation arises when there is correlation between the infectivity of the 'donor' and the susceptibility of the 'receiver' involved in a transmission event. In that case it will be implicitly determined by the dynamics of the generation process how new cases will be distributed with respect to type. In Part II we shall present a systematic analysis based on the spectral theory of positive matrices and operators.

Exercise 2.14 The distribution of new cases, with respect to type, for sampling in real time, may or may not be identical to this distribution for sampling on a generation basis. Can you think of a condition that guarantees that they are identical?

When a population is subdivided into types, the monotonic relation between R_0 and $s(\infty)$ is lost. This is most easily seen by considering first the degenerate limiting case of two decoupled subpopulations: one small and one large. If we give the small population $R_0 = 1$ and the large population $R_0 = 2$, the final size, expressed as a fraction of the total population, will be close to the final size corresponding to $R_0 = 2$. If we give the small population $R_0 = 2$ and the large population $R_0 = 1$, the final size, again expressed as a fraction of the total population, will be rather small. By introducing some coupling, we can arrange that in the first case the coupled population has R_0 slightly less than in the second case. On the other hand, the final size expressed as a fraction of the total population will hardly change by the coupling.

The conclusion is that a small *core group* (see Chapter 10) may have a large impact on initial growth while, not surprisingly, it has little impact on the final size expressed as a fraction of the total population.

Even if we know that averaging introduces errors, we may still choose to do it. Tractable approximations are often to be preferred over scrupulous representations involving a multitude of badly known parameters. After all, modelling is the art of simplification without oversimplification.

So when building a model, a key question is which traits of individuals to take into account and which to neglect. Quite naturally, age and spatial position score highly when it comes to being included in a model.

The assumption that everybody is equally likely to be infected by a particular infectious individual is certainly not warranted when relative spatial position decides about the possibility of having contact. In such a case, $R_0 > 1$ is still a threshold criterion, but r is not always a good indicator of (initial) growth. In large spatial domains one finds wavelike spread with a characteristic velocity, the so-called *asymptotic speed of propagation*. We shall deal with this issue in Chapter 8.

Patterns of human social behaviour and sexual activity correlate with age. The seriousness of many infectious diseases depends on age. Data on the distribution of the random variable 'age at infection' contain information about the prevailing force of infection in an endemic situation. These are three of the reasons to incorporate age as a variable characterising individuals in models for the spread of contagious diseases, notably in human communities. We refer to Chapter 7 for elaboration.

Still other aspects of heterogeneity are the following. Space is neither homogeneous nor isotropic (winds, mountains!). Contact rates change during the year, for example as a result of the school system. And weather conditions influence the transport of aerosols and the survival of a free virus. We shall pay some, but not much, attention to such aspects.

2.3 Heterogeneity: a preliminary conclusion

Individuals of the host population show variation with respect to properties that are relevant for the transmission of the agent. These properties may affect susceptibility, infectivity, or both.

When every individual is equally susceptible, we can, in a deterministic setting (i.e. for large numbers of all categories of individuals involved), simply work with the *mean* infectivity. When the susceptibility differs, but a certain form of independence holds between susceptibility and infectivity, we can still compute the appropriate averages directly. More precisely, this can be done when the properties of two individuals independently influence the probability of contact and transmission. This was demonstrated by way of example in Section 2.2 and will be covered in detail in Section 5.3.

How to handle the case of dependence cannot be explained at this stage. We simply refer to Chapter 5.

3

Dynamics at the demographic time scale

3.1 Repeated outbreaks versus persistence

If population turnover is slow relative to the transmission of infection, we reach almost the final size of the epidemic in a closed population before the gradual inflow of new susceptibles has any effect. When an agent has struck a virgin (or naive) population, the susceptible fraction of the population is then of the order of $\exp(-R_0)$, and it will therefore take a long time before susceptibles will constitute a substantial fraction of the population again. During this period there are so few infectives that demographic stochasticity will lead to extinction of the infective agent. When, after a possibly long time, the population of susceptibles is above threshold again, re-introduction of the agent from outside leads to another epidemic. Thus we expect to see recurrent outbreaks with fade-out with irregular periods in between.

Data about influenza and on Iceland show exactly this pattern (see Cliff & Haggett (1988)). (In some cases it was even possible to trace the ship that carried the infected sailor who triggered a specific measles epidemic!) But data about measles in New York show a different pattern: large outbreaks every two years with low but non-zero incidence in the years between outbreaks.

There are many ways in which Iceland and New York differ. Two relevant ways seem to be (i) the isolation from the 'outside world' and (ii) the population size. How do such factors influence the probability that a virus goes extinct after a large outbreak? Is there a *critical community size* for virus ? (Note that $N \exp(-R_0)$ may still be reasonably large if total population size is large; a low density over a large domain may yield an appreciable number.) Or is geographical expanse the key point? If local epidemics are out of phase then the proneness to global extinction may be much smaller, cf. metapopulation models in ecology (Gilpin & Hanski (1991); Hanski & Gilpin (1997)). Note, in addition, that whether or not we find a 'fade-out followed by re-introduction' in the data, may depend on the somewhat arbitrary geographical division of the public health administration. When trying to analyse the wealth of available measles data (Grenfell & Harwood (1997); Grenfell et al. (1995)),[1] one is

---

[1] B.T. Grenfell & J. Harwood: (Meta)population dynamics of infectious diseases. *TREE*, **12** (1997), 395-399; B.T. Grenfell, B.M. Bolker & A. Kleczkowski: Seasonality, demography and the dynamics of measles in developed countries. In: Mollison (1995), pp. 248-268.

forced to include other complicating factors, such as age structure and seasonality (both the weather and the school system).

The question 'will the agent go extinct after the first outbreak?' cannot be answered within the context of a deterministic description. So we would like to be able to switch back to a stochastic description at the end of the epidemic outbreak. While it is well known how to calculate the probability of extinction from a branching process in a constant environment (where we can work within a generational perspective; see Section 1.2.2), it seems difficult to do so when environmental quality (from the point of view of the agent, i.e. the presence of susceptibles!) is improving linearly at a certain rate. We are not aware of any work in this direction. In fact we know only one paper[2] in which the relevant probability is calculated for a stochastic version of the Kermack-McKendrick ODE model of Exercise 1.21. The calculation is based on approximate solutions of the Fokker-Planck equation and constitutes an ingenious piece of work. It is to be hoped that this will trigger more work in this direction, concentrating on other models and different methods such that in the end a robust picture emerges.

In stochastic models in which the number of individuals cannot grow without bound, the endemic state can only be *quasi-stationary*, which means that it may exist for a long period of time. Ultimately, however, a rare combination of chance events will drive the infective agent to extinction, which is an 'absorbing' state from which return is only possible by the deus ex machina of re-introduction from outside. The expected time until extinction is an important quantity, since it tells us on what time scale the quasi-stationary state is a reasonable description. See Nåsell (1995) and Grasman & van Herwaarden (1999) for efficient methods for the computation of this quantity.[3]

In the next section we will at first simply forget about fade-out and concentrate on the dynamical behaviour near the endemic steady state, in which the inflow of new susceptibles is balanced by the incidence (and by death). We take the rate at which newborn susceptibles are added to the population as a given constant (we ignore immunity derived from maternal antibodies). This means that, on the time scale considered, demography can influence infection dynamics, but not vice versa. In a short heuristic interlude (Interlude 3.7) we shall then comment on the concept of critical community size and its relation with the ratio of the time scales of demography and transmission of infection.

In Section 3.3 we shall investigate the influence of disease on demography: can the infective agent regulate the host population? There we consider a population that, on the time scale considered, grows exponentially in the absence of the agent, and we ask to what extent the growth rate of the host is affected by the agent.

We end this section by noting a paradox for transmission situations in which the standard final-size equation (1.11) applies: for the infective agent, a virgin host population is both a best and a worst case. It is best since it gives the highest initial growth rate. It is worst since the probability of extinction after the outbreak is highest. Indeed, as Figure 1.3 and Exercise 1.19 show, $s(\infty)$ is a decreasing function of R_0 and

[2] O.A. van Herwaarden: Stochastic epidemics: the probability of extinction of an infectious disease at the end of a major outbreak. *J. Math. Biol.*, **35** (1997), 793-813.

[3] I. Nåsell: The threshold concept in stochastic epidemic and endemic models. In: Mollison (1995), pp. 71-83; J. Grasman & O.A. van Herwaarden: *Asymptotic Methods for the Fokker-Planck Equation and the Exit Problem in Applications.* Springer-Verlag, Berlin, 1999.

when only a fraction p of the host population is susceptible, one has to replace R_0 by pR_0 in equation (1.11). We encourage the reader to do some computations to ascertain the quantitative effect, which is quite substantial.

3.2 Fluctuations around the endemic steady state

To describe demographic turnover in the absence of any infection, we use the caricature

$$\frac{dS}{dt} = B - \mu S, \tag{3.1}$$

where S denotes susceptibles, B the population birth rate and μ the per capita death rate. So life expectancy is μ^{-1} and the population stabilises at the size

$$N := B/\mu \tag{3.2}$$

at which inflow and outflow match.

To model the spread of the agent, we use the Kermack-McKendrick ODE model of Exercise 1.21. In combination with (3.1), this yields

$$\begin{aligned}
\frac{dS}{dt} &= B - \beta SI - \mu S, \tag{3.3} \\
\frac{dI}{dt} &= \beta SI - \mu I - \alpha I,
\end{aligned}$$

to which we could add an equation for the removed (which we think of as immunes, such as for relatively innocent children's diseases, in order not to have to model how contact intensity changes with population size):

$$\frac{dR}{dt} = -\mu R + \alpha I.$$

But since system (3.3) is a closed system (i.e. R does not appear on the right-hand side), we can disregard R when doing our analysis.

Exercise 3.1 Show that the 'virgin' (or 'infection-free')state $(S, I) = (N, 0)$ with $N = B/\mu$ is stable if and only if $R_0 < 1$ where

$$R_0 = \frac{\beta N}{\alpha + \mu}. \tag{3.4}$$

Exercise 3.2 Show that in an endemic steady state $(S, I) = (\overline{S}, \overline{I})$ with $\overline{I} > 0$ necessarily

$$\frac{\overline{S}}{N} = \frac{1}{R_0}. \tag{3.5}$$

Does this surprise you? (If so, return to Exercise 1.21-vi and consider that the same argument now applies to minima of I at which S is increasing; in more mathematical terms, verify that $S = \frac{N}{R_0}$ is an isocline (also called nullcline by many authors) and draw a picture of how orbits may proceed through the (S, I) plane.) Reflect upon the possibilities of estimating R_0 from endemic steady-state data.

Exercise 3.3 Show that in an endemic steady state

$$\bar{I} = \frac{\mu}{\beta}(R_0 - 1) \tag{3.6}$$

and that consequently such a state exists if and only if $R_0 > 1$. (Recall the requirement $\bar{I} > 0$!). Note that the equivalent formula

$$\frac{\bar{I}}{N} = \frac{(\alpha + \mu)^{-1}}{\mu^{-1}}\left(1 - \frac{\bar{S}}{N}\right)$$

expresses the relative steady-state incidence in terms of measurable quantities: the life expectancy μ^{-1}, the expected length of the infectious period $(\alpha + \mu)^{-1}$, and the steady-state fraction of susceptibles $\bar{S}/N$.

Existence of a steady state does not guarantee that the balance between inflow of new susceptibles and the combined effect of infection and death is actually exact at every instant. There has to be a balance, but it may be over a longer time interval. Fluctuations around the steady state are not necessarily damped. In order to find out what happens in the present model, we linearise around the steady state. The linearised system has solutions that depend on time through a factor $\exp(\lambda t)$ for special values of λ. When λ is real, this gives information about growth or decay rates. When λ is complex, the real part determines the growth or decay rate, whereas the imaginary part determines the frequency of the oscillations that accompany the growth or decay. The principle of linearised stability guarantees that, provided the real parts of the λs are non-zero, the information about solutions of the linearised system carries over to solutions of the nonlinear system, as long as these stay in a small neighbourhood of the steady state (see e.g. Hirsch & Smale (1974)). The last proviso matters only if we find instability, i.e. growing exponentials; in that case we can only conclude that there are solutions that leave a given neighbourhood of the steady state and not that the distance to the steady state keeps increasing exponentially, since further from the steady state quadratic and higher-order terms matter much).

The linearised system is fully characterised by a matrix M and the λs are precisely the *eigenvalues* of M, which can be found by solving the *characteristic equation* $\det(\lambda I - M) = 0$, which is a polynomial in λ of degree n, where n is the dimension of the system. In the two-dimensional case the characteristic equation reads

$$\lambda^2 - T\lambda + D = 0 \tag{3.7}$$

where T is the *trace*, i.e. the sum of the diagonal elements $T = m_{11} + m_{22}$, and D the *determinant*, i.e. $D = m_{11}m_{22} - m_{12}m_{21}$, of the matrix $M = (m_{ij})_{1 \leq i,j \leq 2}$. It follows at once from the explicit formula

$$\lambda = \frac{T \pm \sqrt{T^2 - 4D}}{2}$$

that

$$T < 0 \text{ and } D > 0 \tag{3.8}$$

is the condition for linearised stability (i.e. decaying exponentials) and that

$$T^2 < 4D \tag{3.9}$$

is the condition for an oscillatory approach to the steady state.

Exercise 3.4 Show that the matrix

$$\begin{pmatrix} -\beta\bar{I} - \mu & -\beta\bar{S} \\ \beta\bar{I} & 0 \end{pmatrix}$$

corresponds to the linearisation of system (3.3) around the endemic steady state. Deduce that the endemic steady state is stable.

Exercise 3.5 Show that the characteristic equation can be written as

$$\left(\frac{\lambda}{\mu}\right)^2 + R_0\frac{\lambda}{\mu} + \left(1 + \frac{\alpha}{\mu}\right)(R_0 - 1) = 0,$$

where each term is dimensionless. Consider the situation where the life expectancy μ^{-1} is much bigger than the expected duration α^{-1} of the infectious period. Show that, unless R_0 is only slightly above 1, the model predicts damped oscillations around the steady state, with relaxation time $2/\mu R_0$ (this means, by definition, that in a time interval of length $2/\mu R_0$ the amplitude diminishes by a factor e^{-1}) and approximate frequency $\sqrt{\mu\alpha(R_0 - 1)}$. The approximate period is $2\pi/\sqrt{\mu\alpha(R_0 - 1)}$. Use once more that $\mu \ll \alpha$, to deduce that the relaxation time is much longer than the period and that, consequently, we can expect to see many oscillations before the steady state is reached.

Exercise 3.6 In steady state the force of infection (recall that this is the probability per unit of time for a susceptible to become infected) is a constant, say Λ. Show that the *mean age at infection* $\bar{a}$ is given by

$$\bar{a} = \frac{1}{\mu + \Lambda} = \frac{1}{\mu R_0}. \tag{3.10}$$

Hint: The safe way is to write out the probability density function for exit from the susceptible state, while conditioning on not dying. A short cut is obtained by arguing that $\bar{a}$ = expected sojourn time in the susceptible state (without any condition), since exits to 'death' and to 'infectious' occur throughout in the same fixed proportion (determined by μ and Λ). Finally, observe that $\Lambda = \beta\bar{I} = \mu(R_0 - 1)$.

Combining the results of the last two exercises, we see that we can give a rather complete description in terms of three *observable* quantities, all with the dimension of time: the life expectancy $L = 1/\mu$, the expected duration of the infectious period $1/\alpha$ and the mean age at infection $\bar{a}$. When $1/\mu \gg 1/\alpha$, the relaxation time for approach to the stable endemic state equals $2\bar{a}$, while the period of oscillation equals $2\pi\sqrt{\bar{a}/\alpha}$ (which involves the geometric mean of the two time scales $\bar{a}$ and $1/\alpha$). Here we assume that R_0 is big enough for the difference between μR_0 and $\mu(R_0 - 1)$ to be negligible.

As an important side-remark we note that (3.10) can be rewritten in the form

$$R_0 = \frac{L}{\bar{a}},$$

which indicates a possibility to estimate R_0 from data (a second possibility for endemic infections; cf. Exercise 3.2). See also Exercise 7.13-ii.

Data about measles from many towns and regions show *sustained*, rather than damped, oscillations. So we are naturally led to the question: what is missing in the present model? Various possibilities present themselves. Stochastic effects may enhance the deterministic fluctuations (Bartlett, 1960). Weather conditions may influence the probability of transmission and make β periodic (see Kuznetsov & Piccardi (1994)[4] and the references given there). Age structure may necessitate the use of a more complex model, including seasonal effects of the school system.[5]

Interlude 3.7. On critical community size and time-scale differences

From Bartlett (1960), p. 66, we quote:

> 'This phenomenon [the time to fade-out of infection increasing rapidly with average weekly notifications] suggests that there will be a critical community size, above which measles should tend to maintain itself, whereas for smaller communities it will die out and require reintroducing from outside before another epidemic can materialise.'

Here we attempt to elaborate this idea in the context of a model, rather than in the context of actual data. The heuristic arguments presented in this interlude were catalysed by discussions with K. Dietz, I. Nåsell, H. Andersson, T. Britton, D. Rand and A. Martin-Löf during a meeting on the Isle of Skye organised by D. Mollison and V. Isham in 1997. Yet they present the personal view of the authors, which is not necessarily shared by (all of) the persons named (or indeed persons not named).

Within reasonable families of stochastic models the agent will go extinct with certainty. So one cannot define critical community size on the basis of the extinction criterion alone, the expected time till extinction has to come in. However, the expected time till extinction of the agent will, other things being equal, be an increasing function of population size N, and this gives no clue.

Somewhat arbitrarily, one may choose a positive number T and a number p between 0 and 1 and declare that the population size is above criticality if the probability of extinction before time T is less than p (of course, this probability depends on the initial condition; as explained in detail in Nåsell (1999)[6], and references therein, it makes sense to represent the initial condition by the so-called quasi-stationary distribution). The number of constants that have to be chosen may be reduced from two to one by considering limits. Limits are always mathematical idealisations, while reality is finite and discrete. Yet limits often provide information and insight.

One would like to be able to define critical community size by concentrating on the limit $N \to \infty$. However, then the expected time till extinction simply tends to infinity, taking astronomical values of the order of $\exp(cN)$ for N large. It is therefore impossible to define 'critical community size' by considering just this limit.

What we do instead is to consider 'going to infinity' in a two-parameter plane, spanned by the N axis and the α/μ axis (ratio of two time scales). We already considered what happens along the N axis. Along the other axis, for α/μ tending

---

[4] Y. Kuznetsov & C. Piccardi: Bifurcation analysis of periodic SEIR and SIR epidemic models. *J. Math. Biol.*, **32** (1994), 109-121.

[5] D. Schenzle: An age-structured model of pre- and post-vaccination measles transmission. *IMA J. Math. Appl. Med. Biol.*, **1** (1984), 169-191.

[6] I. Nåsell: On the time to extinction in recurrent epidemics. *J. Roy. Stat. Soc.*, B, **61** (1999), 309-330.

to infinity, we find totally different behaviour, viz. instantaneous extinction (that is, after the first outbreak). The next idea is to look for a 'phase transition', i.e. a way of approaching infinity in this plane such that the expected time till extinction neither blows up nor goes to zero but stays bounded and bounded away from zero. Our calculations are a heuristic way of determining the paths in the plane that provide the transition: $\frac{\alpha}{\mu}\frac{1}{\sqrt{(N)}}$ should be bounded (see below).

How does one apply this to data? The usual setting is to consider one agent and many different populations (with various sizes N), but in principle one could just as well consider one host population (i.e. one particular N) and ask which agents persist. In both cases, however, the difficulty is that 'bounded' is rather unspecific about the bound. So here it avenges itself in that we cannot really take the limit and we have to make an arbitrary choice for a constant (which is reminiscent of the choice of a constant in the work of I. Nåsell). So we 'define' $\frac{\alpha}{\mu}\frac{1}{\sqrt{(N)}} = C$ as the critical relationship, determine α/μ, choose C to be one and compute N_{crit}.

Our starting point for calculations is the model behind the equations (3.3), which we now, however, write as

$$\frac{dS}{dt} = \mu N - \mu S - \gamma\frac{SI}{N},$$

$$\frac{dI}{dt} = -\mu I + \gamma\frac{SI}{N} - \alpha I.$$

That is, we use (3.2) to write $B = \mu N$ and, thinking now in terms of numbers rather than densities, we have put $\beta = \frac{\gamma}{N}$ to explicitly account for dependence on population size. The steady-state value for I is given by

$$\bar{I} = \frac{\mu(N - \bar{S})N}{\gamma\bar{S}} = \frac{\mu N}{\mu + \alpha}\left(1 - \frac{\bar{S}}{N}\right) = \frac{\mu}{\mu + \alpha}N\left(1 - \frac{1}{R_0}\right),$$

since $\bar{S} = \frac{\mu + \alpha}{\gamma}N = \frac{N}{R_0}$.

For birth-death processes of various kinds, it is a well-established phenomenon that demographic stochasticity leads to fluctuations of the order of $\sqrt{N}$, with N the population size.[7] Assume $R_0 = O(1)$, by which we mean that changes in N do not substantially influence R_0. Also assume that $\frac{\mu}{\mu+\alpha} = O(\frac{1}{\sqrt{N}})$. By this we do not mean that $\frac{\mu}{\mu+\alpha}$ should depend on N, but we interpret the assumption in the context of taking limits in the $\frac{\alpha}{\mu}$ versus N plane, as above, where it serves to keep the extinction time $t_e = t_e(\frac{\alpha}{\mu})$ bounded from above and bounded away from zero. These assumptions were chosen such that they imply that $\bar{I} = O(\sqrt{N})$, and one sees that consequently the average level of infected individuals in the population lies within the range of natural (i.e. not very rarely occurring) fluctuations. Extinction of the agent is therefore to be expected sooner or later (and extinction does not require a rare combination of events that has e^{+vN} (for some v) as its natural time scale). Now when $\mu \ll \alpha$ we have that $\frac{\mu}{\mu+\alpha} \approx \frac{\mu}{\alpha}$ and our assumption boils down to assuming that $\frac{\alpha}{\mu} = O(\sqrt{N})$. Since $R_0 = \frac{\gamma}{\mu+\alpha} = O(1)$, this requires that also $\frac{\gamma}{\mu} = O(\sqrt{N})$.

---

[7] See Nisbet & Gurney (1982), Goel & Richter-Dyn (1974), Taylor & Karlin (1984).

We conclude that there is not so much a critical community *size*, but rather a critical relationship between population size and the ratio of the two time scales involved (that of demography and that of transmission). When both $\frac{1}{\sqrt{N}}\frac{\alpha}{\mu}$ and $\frac{1}{\sqrt{N}}\frac{\gamma}{\mu}$ are very small, we expect a single outbreak; when they are both very large we expect an endemic situation; everything in between could be called critical. Heuristic as these considerations may be, we think they may be enlightening.

A quantitative illustration (a bit dangerous, it is the only one in the book): for measles in only moderately developed countries we may take $\mu = \frac{1}{50}, \alpha = 25 \, and R_0 = 20$ to find

$$\sqrt{N_{crit}} = \frac{1}{1 + \frac{\alpha}{\mu}} N_{crit} \left(1 - \frac{1}{R_0}\right) = \frac{19}{50020} N_{crit},$$

from which we deduce that $N_{crit} \approx 1.65$ million. (A point to note is that we set up the simple caricatural in such a way that $I(t)$ becomes steady. If, for instance due to seasonality, $I(t)$ fluctuated, it is the troughs of the fluctuations that yield relatively high probabilities of fade-out (as pointed out by B. Grenfell). This phenomenon may have a substantial impact on N_{crit}, yet we have ignored it in our calculation.)

See Nåsell (1999), loc. cit., and the references given there for more information. We end by quoting from Nåsell (1999) the approximation formula

$$\bar{t}_{\text{extinction}} \approx \frac{(R_0 - 1)N}{2(\frac{\alpha}{\mu})^2} \frac{1}{\mu}$$

for the expected time until extinction under critical conditions. Note that the right-hand side is $O(\frac{1}{\mu})$ when $\frac{\alpha}{\mu}$ is $O(\sqrt{N})$, which is completely in line with our conclusions above.

End of Interlude 3.7

We conclude this section with several exercises dealing with generalisations, variations on the theme, different aspects, etc. The first intends to demonstrate quantitatively that our estimate of the period is in fact not bad at all.

Exercise 3.8 For measles in pre-vaccination Western Europe and the USA one estimates $\bar{a}$ as somewhere between 4 and 5 years, while $1/\alpha$ is approximately 12 days. Compute $2\pi\sqrt{\bar{a}/\alpha}$ and compare the result with the observed period of two years.

Our next two exercises are intended to test the robustness of our conclusions by investigating the influence of minor modifications to the basic model.

Exercise 3.9 Consider the SEIR system (cf. Exercise 2.2) with demographic turnover described by

$$\begin{aligned}
\frac{dS}{dt} &= B - \beta SI - \mu S, \\
\frac{dE}{dt} &= \beta SI - \mu E - \theta E, \\
\frac{dI}{dt} &= -\mu I + \theta E - \alpha I.
\end{aligned} \tag{3.11}$$

i) Recall that now $R_0 = \frac{\theta}{\theta+\mu} \frac{\beta N}{\alpha+\mu}$ (Exercise 2.4-iv).

ii) Show that in the endemic steady state $\overline{S}/N = 1/R_0$, $\overline{I} = \mu(R_0 - 1)/\beta$ and $\overline{E} = \frac{\alpha+\mu}{\theta}\overline{I}$.

iii) The linearised system is now described by the 3×3 matrix

$$\begin{pmatrix} -(\beta\overline{I} + \mu) & 0 & -\beta\overline{S} \\ \beta\overline{I} & -(\mu+\theta) & \beta\overline{S} \\ 0 & \theta & -(\mu+\alpha) \end{pmatrix}$$

and the eigenvalues are the roots of the characteristic equation

$$\lambda^3 + (\mu R_0 + 2\mu + \alpha + \theta)\lambda^2 + \mu R_0(\alpha + 2\mu + \theta)\lambda + \mu(R_0 - 1)(\alpha + \mu)(\theta + \mu) = 0.$$

Algebraically inclined readers are invited to check these statements, while others are asked to believe them.

iv) When both α and θ are large relative to both μ and μR_0, roots of the characteristic equation should lie close to roots of the simplified equation

$$\lambda^3 + (\alpha + \theta)\lambda^2 + \mu R_0(\alpha + \theta)\lambda + \mu(R_0 - 1)\alpha\theta = 0,$$

which we can rewrite as

$$\lambda^3 + (\alpha + \theta)\left(\lambda^2 + \mu R_0\lambda + \mu(R_0 - 1)\frac{\alpha\theta}{\alpha + \theta}\right) = 0.$$

In Anderson & May (1991), Appendix C, p. 668, it is concluded that this cubic equation has one root $\lambda \approx -(\alpha + \theta)$ (corresponding to perturbations that decay rapidly) and two other roots given approximately by the roots of the quadratic equation between braces. Thus one finds that now the period of the oscillations is given by $2\pi\sqrt{\overline{a}\frac{\alpha+\theta}{\alpha\theta}}$, or, in other words, that $1/\alpha$ has to be replaced by $\frac{\alpha+\theta}{\alpha\theta} = \frac{1}{\alpha} + \frac{1}{\theta}$, which is still the expected duration of 'infection', in the sense of the period between being infected and becoming immune.

In order to sustain this claim by formal asymptotics, one has to consider the limit $\mu \to 0$. Readers who like to do asymptotic calculations are invited to derive these approximations for the roots.

Exercise 3.10 Let us now consider a model in which expected infectivity is described by a general integral kernel A. We still take

$$\frac{dS}{dt} = B - \mu S - \Lambda S,$$

with $\Lambda(t)$ the force of infection at time t, but we assume that

$$\Lambda(t) = \frac{c}{N} \int_0^\infty A(\tau)\Lambda(t - \tau)S(t - \tau)\, d\tau$$

(to understand this expression, note that $\Lambda(t-\tau)S(t-\tau)$ is the incidence τ units of time before the current time t, so that the individuals infected then, now have infection-age τ).

i) Recall that $R_0 = c \int_0^\infty A(\tau) \, d\tau$ (see Section 2.1).

ii) Use the equation for Λ to deduce that in an endemic steady state necessarily $\overline{S}/N = 1/R_0$.

iii) Use the variation-of-constants formula (see any book on ODE, for instance Hale (1969), or the first paragraph of the elaboration) to rewrite the differential equation for S as the integral equation

$$S(t) = B \int_{-\infty}^t e^{-\mu(t-\tau)} \exp\left(- \int_\tau^t \Lambda(\sigma) \, d\sigma\right) d\tau$$

or, equivalently,

$$S(t) = B \int_0^\infty e^{-\mu\sigma} \exp\left(- \int_{t-\sigma}^t \Lambda(s) \, ds\right) d\sigma.$$

iv) Show that in an endemic steady state $\overline{\Lambda} = \mu(R_0 - 1)$ and compare this expression with the expression (3.6) for $\overline{I}$.

v) Write $S(t) = \overline{S} + x(t)$ and $\Lambda(t) = \overline{\Lambda} + y(t)$ and derive the linearised system of integral equations

$$
\begin{aligned}
x(t) &= -\frac{N}{R_0} \int_0^\infty e^{-\mu R_0 \sigma} y(t - \sigma) \, d\sigma, \\
y(t) &= \frac{c}{R_0} \int_0^\infty A(\tau) y(t - \tau) \, d\tau + \frac{c}{N} \mu(R_0 - 1) \int_0^\infty A(\tau) x(t - \tau) \, d\tau.
\end{aligned}
$$

Verify that this system has solutions of the form

$$\begin{pmatrix} x(t) \\ y(t) \end{pmatrix} = e^{\lambda t} \begin{pmatrix} x_0 \\ y_0 \end{pmatrix}$$

if and only if λ is a root of the characteristic equation

$$1 = c \frac{\lambda + \mu}{R_0(\mu R_0 + \lambda)} \overline{A}(\lambda),$$

where

$$\overline{A}(\lambda) := \int_0^\infty e^{-\lambda \tau} A(\tau) \, d\tau$$

or, in words, $\overline{A}$ is the Laplace transform of A.

vi) Consider the characteristic equation for $R_0 = 1$. Show that $\lambda = 0$ is a root and that all other roots lie in the left half-plane, i.e. have negative real part. Hint: Use the non-negativity of A.

vii) Use the fact that the roots depend continuously on parameters to deduce that for R_0 slightly bigger than 1, all roots lie in the left half-plane.

viii) Convince yourself that roots can only enter the right half-plane by crossing the imaginary axis. In other words, make plausible that roots cannot enter the right half plane at infinity.

ix) Show that for $R_0 > 1$ the characteristic equation cannot have a root on the imaginary axis. Hint: Take the modulus of both sides of the characteristic equation and use that $c|\overline{A}(i\omega)| \le R_0$ and $|(i\omega + \mu)/(i\omega + \mu R_0)| < 1$ for $R_0 > 1$.

x) Accepting that the principle of linearised stability holds for these systems of integral equations and that the growth or decay of solutions of the linear system is completely determined by the position of the roots relative to the imaginary axis, conclude that the endemic steady state is locally asymptotically stable for *every* non-negative and integrable kernel A. (See Diekmann et al. (1995)[8] for justification of the assumptions about linearised stability and about the exponential decay of all solutions of the linearised problem when all roots of the characteristic equation lie in the left half-plane.) We conclude that the model predicts *damped* oscillations and that therefore other mechanisms are responsible for the observed sustained oscillations.

Exercise 3.11 Returning to the basic model (3.3), show that the endemic steady state is in fact *globally* asymptotically stable.

Hint: consider the Lyapunov function

$$V(S, I) = S - \overline{S} \ln S + I - \overline{I} \ln I.$$

Verify that

$$\frac{dV}{dt} = \frac{\partial V}{\partial S} \frac{dS}{dt} + \frac{\partial V}{\partial I} \frac{dI}{dt} = -\frac{\mu N}{S \frac{N}{R_0}} \left(S - \frac{N}{R_0} \right)^2.$$

Check that on the line $S = \frac{N}{R_0}$, at which $\frac{dV}{dt} = 0$, the maximal invariant subset is precisely the endemic steady state. Conclude that all orbits converge to this point. See Hale (1969), Chapter X for background information.

Exercise 3.12 The so-called *microcosm principle* (cf. Mollison (1995)[9]) asserts that for a quite general population process in steady state, the fraction of the population π_j in state j is proportional to the mean time τ_j an individual spends in that state. Hence $\pi_j = \tau_j/L$, where L denotes life expectancy, so $L = 1/\mu$. Apply this principle to both the basic model described by (3.3) and the SEIR model of Exercise 3.8, and verify the correctness.

Exercise 3.13 Check that for the basic model the identity

$$\frac{\text{incidence}}{\text{prevalence}} = \frac{1}{\text{infectious period}}$$

holds for the endemic steady state. Reflect upon the possibility to estimate the infectious period from data at the population level.

Exercise 3.14 Suppose we vaccinate a fraction q of all newborns, i.e. we modify the inflow term in the equation for dS/dt to $(1 - q)B$. Assume that q is too small

---

[8] O. Diekmann, S.A. van Gils, S.M. Verduyn Lunel & H.-O. Walther: *Delay Equations: Functional-, Complex-, and Nonlinear Analysis.* Springer-Verlag, Berlin, 1995.

[9] D. Mollison: The structure of epidemic models. In: Mollison (1995), pp. 17-33.

to achieve eradication (i.e. $q < 1 - \frac{1}{R_0}$). What happens to $\overline{S}/N$? What happens to the force of infection and the mean age at first infection? (What are the repercussions for a disease like rubella, which is rather innocent except when contracted during pregnancy.) What happens to the relaxation time and the period of the oscillations near the stable endemic steady state?

Exercise 3.15 If a disease is lethal with high probability, would a model with *constant* inflow B of new susceptibles make any sense?

(The question is perhaps less rhetoric than it may seem at first sight: think of HIV and the homosexual subpopulation. In that case, however, one may want to incorporate behavioural reactions to the level of prevalence.) (Also see Exercise 3.20).

Exercise 3.16 Formulate and analyse a model in which births are all concentrated in a very short period once a year, which we describe somewhat caricaturally as an event taking place at the integer values of time t. In a sense the model is hybrid, in that it combines continuous-time and discrete-time features. To investigate the analogue of the endemic steady state, we have to adopt a stroboscopic way of monitoring the population; that is, we look only at the values of S, I, etc. immediately after (or, if you prefer, before) each birth pulse.

Exercise 3.17 Formulate and analyse a simple SIS model for a closed population. For a disease like gonorrhea, in which one returns to the class of susceptibles immediately after treatment, one can assume that those leaving I go to S. For other diseases an SIRS formulation in which immunes have a certain fixed probability per unit of time of losing immunity may be more appropriate.

Exercise 3.18 Consider a disease for which *vertical transmission* (i.e. from mother to foetus) cannot be ignored. How would you modify the basic model (3.3)? Analyse the modified system. (Among other things, calculate R_0.)

Exercise 3.19 Consider a host population in which infective agent 1 is in steady state. Now let infective agent 2 enter the population. Assume cross-immunity (i.e. hosts that have ever been infected by agent 1 cannot become infected by agent 2, and vice versa). Show that agent 2 will increase if and only if $R_0^2 > R_0^1$ and conclude that, given our assumptions, natural selection will tend to increase the R_0 of the agent.

In general, one expects a trade-off between infectiousness (a component of β) and the length of the infectious period (as determined by α). A very virulent agent incurs a high death rate of the host or at least a strong reaction of the immune system. To model this, we consider β as a 'free' parameter under natural selection, while α is constrained as depicted in Figure 3.1.

Devise a graphical procedure to find the 'uninvadable' value of β, i.e. the value of β at which R_0 is maximal. See if you agree with the following statement: intermediate virulence is favoured by natural selection, since the price for high infectiousness is a very short infectious period.

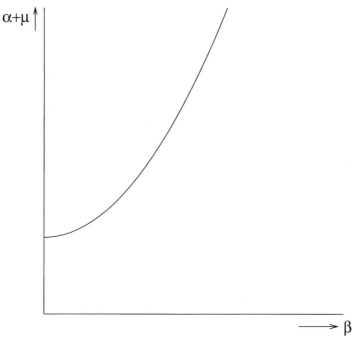

Figure 3.1

Exercise 3.20 Greenwood and co-workers $(1936)^{10}$ carried out experiments with mice infected with *Pasteurella muris*. Some mice die from the clinical effects of this infection and some recover to become temporarily immune. The parameter B was experimentally tuned to 6, 3, 2, 1 or $\frac{1}{2}$ mice being added every day (the last value meaning, of course, one mouse every two days). Every day, except on Sunday, the mice were transferred to sterilised and cleaned cages while dead mice were removed. When transferring the mice, the number of cages was adjusted so as to keep the number of mice per unit area constant. All cages were connected so as to form one large cage. Reproduction did not take place under the circumstances. When B was held fixed at a certain value, the total number N of mice settled to a rather constant level after a while. The value of N was monitored as a function of B.

In this exercise we want to address two related questions: how should we model this situation and what information can we deduce from the experimental data?

i) Would you model the incidence term by βSI or by $\gamma \frac{SI}{N}$? To compare the implications of each, we shall analyse both. To make the calculations relatively easy, we shall ignore the possible loss of immunity.

ii) Show that for the system

$$\frac{dS}{dt} = B - \beta SI - \mu S,$$

---

[10] M. Greenwood, A.T. Bradford Hill, W.W.C. Topley & J. Wilson: *Experimental Epidemiology.* MRC Special Report Series **209**, HMSO, London, 1936.

$$\frac{dI}{dt} = \beta SI - \mu I - \alpha I,$$

$$\frac{dR}{dt} = -\mu R + f\alpha I,$$

we have

$$\overline{N} = \overline{S} + \overline{I} + \overline{R} = \frac{1 + \frac{f\alpha}{\mu}}{\mu + \alpha}B + (1 - f)\frac{\alpha}{\beta}$$

which is, as a function of B, a straight line with positive intercept of the N axis.

(Digest the following statement as a test for your understanding: for $B < \frac{\mu(\alpha+\mu)}{\beta}$ we have $R_0 < 1$ and we expect to find experimentally that N settles at the infection-free value B/μ. So in that range the expression above refers to a mathematical extrapolation, not to actual measurements).

iii) Show that for the system

$$\frac{dS}{dt} = B - \gamma\frac{SI}{N} - \mu S,$$

$$\frac{dI}{dt} = \gamma\frac{SI}{N} - \mu I - \alpha I,$$

$$\frac{dR}{dt} = -\mu R + f\alpha I,$$

with $N = S + I + R$, we have in steady state that

$$\overline{N} = \frac{1 + \frac{f\alpha}{\mu}}{(1 - \frac{1-f}{\gamma}\alpha)(\mu + \alpha)}B,$$

which is a straight line through the origin.

iv) Conclude that in principle the data could be used to falsify one of the alternatives, by fitting a straight line and finding whether or not it passes through the origin. Unfortunately this test was inconclusive. See de Jong et al. (1995).[11] There, a slightly different variant is analysed that, in particular, includes return to the susceptible class by loss of immunity. In that reference one can also find other data that point more clearly to the system of iii) as the appropriate description.

Exercise 3.21 What vaccination effort is required to eradicate an infectious disease? How does this effort depend on the vaccination strategy? The aim of this exercise is to compare the strategy of successfully vaccinating (i.e. resulting in complete immunity) a certain fraction q of all newborns (cf. Exercise 1.36) with the so-called *pulse vaccination* strategy of vaccinating periodically a certain fraction p of all susceptible individuals (which presupposes that the public health system is capable of knowing who these are).

---

[11] M.C.M. de Jong, O. Diekmann & J.A.P. Heesterbeek: How does transmission of infection depend on population size? In: Mollison (1995), pp. 84-94.

When we have a constant population birth rate B and a constant per capita death rate μ, the population density stabilises at the level

$$N = \frac{B}{\mu}.$$

When vaccination is periodic, say with period T, the susceptible density may nevertheless fluctuate with that period.

If S is a given positive T-periodic function then the zero steady state of the linear differential equation

$$\frac{dI}{dt} = (\beta S - \mu - \alpha)I$$

is determined by whether the dominant Floquet multiplier (see Section 9.4, and Hirsch & Smale (1974))

$$e^{\beta \int_0^T S(t)\,dt - \mu T - \alpha T}$$

is greater or less than one. Hence it is determined by the sign of

$$\beta \int_0^T S(t)\,dt - \mu T - \alpha T.$$

To achieve stability of the infection-free steady state, we should have that

$$\frac{1}{T} \int_0^T S(t)\,dt < \frac{\mu + \alpha}{\beta} = \frac{N}{R_0}$$

i.e. the average value of S over the period should be below the inverse of the basic reproduction ratio. Define the vaccination effort as the average number of vaccinations per unit of time. For pulse vaccination at time intervals of length T, how does the effort required for elimination depend on T?

3.3 Regulation of host populations

From the point of view of the infective agent, the host population constitutes a renewable resource. Depending on the time scale considered and on the, perhaps implicit, incorporation of other factors, we can consider the birth term of the host population as zero, a constant or a per capita constant. The first is adopted when we want to model and understand an epidemic outbreak, the second when we want to study the dynamical balance of supply and 'consumption' of susceptibles and the third when we want to investigate the possible long-term influence of infectious diseases on population growth. So our own objectives matter. In this section we address the regulation problem and ask a number of questions. Under what conditions on the parameters does the infective agent: go extinct, grow but at a slower rate than the host, grow at the same rate as the host (and if so, how much is that rate reduced relative to the infection-free growth), or induce a steady host state (i.e. stop population growth), turn exponential host growth into exponential decline and eventual extinction?

But what does host population growth mean? In any case it means that numbers grow. But does it also entail that the density increases? Or is simultaneously the occupied area increased such that the density remains roughly constant (as in an expanding city). The key issue is of course whether or not the number of contacts per unit of time per individual increases or not. And if it increases, in what manner?

We shall start to consider the situation in which the per capita contact rate does not change. That is, we imagine a population that expands while growing so as to keep the density constant. The system

$$
\begin{aligned}
\frac{dS}{dt} &= bS + bR - \mu S - \gamma \frac{SI}{N}, \\
\frac{dI}{dt} &= -\mu I + \gamma \frac{SI}{N} - \alpha I, \\
\frac{dR}{dt} &= -\mu R \quad\quad + f\alpha I
\end{aligned}
\tag{3.12}
$$

with $N := S + I + R$, incorporates in addition the assumption that S and R individuals have a per capita birth rate b, while I individuals do not reproduce successfully. Note that the system is first-order homogeneous, i.e. if (S, I, R) is a solution then so is (kS, kI, kR) for any constant k. This makes exponential solutions feasible even though the system is nonlinear.

Quite in general it is worth the effort to change to relative quantities, i.e. fractions, when studying regulation problems. Therefore we define

$$
y = I/N, \quad z = R/N. \tag{3.13}
$$

Exercise 3.22 Verify that (3.12) decouples into the two-dimensional system

$$
\begin{aligned}
y' &= y\{\gamma(1 - y - z) - \alpha + \alpha(1 - f)y + b(y - 1)\}, \\
z' &= y(f\alpha + (1 - f)\alpha z) - bz(1 - y),
\end{aligned}
\tag{3.14}
$$

with a scalar equation for N appended,

$$
N' = \{b - \mu - (b + \alpha(1 - f))y\}N.
$$

We assume that $b > \mu$, so in the infection-free situation N grows exponentially with rate $b - \mu$. Before going into the analysis, we summarise the conclusions by plotting in Figure 3.2 the rate of growth of both host (full line) and parasite (dash-dotted line) as a function of the parameter γ which measures the infectivity.

At $\gamma = \gamma_0 = \alpha + \mu$ we have $R_0 = 1$. For $\gamma_0 < \gamma < \gamma_1$ the agent grows exponentially with rate $\gamma - \alpha - \mu$, but the host grows with the larger rate $b - \mu$, and the 'dilution' effect is that the proportion y of hosts infected still remains zero. At $\gamma = \gamma_1$ the two rates become equal, i.e. $\gamma_1 - \alpha - \mu = b - \mu$, so $\gamma_1 = b + \alpha$.

An alternative way to characterise γ_1 has the advantage that it carries over to more complicated situations. The idea is to compute the expected number of secondary cases, while discounting for the host population growth by using the proportion of the population as the 'currency', rather than absolute numbers. (Once again and in more detail: Consider a population of size $N(t) = N_0 e^{rt}$. Consider one individual infected at time zero. It constitutes a fraction N_0^{-1} of the population. Any secondary case produced by this individual after a time interval of length τ constitutes a fraction

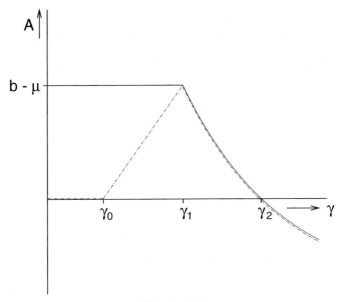

Figure 3.2

$N_0^{-1}e^{-r\tau}$ of the population. We want to determine whether or not the *fraction* of infecteds will grow. So we compute the ratio of fractions, which is $e^{-r\tau}$, and consider this as the appropriate weighing factor for the secondary case. Finally, integrate the product of the rate at which secondary cases are produced and this weighing factor with respect to τ and you get a number with threshold value one (for *relative* growth).) Thus we define

$$R_{0,\text{relative}} = \gamma \int_0^\infty e^{-(\alpha+\mu)\tau} e^{-(b-\mu)\tau} d\tau = \frac{\gamma}{b+\alpha}$$

(note that the first factor of the integrand is the probability to be still alive and infectious at infection-age τ, while the second derives from the fact that the host population has meanwhile increased by a factor $\exp(b-\mu)\tau$). At $\gamma = \gamma_1$ the $R_{0,\text{relative}}$ passes the threshold value one. For $\gamma > \gamma_1$ the relative prevalence y is positive; host and parasite grow at the same reduced rate $b - \mu - (b + \alpha(1 - f))\bar{y}$. Whether or not a further increase of γ leads to a reduction of the growth rate to below zero is determined by another quantity admitting a biological interpretation (introduced by Viggo Andreasen[12]). The expected number of offspring produced by an individual that is infected right at birth equals

$$\frac{\alpha}{\alpha + \mu} f \frac{b}{\mu}.$$

If this quantity is greater than one, the host population (and therefore the parasite population) will increase exponentially no matter how prevalent the infection is. If the quantity is less than one, increasing γ will inevitably lead to a situation in which the population will decline. Figure 3.2 is for the latter case.

---

[12] V. Andreasen: Disease regulation of age-structured host populations. *Theor. Pop. Biol.*, **36** (1989), 214-239.

The mathematical arguments underpinning these conclusions derive from a phase-plane analysis of the (y, z) system. For $R_{0,\text{relative}} < 1$, all solutions in the positive quadrant converge to the origin. For $R_{0,\text{relative}} > 1$, there exists a locally stable non-trivial steady state and the y coordinate $\bar{y}$ of this steady state increases monotonically with the parameter γ, with limit

$$\bar{y}_\infty = \frac{\alpha + 2b - \sqrt{\alpha^2 + 4b\alpha f}}{2(b + (1 - f)\alpha)}$$

for $\gamma \to \infty$. So the population growth rate reduces from $b - \mu$ to the value $\frac{1}{2}\sqrt{\alpha^2 + 4b\alpha f} - \mu - \frac{1}{2}\alpha$, which is negative if and only if $b\alpha f/(\alpha + \mu)\mu < 1$. The critical value γ_c at which the growth rate changes sign is given explicitly by

$$(\alpha + \mu)(b + \alpha(1 - f))(\alpha + \mu - b\alpha f\mu^{-1})^{-1}.$$

Exercise 3.23 Perform the phase-plane analysis and check the calculations.

The conclusion that the host is possibly driven to extinction for high values of γ hinges upon the homogeneous expression for the incidence. For low values of N this form is in fact debatable (for instance, for certain species of whale, the number of animals in parts of their habitat is so low that finding a mate is a problem). The next exercises discuss an alternative, which is only asymptotically (for large population sizes) homogeneous.

Exercise 3.24 Consider the system

$$\frac{dS}{dt} = b_1 S + b_2 I - \mu S - \frac{\gamma SI}{K + N},$$

$$\frac{dI}{dt} = -\mu I + \frac{\gamma SI}{K + N} - \alpha I,$$

where now $N = S + I$. (To keep things simple, we have 'eliminated' the removed; you may think of a lethal disease.) Check that the behaviour as a function of γ shows exactly the same pattern as before, but that now a (locally stable) steady state exists for $\gamma > \gamma_c$. Derive the condition for the existence of γ_c and check its interpretation.

Exercise 3.25 Repeat the analysis for the system of the preceding exercise, but with the incidence term replaced by

$$\gamma C(N)\frac{SI}{N}.$$

What properties should the function $C(N)$ have? What is its interpretation? (We will return to this incidence in Section 10.2.)

Exercise 3.26 Consider the system

$$\frac{dS}{dt} = b_1 S + b_2 I - \mu S - \beta SI,$$

$$\frac{dI}{dt} = -\mu I + \beta SI - \alpha I.$$

i) Show that a steady state exists provided that $b_2/(\mu + \alpha) < 1$ (a condition which we hope now has a familiar interpretation). Show that this steady state is locally stable (at least).

ii) What happens when the condition does not hold?

Hint: Prove first that $S(t) \geq b_2/\beta$ for large t, next that $I(t) \to \infty$ for $t \to \infty$, and subsequently that $S(t) \to b_2/\beta$ for $t \to \infty$. Conclude that the agent reduces the host growth rate to $b_2 - \mu - \alpha$ and that, essentially, everybody spends their whole life in the class of infecteds.

We conclude that the parameter space is divided into regions according to three threshold criteria: one for the absolute growth of the agent, one for relative growth and one for stopping host growth. We have emphasised the biological interpretation of the various criteria, since these are robust, while explicit expressions are usually out of the question for more complicated models.

3.4   Beyond a single outbreak: summary

This chapter has focused on the influence of the demography on the persistence of an infective agent and, vice versa, on the influence of the agent on host population growth and persistence.

When new 'fuel' is provided through the replacement of immune individuals by newborn susceptibles, the infective agent may strike again at a later time or even persist and become endemic. Whether we observe repeated outbreaks or relatively small fluctuations around a steady endemic level depends on the temporal and spatial scale that we, the investigators of the system, choose to monitor and/or to model. And likewise it depends on the degree of isolation of the (sub)population on which we focus (e.g. measles on Iceland showed repeated outbreaks). Data on measles provide a rich source to study the above phenomena. For aspects of data collection and analysis in relation to measles, we refer to Cliff, Haggett & Smallman-Raynor (1993), Keeling (1997) and especially Bryan Grenfell and co-workers, reviewed in Grenfell & Harwood (1997), and the references given there.[13]

In Section 3.1, we suggested that two parameters, rather than just one, are needed to distinguish between a single outbreak (which may much later re-occur by re-introduction of the agent from the outside) and the endemic situation. In addition to population size N, we consider the ratio α/μ of the time scale of life expectancy and of transmission of the agent. When $N \to \infty$ for fixed α/μ, we expect an astronomical time until extinction (of the order e^{+vN} for some positive constant v). When $\alpha/\mu \to \infty$ for fixed N, we expect immediate extinction after an outbreak. The 'phase transition' occurs when both N and α/μ tend to infinity such that $\alpha/\mu = O(\sqrt{N})$ and so, in the area of parameter space that is somewhat imprecisely characterised by this relation, we expect extinction on the demographic time scale $1/\mu$. Therefore we propose to replace

---

[13] A.D. Cliff, P. Haggett & M. Smallman-Raynor: *Measles: an Historical Geography of a Major Human Viral Disease from Global Expansion to Local Retreat, 1840-1990*. Blackwell, London, 1993; B.T. Grenfell & J. Harwood: (Meta)population dynamics of infectious diseases. *TREE*, **12** (1997), 395-399; M.J. Keeling: Modelling the persistence of measles. *Trends Microbiol.*, **5** (1997), 513-518.

the notion of 'critical community size' by 'critical relationship between population size and time-scale ratio' (we admit some work could be done on the name). The work of Nåsell (loc. cit.) comes very close to underpinning our hand-waving arguments.

In Section 3.2 we took the population birth rate B as a constant and found a stable steady endemic state, characterised by damped fluctuations. As the damping requires many periods, the oscillations should be visible in data. But in fact one can easily imagine that periodic temporal heterogeneity (e.g. seasons and the school system) drives the system to truly periodic motion. This is indeed what Schenzle (1984)[14] and others found (and more: even chaotic fluctuations).

Even when analysing caricaturally simplified models, it pays to write expressions for quantities like the level of prevalence, the time scale at which disturbances decay, the period of oscillations et cetera, in terms of observable quantities such as the mean age $\bar{a}$ at infection. First of all, this helps to estimate parameters from data. But, secondly, it makes the conclusions more robust, that is, usable perhaps as rules of thumb in situations where clearly the caricature is an inappropriate description. We emphasise this as a general principle (without repeating the examples we gave in Section 3.2).

On a longer time scale, the spread of an infective agent may have its repercussions on population growth. Indeed, in Section 3.3 we considered a host population that grows exponentially and found that an infective agent may reduce the population growth rate. In fact, it may reduce it to the extent that growth is stopped completely.

The competition between very similar agents for one and the same host is a topic we only very briefly touched upon in Exercise 3.19, and will make some remarks about in the next section. We think that this is a very important topic, in particular when put into the invasibility framework of adaptive dynamics.

3.5 Some evolutionary considerations about virulence

This final section touches upon only the tip of an enormous iceberg of largely open questions: can we understand how evolution by natural selection has shaped the various ways in which infective agents exploit their hosts (and travel from one host to the next)?

The adage 'optimal adaptation to the environment' presupposes that the environment is somehow given and fixed. Yet if ever it is clear that individuals interact by feedback to components of the environment (which are therefore called 'environmental interaction variables'), it is when we consider parasitism (think of both the availability of susceptible hosts, the provoked immune response within one host and the death of a host). So the key issue is *invasibility*: is a new type, the mutant or rival, able to increase in number in the environment as set by the current type, the resident? Compare Exercise 3.19. For background reading on the rapidly growing theory of *adaptive dynamics* (or, as it is called when genetics is properly dealt with, *evolutionary dynamics*) we refer to the literature.[15] An inspiring study of the

---

[14] D. Schenzle: An age-structured model of pre- and post-vaccination measles transmission. *IMA J. Math. Appl. Med. Biol.*, **1** (1984), 169-191.

[15] O. Diekmann, F.B. Christiansen & R. Law (guest eds.) (1996): *Evolutionary Dynamics, J. Math. Biol.*, **34**, issue 5/6, 483-688. U. Dieckmann & J.A.J. Metz (eds.): *Elements of Adaptive Dynamics*. Cambridge University Press, Cambridge, to appear 2000.

evolution of infectious disease can be found in the book by P.W. Ewald[16] (see also van Baalen & Sabelis (1995)[17]). The general conclusion that emerges is that parasites are 'forced' to deal more carefully with their hosts when transmission is more difficult. So the virulence of the parasite depends on the host contact process and structure, on which transmission is superimposed. For a recent review of the evolutionary dynamics of virulence see Pugliese (2000).[18]

A metapopulation is a population consisting of many local populations (or colonies). When reproduction is both local (within the colony) and global (the founding of a new colony), it is often the second process that matters for evolution by natural selection, in particular when colonies consist of individuals with a high degree of genetic correlation (and perhaps even of individuals that are genetically identical). The problem that the organism has to face is to tune the reproduction within the colony in such a way that reproduction at the colony level cannot be improved (i.e. no mutant competitor can do any better). When a rival mutant cannot invade an existing colony, this is essentially the problem of optimal exploitation of a renewable resource. But it becomes a more subtle game when a patch of habitat consisting of a renewable resource has to be shared with the rival, once that rival has arrived at the patch. Indeed, in such a situation the economically attractive strategy of prudent exploitation of the resource (the so-called 'milker strategy') may be invadable since a more aggressive exploitation strategy (the so-called 'killer strategy') will secure a larger part of the resource. The dilemma is whether one should gain a lot at a slow pace or quickly take what can be gotten right away? The answer certainly depends on the likelihood of arrival of potential competitors.

Microparasites form a metapopulation. Indeed, from their point of view, the world consists of inhabitable patches, viz. hosts. Virulence is strongly coupled to reproduction (usually leading to genetically identical copies; but see below) within the host, while transmission corresponds to reproduction at the patch level. In our basic models, we ignore *superinfection* (i.e. the arrival, within the host, of yet another microparasite from outside the host) since reproduction within the host is fast and enormous. However, as we have just argued, for evolutionary considerations superinfection is crucial.[19]

A feature specific to the evolution of infective agents is the interaction with the immune system of the host. The host is not just a (renewable) resource, it also actively defends itself to prevent or stop its exploitation. Acquired immunity assures that a host cannot be colonised repeatedly. When evaluating the fitness (in the sense of Metz et al. (1992)[20]) of a mutant rival, the issue of cross-immunity plays a crucial role. We think that the related aspects of superinfection and cross-immunity make the evolution of virulence such a complicated process to describe and analyse. (Note the paradox

---

[16] P.W. Ewald: *Evolution of Infectious Disease.* Oxford University Press, Oxford, 1994.

[17] M. van Baalen & M.W. Sabelis: The scope for virulence management—a comment on Ewald's view on the evolution of virulence. *Trends Microbiol.*, **3** (1995), 414-415.

[18] A. Pugliese: Evolutionary dynamics of virulence. In: U. Dieckmann & J.A.J. Metz (eds.), loc. cit., 2000, to appear.

[19] M. van Baalen & M.W. Sabelis: The dynamics of multiple infection and the evolution of virulence. *Am. Nat.*, **146** (1995), 881-910.

[20] J.A.J. Metz, R.M. Nisbet & S.A.H. Geritz: How should we define 'fitness' for general ecological scenarios? *TREE*, **7** (1992), 198-202.

that the fact that for many current microparasites superinfection is irrelevant may
be due to the key role of superinfection in shaping these microparasites by natural
selection.)

A special evolutionary puzzle is posed by HIV. Within any particular host, the
agent evolves by mutation and selective force exerted by the immune system, as well
as, nowadays, (cocktails) of drugs. An understanding of the adaptive dynamics of the
quasi-species (i.e. coexisting collections of individuals with only marginally different
traits that are almost faithfully propagated from generation to generation) of HIV
within the microcosm of one host is emerging. However, whether (and, if so, how) there
is a concerted influence on adaptive dynamics at the metapopulation level remains to
be seen/investigated.

This section is short. The reasons are simple: i) the authors themselves have never
worked on the subject; ii) every day only has 24 hours; iii) not that much is known
about the subject; iv) there is a dazzling array of potentially relevant assumptions (for
some agent-host combination), all leading to somewhat different problem formulations.
We have tried to sketch some key features. We are aware that we have missed out on
others (e.g. coevolution of the host and parasite or the battle between the host's
immune system and the parasite). We have not even attempted to present insights or
conclusions. We hope that our readers interpret all this as an invitation to work on
the subject.

We end this section with an exercise, but it is advisable to postpone working through
it until you have read Chapter 5.

Exercise 3.27 Assume that, in the presence of just one infective agent, the
combination of host demography and the transmission of the agent is described
by

$$\frac{dS}{dt} = B - \mu S - \beta SI,$$
$$\frac{dI}{dt} = -\mu I + \beta SI - \alpha I, \qquad (3.15)$$
$$\frac{dR}{dt} = -\mu R \quad + \alpha I.$$

Then, as derived in Exercise 3.1/3.2, an endemic steady state will obtain
provided $R_0 = \frac{\beta}{\alpha + \mu} \frac{B}{\mu} > 1$.

Assume that a rival infective agent is introduced at very low density into this
population in endemic steady state. Assume that this rival can infect all host
individuals.

i) What are the types at birth with respect to infection by the rival agent?

ii) Formulate more detailed assumptions such that the resulting next-generation
matrix has a one-dimensional range.

iii) Derive a formula for the basic reproduction ratio of the rival agent in
the structured (by the residential agent) host population. What can you
conclude from the condition of invasibility of the rival agent? What about
mutual invasibility? Critically examine the influence of the assumptions on the
conclusions.

Part II

Structured populations

4

The concept of state

4.1 i-states

The basic idea of dynamic structured population models is to distinguish individuals from one another according to characteristics that determine the birth, death and resource consumption rates—more generally, the interaction with the environment—and to describe the rates with which an individual's characteristics change themselves. Since we are mainly interested in infectious diseases, we limit ourselves to those characteristics that influence the force of infection of a given infectious agent, i.e. those traits that influence the rate with which susceptible individuals become infected (encompassing both infectivity, susceptibility and contact pattern).

The first step in building a structured population model to investigate a concrete question is to choose those characteristics that are deemed relevant to the problem one is interested in. In mathematical jargon, this is called choosing the i-state, where 'i' denotes 'individual'. The i-state of an individual at some point in time t is therefore the set of values for the chosen traits for that individual at time t. For example, if we choose age as the only relevant characteristic then the i-state of an individual at any time t is simply the individual's age at t. If the individual is born at some time t_0 then its age at some time $t > t_0$ will be $t - t_0$. Another example is to take sex, partnership status (i.e. single or with steady partner) and age (in discrete classes) simultaneously as individual characteristics. A possible value of this i-state is (male, single, 34).

The concept of 'state' has a more fundamental content than just any collection of individual characteristics. Informally speaking, the *state* of a system—in our case an individual—is the set of precisely that information about the system that is relevant to predict the system's future development/behaviour. In the context of epidemic models, the future 'behaviour' we wish to characterise often encompasses the expected infectious output of an infected individual as a function of its i-state; we wish to determine how infectious this individual is to others depending on its characteristics. The state could change according to, for example, a collection of laws describing the individual's life, interaction with other individuals, the transmission of infection, the time-course of the disease inside the individual (if infected), or changes in a, broadly defined, environment.

It is clear from the first example that the following can hold for an individual's i-state. Given the state, say $x(t_0)$, at some time t_0, the state at time $t_1 + t_0$ is determined by

$$x(t_1 + t_0) = \mathcal{T}(t_1)x(t_0).$$

Here $\mathcal{T}(t_1)$ denotes an operator that maps states into states and that has the semigroup property

$$
\begin{aligned}
\mathcal{T}(0) &= I, \\
\mathcal{T}(t_2)\mathcal{T}(t_1) &= \mathcal{T}(t_2 + t_1), \quad t_2, t_1 \geq 0.
\end{aligned}
$$

In words, states have the following property. If we start at time 0 and want to know the state at some future time $t_2 + t_1$, it does not matter whether we go from 0 to $t_2 + t_1$ immediately, $x(t_2 + t_1) = \mathcal{T}(t_2 + t_1)x(0)$, or whether we take the intermediate step of first going from 0 to t_1, and then taking $x(t_1)$ as our new starting value to go a time t_2 further: $x(t_2 + t_1) = \mathcal{T}(t_2)\mathcal{T}(t_1)x(0)$.

So, whatever the dynamics are in the time interval $[0, t_1]$, if we want to predict the future after t_1, we need only look at the state of the system at time t_1 and can disregard the precise history of the evolution between 0 and t_1. In a way, the state at t_1 carries with it a 'memory' of what happened to the system before t_1, at least of those aspects that are relevant to predicting the future course of the system.

In the context of epidemic models, choosing the i-state ingredients is a double task since we have to deal with both population dynamics and infection transmission, and we shall accordingly refer to those components of the i-state that describe the development of the infection within an individual as the d-state—where 'd' denotes 'disease' (it would be more appropriate to use the 'i' of 'infection', but we have already used 'i' to denote 'individual'). More precisely, the d-state is that part of the i-state that describes the difference between infected and susceptible individuals. From the point of view of the infectious agent, the rest of the i-state reflects heterogeneity in the population. Accordingly, we will call this part of the i-state the h-state. We will usually call h-state values *types*.

In the presentation above, we have concentrated on so-called 'autonomous' systems, characterised by the absence of time-dependent input. This allows us to work with time as a relative quantity ('the time elapsed since ...'). If, for instance, seasonal weather conditions do have an impact on the success of transmission or if contact patterns are time-dependent (think of the school system), we should take absolute time into account. This then is an additional technical complication in the bookkeeping, but it does not in any way influence the state concept. The state should contain all information relevant for predicting the future development/behaviour, given the environmental input in the intervening period.

4.1.1   d-states

The d-state determines the infectious output of an infected individual. The course of an individual infection is a stochastic process, reflecting among other things the status of the immune system and its 'battle' with the infectious agent. We would like to avoid modelling this complex process in detail. From a system-theoretic point of view, there are but two kinds of d-state: those where the input is a unique event and those where the input is a repeated process.

- *Infection-age:* Suppose that after infection the disease develops as an autonomous process within the infected individual, and that superinfections (re-infections of an already-infected individual) therefore play no role. We have in mind that the invading organism reproduces within the host at such a rate that further infections with the same agent can be neglected; examples include measles, influenza, rabies and HIV, and fall under the general heading of *micro*parasites. Usually they are viruses or bacteria, hence the name. Irrespective of the precise biological-chemical-physical interpretation of the d-state we can then describe disease-progress—and with it, morbidity, mortality and infectious output—by an infection-age representation (d-age). We act as if a clock starts ticking the moment the individual becomes infected. It is convenient in this respect to refer to the h-state at d-age zero as the *state at birth* or *type at birth* of the individual - birth meaning recruitment into the infected population.

We will usually denote infection-age by the variable τ. In Section 2.1 we gave examples of how to compute the expected infectivity $A(\tau)$ from submodels for the dynamics of an underlying d-state that can assume finitely many 'values'. See in particular expression (2.3).

- *Infection-degree:* Here infection is not a unique event but rather a repeated process. Examples include schistosomiasis and other diseases caused by helminths and they fall under the general heading of *macro*parasites. As a rule, the parasites within a host in this class can be counted, in contrast to microparasites. Sexual reproduction or cloning of a particular life stage of the parasite within the host is allowed to take place, but the full life cycle of the parasite usually involves one or more stages outside the host (possibly in another host species). Since in these cases the morbidity, mortality and infectious output of an infected host depend on the level of infection, the d-state is represented by the number of parasites a host harbours (infection-degree or d-degree).

While these seem to be exclusive categories, there is at least one important class of infectious agents, the protozoan parasites (among these are the causal agents of malaria), that belong to both. Superficially speaking, protozoan infections would belong to the first category because protozoans in relevant stages of the life cycle multiply very rapidly within the host. However, the phenomenon of acquired immunity occurs in many protozoan infections. The more additional infections with the parasite (possibly different strains of the same species) that an individual acquires, the higher its level of immunity will rise (leaving aside intricacies that concern the required length of the time period between successive infections). The immunity usually does not protect against re-infection, but individuals with a high level of immunity do not experience the severe disease symptoms. Immune individuals can still be infectious to others for a number of protozoan infections, but usually this infectivity is much reduced compared to that of nonimmunes. This phenomenon of acquired immunity through superinfection places the protozoan infections in the second category.

In most of what follows we restrict ourselves to systems allowing for infection-*age* as the state representation. Models incorporating infection-*degree* will be treated briefly in Chapter 9.

There is a type of distinction in transmission opportunity for which both of the above descriptions of d-state are inadequate, or, more precisely, incomplete. This concerns sexually transmitted infections, for which one should often explicitly take the formation of long-term monogamous partnerships into account. In that case a susceptible can only become infected if its current partner is infected and an infected partner can cause at most one new infection while the partnership lasts. As a consequence, all contacts between the partners are 'wasted' from the point of view of the infectious agent, once both are infected. Potentially, the infected individual would be capable of causing more infections with the same infectious output. Because of the contact structure, however, these infections are not realised. This implies that the description with d-age, which would be appropriate in the case of, for example, random contacts, is not a good indicator of transmission ability. What we have to take into account is the survival of the partner of the infected individual. Only when the partner dies, or the partnership is dissolved for other reasons, can the infected individual cause new infections. We cannot describe this situation by simply looking at the d-age of the infected, since that quantity does not describe the status of the partner. In Section 5.8 we will show how our general methodology for calculating R_0 easily extends to cover this kind of situation, involving 'super-individuals', viz. pairs of individuals.

4.1.2   h-states

We now turn to the possible h-states. Characteristics on which these are based can, for a given individual, be *static* (like sex, genetic composition) or *dynamic* (suffering from another disease, stage of development), and they can take *discrete* values (like sexual orientation, partnership status) or *continuous* values (like spatial position of a plant, or age). In particular cases h-states can be very complicated, and contain simultaneously continuous and discrete, static and dynamic components. We will denote by Ω the state space of all possible values of the chosen set of characteristics (i.e. the h-state space.

If the h-component of the i-state has more than one possible value, and the d-state has a d-age representation, we have to take into account that the expected infectivity function A may depend on both the h-state of the susceptible and the h-state of the infected individual taking part in a contact. The major modelling effort involved in addressing a question pertaining to a specific host/infection system is to make precise how A depends on these states. The dependence of A of a given infected individual on the h-state of susceptibles is primarily through the frequency of contacts (for example, if the h-state denotes sex, we could specify different contact rates for homosexual and heterosexual contacts). In Chapter 5, we give one method to obtain A expressed in the parameters (of submodels) that govern the changes in individual state, the infection transmission and contact pattern, illustrated by a number of extended examples. As is typical for structured population models, the process of obtaining A involves detailed stochastic modelling of events at the individual level (fortunately, on this level one often has possibilities to experimentally measure or estimate parameters). By assuming that our population consists of many individuals, we can then invoke a law-of-large-numbers argument that allows us to describe the changes at the population level by deterministic equations (we will come back to this in the Appendix). The theory of Markov chains is tailor-made to compute A from submodels. The Markov property states that the conditional probability of a given event only depends on the present state of the system and not on the manner in which the present state was reached. This is exactly the property that characterises i-states.

4.2   p-states

The population state (p-state) is nothing more than the *distribution* of individuals over the i-state space. Changes on the individual level give rise to changes in the composition of the population. The equation for p-state change is obtained basically by bookkeeping, once the dynamics of i-states has been described. We do not go into this issue here, but refer for general background reading on structured population models to Metz & Diekmann (1986), Tuljapurkar & Caswell (1997), Cushing (1998), Diekmann et al. (1998) and Diekmann (1999).[1]

In view of the next chapter, we now focus on the h-state distribution when explaining some concepts and some notation.

When there are finitely many h-states and finitely many individuals, it is simple what we mean by 'distribution': for each h-state we specify the number of individuals that happen to have that state. Or, alternatively, the fraction of the population with that state (in which case we complement the information by one additional number, the total population size).

When we talk about people aged 67, we mean those people who were born between 67 and 68 years ago. True age, in the sense of 'time elapsed since birth', is actually a continuous variable that can take uncountably many different values. But we are used to subdividing the age axis in one-year intervals and to count accordingly. Other subdivisions are conceivable (and sometimes useful) and they can be taken as a basis for counting.

The object that assigns to every reasonable subset of the age axis the number/fraction of individuals that have their 'true age' in that subset is called a *measure* (the mathematical jargon for 'reasonable' is 'measurable', and there is a precise technical definition for it). We shall, as much as we can, denote measures by the letter m.

When the measure m is used to describe a population age distribution and ω is an age interval (or union of such intervals, or just a measurable subset) then $m(\omega)$ is the number/fraction of individuals with age in ω. We also write

$$m(\omega) = \int_\omega m(da)$$

where the right-hand side is called an integral and where the notation symbolises the (mathematical idealisation of the) process of adding/counting all individuals with age in ω.

If, for instance, individuals with age a have contact intensity $c(a)$ then the mean contact intensity in the population is given by

$$\bar{c} := \frac{\int_\Omega c(a)m(da)}{\int_\Omega m(da)},$$

---

[1] O. Diekmann, M. Gyllenberg, J.A.J. Metz & H.R. Thieme: On the formulation and analysis of general structured population models. I: Linear theory. *J. Math. Biol.*, **36** (1998), 349-388; O. Diekmann: Modelling and analysing physiologically structured populations. In: V. Capasso, O. Diekmann (eds.), *Mathematics Inspired by Biology*. Lect. Notes in Math. vol. 1714, Springer-Verlag, Berlin, 1999.

where $\Omega = [0, \infty)$. Whenever $m(\Omega) = \int_\Omega m(da) = 1$, we call m a *probability measure*, and we can omit the denominator. Such is the case when m describes fractions. The terminology refers to the fact that the probability that a randomly chosen individual belongs to some specified subgroup equals the fraction of the total population that that subgroup constitutes. The numerator is called the integral of c with respect to the measure m. The operation as a whole corresponds precisely to giving individuals of age a the weight $c(a)$ and then computing the average.

Hidden in the notation are certain mathematical subtleties related to the fact that $[0, \infty)$ is uncountable. For instance, imagine entering a classroom and try to guess what the probability is that one of the students will have true age 19 years, 3 months, 1 week, 2 days, 5 hours, 10 minutes, 57 seconds, 817 milliseconds, ...? Probably you will quickly conclude that already without the '...' being specified, the probability will be low, but that it will be zero if we could specify 'age' with infinite precision. Thus the theory of measure and integration constitutes a highly non-trivial mathematical challenge. In this book, however, all we need is the small conceptual and notational part sketched above and below.

Sometimes the integral with respect to a measure boils down to the integral that is familiar from calculus. We say that the measure m has a *density* n when, symbolically, $m(da) = n(a)da$, by which we mean that, for all (measurable) ω,

$$m(\omega) = \int_\omega m(da) = \int_\omega n(a)\, da.$$

For instance, when there is a constant population birth rate b, and of all newborns a fraction $\mathcal{F}(a)$ survives until at least age a, then the above holds with

$$n(a) = b\mathcal{F}(a).$$

The point is that the density gives complete information about the measure and, because it is a function for which one can draw a graph, it is a more familiar object. There are, however, at least two points in favour of measures:

- not all measures have a density;
- densities are not directly interpretable in terms of numbers, which makes it more dangerous to employ intuitive arguments based on the interpretation.

Returning from age to general h-state, we add two final remarks. We always assume that Ω is measurable, i.e. that one can define measures on Ω. Certainly this is the case when Ω is a nice subset of $\mathbb{R}^k$ for some $k \geq 1$. In that case, one can also introduce the notion of density on the basis of the standard n-dimensional integral. Just as in the case of age $(k = 1)$, one requires that, for all ω,

$$\int_\omega m(d\xi) = \int_\omega n(\xi)\, d\xi.$$

For a gentle introduction to the theory of measures see e.g. Kolmogorov & Fomin (1975) or Rudin (1974).

4.3   Recapitulation, problem formulation and outlook

Suppose individuals differ from each other with respect to traits that are relevant for the transmission of an infectious agent. How do we describe the spread of the agent? How do we quantify the infectivity? What happens in the initial phase? Can we characterise the final size?

Examples of the 'traits' we have in mind are age, sex, sexual activity level, sexual disposition and spatial position. So a trait may be *static* or *dynamic*, it may be *discrete* or *continuous*. Often the modeller's subjective and pragmatic striving for manageable problems will suggest to take a trait such as, for example, 'sexual disposition' as static and discrete, while one may rightfully wonder whether reality isn't more polymorphic and changeable than that.

Despite such doubts, we shall consider the traits as i-states, where 'i' means 'individual' and where 'state' signifies that the current value together with the environmental input in the intervening period completely determines future behaviour (although probably in a stochastic sense).

Thus we classify the heterogeneity of individuals in terms of a component, h-state, of their i-state, while the other component, d-state, summarises all relevant information about output of infectious material. A population of such individuals is no longer characterised by one number, the population size. We need, in addition, to know the composition of the population, i.e. the distribution with respect to i-state. We are dealing with a *structured* population.

Our approach is *top down*. This means that we start with abstract general principles and then gradually become more concrete by being more specific and quantitative. Some readers might conclude that we stay abstract throughout the book, but that is partly a matter of taste and background (determining what the words 'abstract' and 'concrete' mean to someone). We shall sketch computational schemes, but not perform actual computations.

The next chapter concentrates on the definition and the computation of the basic reproduction ratio R_0 in the context of structured population models. In Chapter 6 we deal much more briefly with the real-time growth rate r, the final size of an epidemic and the probability of a minor outbreak. In particular, we show how the assumption of *separable mixing* allows one to reduce all computations to the situation where there is only a one-dimensional unknown. In Chapter 7 we elaborate on the special case of age-structure, while in Chapter 8 we concentrate on the *asymptotic speed of propagation* c_0 as an important indicator of the spread at the population level in a spatially structured population.

5

The basic reproduction ratio

5.1 The definition of R_0

The following simple but helpful idea allows us to postpone a discussion of the dynamics of h-states: we characterise infected individuals by the h-state at the moment of becoming infected, rather than by their current h-state. So, whenever we say that an infected individual has h-state j, we mean that at the moment this individual became infected it had h-state j. The h-state at the moment of becoming infected will be called the *type at birth* (or the *state at birth*) of the infected individual, since the individual is 'born' from an epidemiological point of view at that moment.

In a similar spirit one can mask the distinction between a discrete h-state and a continuous h-state by working with measures, but, as this may look rather unfamiliar to most readers, we choose not to do so from the beginning. In fact, we start by considering the situation in which there are only finitely many h-states $1, 2, ..., n$. In this chapter we address the 'introduction' issue: given a population in demographic steady state, with no history of a given infection, will the introduction of the infectious agent cause an outbreak? We adopt a deterministic point of view— that is, we only consider expected values—and we linearise—that is, we neglect that the agent itself diminishes the availability of susceptibles. The assumption of a demographic steady state means that the agent experiences the world as constant, i.e. environmental conditions do not vary. This allows us to judge growth by adopting a generation perspective.

Define k_{ij} to be the expected number of new cases that have h-state i at the moment they become infected, caused by one individual that was itself infected while having h-state j, during the entire period of infectiousness. (So, implicitly, k_{ij} is the integral with respect to τ from 0 to ∞ of a more basic model ingredient, which we might denote by $A_{ij}(\tau)$. As all subsequent considerations in this section concern k_{ij}, we choose not to introduce $A_{ij}(\tau)$ explicitly here.) With this definition of the k_{ij}, we obtain n^2 non-negative numbers. The question is how these numbers should be averaged. More precisely, we look for a single summarising number that has two desired properties: (i) the introduction is successful if and only if the number is greater than 1, and (ii) the number has almost the same biological interpretation as in the homogeneous case.

Example 5.1 (cf. Section 2.2) Consider the matrix

$$K = (k_{ij}) = \begin{pmatrix} 0 & 100 \\ 10 & 0 \end{pmatrix},$$

which could, for example, relate to a heterosexually transmitted infection with female and male as h-state values (or to host-vector transmission). The multiplication factors 10 and 100 alternate because, starting from, for example, a female (host), the infectious agent has to 'pass through' a male (vector) before it can enter a new female (host). In other words, it takes two generations to get back and every two generations numbers are multiplied by $10 \times 100 = 1000$. The average per generation multiplication factor is therefore $\sqrt{1000}$. How do we arrive at such a number for general matrices? How does one measure the 'size' of a matrix with non-negative entries?

We regard *generations* of infected individuals, described by vectors (note that now we use 'vector' in the mathematical sense, whereas above we used the same word to denote a biological carrier; the etymology is the same). The jth component of the vector equals, by definition, the number of cases with h-state j, in that particular generation. The vector describing the next generation is obtained from the vector describing the current generation by applying the matrix K to it, as in the example above:

$$\phi_i^{\text{new}} = \sum_{j=1}^{n} k_{ij} \phi_j^{\text{old}} \tag{5.1}$$

being the number of susceptibles with state i that are infected by an infective with state j, summed over all states j. In short,

$$\phi^{\text{new}} = K \phi^{\text{old}}. \tag{5.2}$$

Let us use a superscript to number the generations and write

$$\phi^{n+1} = K \phi^n. \tag{5.3}$$

We call K the *next-generation matrix* and observe that the generation process is described by iteratively applying K. If we use a superscript for a matrix to denote powers with respect to matrix multiplication, e.g. $K^2 = KK$, we can write

$$\phi^n = K\phi^{n-1} = K^2\phi^{n-2} = \cdots = K^n\phi^0. \tag{5.4}$$

We note that K is a *positive* matrix in that all its elements are necessarily non-negative (we also use 'non-negative matrix' at times). We will use the shorthand notation $K \geq 0$, to indicate that all elements $k_{ij} \geq 0$, $1 \leq i, j \leq n$. Likewise, we speak of positive vectors when all components are non-negative and write $\phi \geq 0$.

Exercise 5.2 Let K be a 2×2 matrix that has two distinct real eigenvalues λ_1 and λ_2, with $\lambda_1 > 0$ and $\lambda_1 > |\lambda_2|$. Let $\psi^{(1)}$ and $\psi^{(2)}$ be the associated (right) eigenvectors. Then any vector $x \in \mathbb{R}^2$ can be written as a linear combination of these eigenvectors: $x = c_1 \psi^{(1)} + c_2 \psi^{(2)}$.

(i) Express $K^n x$ as a combination of the eigenvectors.

(ii) By rewriting the expression for $K^n x$, conclude that, after many generations, the influence of the $\psi^{(2)}$ component on the growth of $K^n x$ will become negligible and that we have asymptotically for the number of generations $n \to \infty$

$$K^n x \sim \lambda_1^n c_1 \psi^{(1)}. \tag{5.5}$$

Where is the influence of the 0th generation expressed? Conclude that λ_1 has the desired threshold property to determine growth or decline (on a generation basis) of the infective population.

In the present context a natural *norm* for vectors is

$$\| \phi \| = \sum_{j=1}^{n} | \phi_j |, \tag{5.6}$$

where the absolute value is not 'active' in our setting of positive vectors and $\| \phi \|$ is the total number of cases in the generation that is described by ϕ. Now that a norm for vectors has been specified, we norm matrices by

$$\| K \| = \sup_{\|\phi\|\neq 0} \frac{\| K\phi \|}{\| \phi \|} = \sup_{\|\phi\|=1} \| K\phi \|. \tag{5.7}$$

(For the matrix in Example 5.1 we find $\| K \| = 100$.) In the epidemiological setting we can interpret $\| K \|$ as the maximum multiplication number for the total number of cases, when we allow the distribution with respect to h-state to take all possible forms.

The point is now that we can choose ϕ^0 quite arbitrarily, but that subsequently the distribution will be determined by the dynamics itself. In Example 5.1, if $\phi^0 = (0, 1)^\top$ (where $\top$ again denotes 'transpose') then $\phi^1 = (100, 0)^\top$ and indeed we find the number of cases multiplied by 100. In the next generation, however, the multiplication factor is only 10. Consequently, the norm of K is too coarse a measure for generation growth. This motivates us to look at multiplication in n generations, but on a 'per generation' basis. In other words, we look at the $\frac{1}{n}$th power of the growth in n generations:

$$\| K^n \|^{1/n}. \tag{5.8}$$

In general, we have to be patient and consider the limit for $n \to \infty$ of this quantity. (Which exists! This requires proof of course, but is not really difficult.) This limit is, for reasons explained below, called the *spectral radius* of K. As it is by definition the long-term average per generation multiplication number, we feel confident to denote it by R_0:

$$R_0 := \lim_{n\to\infty} \| K^n \|^{1/n}. \tag{5.9}$$

This may be a mathematically natural definition of R_0, but should we also use this as an algorithm to compute it? Or are there other characterisations that lend themselves more readily to computation? The answer to the latter question is yes. We can in addition obtain more detailed information about the (linearised) generation process, in particular about the asymptotic behaviour for $n \to \infty$. Here the fact that $K \geq 0$ plays an important part.

Theorem 5.3 *Let* $K \geq 0$. *Then the spectral radius* R_0 *is an eigenvalue of* K, *which we call the* dominant *eigenvalue since* $| \lambda | \leq R_0$ *for all other eigenvalues* λ *of* K. *The eigenvector* ψ^d *('d' denotes 'dominant') can be chosen in such a way that all its components are non-negative and their sum equals one (we then call* ψ^d *normalised).*

We next assume that R_0 is *strictly* dominant in the sense that $| \lambda | < R_0$ for all other eigenvalues λ of K and that R_0 is an algebraically simple eigenvalue (i.e. there is only one factor $\lambda - R_0$ in the characteristic equation for K). Then one can prove, along the lines of Exercise 5.2, that

$$K^n \phi^0 = c(\phi^0) R_0^n \psi^d + o(R_0^n) \quad \text{for } n \to \infty. \tag{5.10}$$

Here $o(R_0^n)$ summarises the transient behaviour in as crude a form as possible, it signifies that $[K^n \phi^0 - c(\phi^0) R_0^n \psi^d]/R_0^n \to 0$ as $n \to \infty$.

The dominant eigenvector ψ^d (when normalised) describes the *stable distribution*. This terminology combines two properties: invariance and attraction. We now elucidate both of these. If we take $\phi^0 = \psi^d$, we have exactly

$$K^n \psi^d = R_0^n \psi^d, \tag{5.11}$$

or, in words, the distribution remains unchanged from generation to generation while numbers are multiplied by R_0. For general initial-generation vectors the situation will more and more resemble this special case in the course of the generations. Only one aspect of the initial situation remains manifest, in the scalar quantity $c(\phi^0)$. There exists a slightly technical procedure to compute $c(\phi^0)$ from a given ϕ^0, and for the interpretation this procedure does not matter much.[1]

So when R_0 is strictly dominant and algebraically simple the situation is very clear: if we apply the next-generation matrix K repeatedly, the (normalised) distribution converges to the stable distribution ψ^d and the per generation multiplication number converges to R_0. The (normalised) stable distribution can be interpreted as the probability distribution for h-state at the moment of infection.

Can we determine whether or not R_0 is strictly dominant and algebraically simple without computing all eigenvalues? The following collection of definitions and results shows that indeed we can. Together with Theorem 5.3 above, this is collectively called *Perron-Frobenius theory* of positive matrices (see e.g. Minc (1988)).

Definition 5.4 *The non-negative matrix K is called* irreducible *if for every index pair i, j there exists an integer $n = n(i, j) > 0$ such that $(K^n)_{ij} > 0$. And K is called* primitive *if one can choose one n for all i, j, that is, if there exists n such that K^n has all its entries strictly positive.*

Exercise 5.5 i) Start with one case that has h-state j. Let K be irreducible. Show that eventually there will be cases with h-state k, no matter what combination of k and j we consider. What do we mean by 'eventually'? And how could we interpret 'eventually' when K would be primitive?

ii) The property discussed in i) can serve to define 'irreducible' and 'primitive'. Do you agree?

iii) Consider the example

$$K = \begin{pmatrix} 0 & 100 \\ 10 & 0 \end{pmatrix}.$$

Is K irreducible? Is K primitive? Compute $\| K^n \|^{1/n}$ for $n = 1, 2, ..., 10$.

iv) Consider a sexually transmitted disease in a population composed of homosexual and heterosexual males and females. Argue that in the absence of bisexuality we have a reducible situation. What does 'reducible' mean?

* * *

[1] For completeness, we describe it. Let ψ^{d*} denote an eigenvector of the transposed matrix $K^\top$ corresponding to the eigenvalue R_0. Normalise it such that $\psi^{d*} \cdot \psi^d = \sum_{j=1}^n (\psi^{d*})_j (\psi^d)_j = 1$. Then $c(\phi^0) = \sum_{j=1}^n (\psi^{d*})_j \phi_j^0$.

Theorem 5.6 *Let K be primitive. Then*

- R_0 *is strictly dominant;*
- ψ^d *and ψ^{d*} have strictly positive components;*
- R_0 *is an algebraically simple eigenvalue;*
- *no other eigenvalue has a positive eigenvector.*

When K is merely irreducible, the last three properties still hold, but on the circle of radius R_0 in the complex plane there are other eigenvalues. These have to be roots of the equation $\lambda^m = R_0^m$ for some m, i.e. they have to be of the form $\lambda = R_0 \exp(i\frac{2\pi l}{m})$, $l = 1, ..., m - 1$.

Exercise 5.7 Consider the situation described in Section 2.2, but now assume that a fraction p_1 of the contacts of 1-individuals is with 1-individuals and a fraction p_2 of the contacts of 2-individuals is with 2-individuals. Consistency requires that $(1 - p_1)c_1 N_1 = (1 - p_2)c_2 N_2$, so for given c_i and N_i only one of the two ps is a free parameter. For which combination of ps is R_0 maximal, respectively minimal?

5.2   General h-state

When the h-state is not purely discrete-valued but is continuous, or has, in case it consists of several traits, continuous elements, all we have to do is to replace summation by integration in the previous section. Let Ω denote the h-state space. One can think of some suitably nice subset of $\mathbb{R}^n$ for some $n \geq 1$.

Let

$$k(\xi, \eta) \quad = \quad \text{expected number (per unit of h-state space) of newcases}$$
$$\text{with h-state } \xi \text{ (at the moment of becoming infected)}$$
$$\text{caused by one individual that was itself infected while}$$
$$\text{having h-state } \eta, \text{ during the entire period ofinfectiousness.}$$

Again, as with k_{ij} in Section 5.1, one can have an identity $k(\xi, \eta) = \int_0^\infty A(\tau, \xi, \eta)\, d\tau$ in mind, but this is not essential. See Section 5.4.

The generations are now described by functions ϕ on Ω such that

$$\int_\omega \phi(\eta)\, d\eta \tag{5.12}$$

equals the expected number of cases with h-state belonging to the subset $\omega \subset \Omega$. The *next-generation operator*, again denoted by K, tells us how the situation changes from one generation to the next:

$$(K\phi)(\xi) = \int_\Omega k(\xi, \eta)\phi(\eta)\, d\eta. \tag{5.13}$$

The function $k(\xi, \eta)$ is called the *kernel* of K; kernels are our main modelling ingredient.

The norm of ϕ is the so-called L_1 norm:

$$\| \phi \| = \int_\Omega | \phi(\eta) | \, d\eta \tag{5.14}$$

and the norm of K is, exactly as before, defined by

$$\| K \| = \sup_{\|\phi\|\neq 0} \frac{\| K\phi \|}{\| \phi \|} = \sup_{\|\phi\|=1} \| K\phi \| . \tag{5.15}$$

By definition, R_0 is the spectral radius of K:

$$R_0 = \lim_{n\to\infty} \| K^n \|^{1/n} . \tag{5.16}$$

(We remark that the product of two operators K and M with kernels k and m is an operator with kernel $\int_\Omega k(\xi,\zeta) \, m(\zeta,\eta) \, d\zeta$, as can be seen from the identity

$$((KM)\phi)(\xi) = \int_\Omega \left(\int_\Omega k(\xi,\eta) \, m(\eta,\zeta) \, d\eta \right) \phi(\zeta) \, d\zeta, \tag{5.17}$$

which follows directly from the definitions and an interchange of the order of the two integrations.)

Often (but not always) R_0 is an eigenvalue. The corresponding (normalised) eigenvector ψ^d is called the *stable distribution* and, provided R_0 is strictly dominant and algebraically simple, the asymptotic behaviour for $n \to \infty$ is completely analogous to that in the finite dimensional setting. Technical conditions for this to be the case often involve the notion of compactness (of both Ω and K).

We now present a third variant of the basic ideas that unifies the preceding two. Let ω denote a subset of Ω (we have a measurable subset in mind, but our presentation will emphasise the interpretation and intuition while being deliberately sloppy about technical aspects). Then $\int_\omega k(\xi,\eta) \, d\xi$ is the expected number of cases with h-state in ω caused by one individual that was itself infected while having h-state η, during the entire period of infectiousness. (Note that, strictly speaking, k itself is not a number, but that we obtain a number only after integration with respect to ξ.) The third variant takes these quantities, with arbitrary η and ω, as its starting point (and forgets about k). We denote the quantities by $\Lambda(\eta)(\omega)$. So, rather than using k, we now take as the main modelling ingredient

$\Lambda(\eta)(\omega)$: = the expected number of cases with h-state in ω caused by one
individual that was itself infected while having h-state η,
during the entire period of infectiousness.

Likewise, we describe the generations by measures m, which assign to every $\omega \subset \Omega$ the number of cases with h-state in ω. (If one can alternatively describe the generations by functions ϕ, the relation between ϕ and m is given by

$$m(\omega) = \int_\omega \phi(\eta) \, d\eta. \tag{5.18}$$

Since measures cannot always be related to functions, this third formulation is more general.) The next-generation operator is now given by

$$(Km)(\omega) = \int_\Omega \Lambda(\eta)(\omega)\, m(d\eta). \tag{5.19}$$

Concerning norms, we confine ourselves to the remark that for non-negative measures $\| m \| = m(\Omega)$. As before we define R_0 to be the spectral radius of K. The handiness of formulations in terms of measures will be demonstrated by (classes of) examples in Sections 5.3.2 and 5.3.3 below.

We end our presentation of the definition of R_0 with a cautionary remark. On the one hand, we linearise (i.e. we neglect the diminishing of susceptibles), which we justify by saying that we consider the initial phase only. On the other hand, we define R_0 by looking after many generations, arguing that the details of how precisely the agent is introduced affect only the transient behaviour in a relatively short period. The upshot of these somewhat contradictory arguments is that one is not sure that indeed data, when organised on a generation basis, should exhibit multiplication by R_0 in the initial phase. Since looking at data in a generation perspective is somewhat artificial anyhow, we emphasise that the remark carries over to the real-time growth rate r. Whether or not we can observe r in data will depend on such factors as the difference between r and the next eigenvalue (ordered according to real part), and the quantitative aspects of irreducibility (measured, for instance, by the ratio of the maximum and the minimum component of the dominant eigenvector). This admonition detracts nothing from the key property of R_0:

$$\{\text{the infective agent will be able to grow}\} \Leftrightarrow R_0 > 1.$$

Exercise 5.8 The aim of this exercise is to illuminate the remark about 'quantitative aspects of irreducibility' above. Consider the reducible matrix

$$\begin{pmatrix} 10^3 & 0 \\ 0 & 1 \end{pmatrix}$$

with dominant eigenvalue 10^3 and corresponding eigenvector $\binom{1}{0}$. If we make the anti-diagonal terms slightly positive, the matrix becomes irreducible (even primitive). Use perturbation analysis to show that the second component of the normalised eigenvector is of the same order of magnitude as the coupling coefficients in the matrix. Now explain the remark above (formulate your conclusions in terms of the interpretation rather than in mathematical terms).

5.3   On conditions that simplify the computation of R_0

There are efficient numerical methods for the computation of the dominant eigenvalue of a non-negative matrix, but that is not what will concern us here. Instead we try to establish analytical procedures. Quite naturally these require extra conditions. In this section we discuss three of these conditions (and their interpretation) that facilitate the computation of R_0.

5.3.1   One-dimensional range

When the range of an operator is one-dimensional, no matter what operator you consider, there is only one candidate eigenvector and a formula for the one and only non-zero eigenvalue follows at once. Following the three variants of the formulation of the next-generation operator presented in the previous section, the assumptions leading to a one-dimensional range and the resulting formulae for R_0 take the following form:

$$k_{ij} \;=\; a_i b_j \quad \Rightarrow \quad R_0 = \sum_{j=1}^{n} b_j a_j, \tag{5.20}$$

$$k(\xi, \eta) \;=\; a(\xi) b(\eta) \quad \Rightarrow \quad R_0 = \int_{\Omega} b(\eta) a(\eta) \, d\eta, \tag{5.21}$$

$$\Lambda(\eta)(\omega) \;=\; \alpha(\omega) b(\eta) \quad \Rightarrow \quad R_0 = \int_{\Omega} b(\eta) \alpha(d\eta). \tag{5.22}$$

Exercise 5.9 Verify at least one of the expressions for R_0. Hint: Write down the respective eigenvalue problems for the special kernels, remark that there is only one candidate eigenvector and find the only eigenvalue.

Exercise 5.10 This exercise extends and continues the example introduced in Section 2.2 and it gives an introduction to 'sexual activity'-based models as presented at the end of the present section. Assume that $k_{ij} = a_i b_j$ with $b_j = c_j$ and $a_i = c_i N_i / (\sum_l c_l N_l)$ (which guarantees that $k_{ij} N_j = k_{ji} N_i$, which is a consistency condition required by a particular context as explained in Section 2.2). Show that

$$R_0 = \text{mean} + \frac{\text{variance}}{\text{mean}},$$

where the right-hand side refers to the activity c.

But what is the interpretation of the assumptions (5.20)-(5.22)? Can we understand when and why they are reasonable?

The assumptions express that the probability distribution for the h-state of a newly infected individual is *independent* of the h-state of the individual that is responsible for the infection. So in a stochastic sense, there is just one possible h-state at infection. Stated in yet another manner: the h-states of the two individuals involved influence transmission independently. We argued before (Section 2.2) that straightforward averaging works unless there is correlation between infectivity and susceptibility and that even if such correlation exists within one individual, we can still average (in the right way as expressed by an explicit formula for R_0) provided there is no correlation between the infectivity of the 'donor' and the susceptibility of the 'receiver'.

The condition is called *separable mixing*. When a is proportional to b (or α to the integral of b) we speak of *proportionate mixing*, which is in fact a weighted form of random mixing. Proportionate mixing was introduced in Barbour (1978).[2]

---

[2] A.D. Barbour: MacDonald's model and the transmission of bilharzia. *Trans. Roy. Soc. Trop. Med. Hyg.*, **72** (1978), 6-15.

Let us discuss an example. Let ξ be some indicator of sexual activity and consider an STD (sexually transmitted disease). It is reasonable to assume that both susceptibility and infectivity are proportional to ξ, since we have in mind that the number of contacts per unit of time is proportional to ξ. We consider a homosexual population (but see Section 5.3.3 below for the slightly more complicated heterosexual case). We take $\Omega = [0, \infty)$ and introduce a measure m on Ω to describe the population composition. In other words, for $\omega \subset \Omega$ the probability that a randomly chosen individual has $\xi \in \omega$ is $m(\omega)$. Let $\bar{\xi} = \int_\Omega \xi m(d\xi)$ denote the mean level of sexual activity and $\bar{\bar{\xi}} = \int_\Omega \xi^2 m(d\xi) - \bar{\xi}^2$ the variance. One could think that R_0 is simply proportional to $\bar{\xi}$, but this is the wrong way of taking averages. As we shall show, the formula above leads to the conclusion that R_0 is proportional to

$$\bar{\xi} + \bar{\bar{\xi}}/\bar{\xi}.$$

When the variance is large (which it is whenever there are some very active and some very inactive individuals) the second term contributes a lot and the wrong way of taking averages leads to a much too optimistic estimate of R_0.

If we sample individuals by choosing a *contact* at random, we will not observe the distribution m but the distribution obtained from m by applying a weight ξ. We denote this distribution by α. So,

$$\alpha(\omega) = \left(\int_\omega \xi \, m(d\xi) \right) / \bar{\xi}$$

(the factor $1/\bar{\xi}$ serves to turn α into a probability distribution, i.e. to ensure that $\alpha(\Omega) = 1$. We *assume* that the partners of any individual follow the distribution α, irrespective of the sexual activity level of the individual itself. And we take $b(\xi) = \pi\xi$, where the constant π involves the probability of transmission, given a contact between a susceptible and an infective. The formula then yields

$$R_0 = \frac{\pi}{\bar{\xi}} \int_\Omega \xi^2 \, m(d\xi) = \pi(\bar{\xi} + \bar{\bar{\xi}}/\bar{\xi}) \tag{5.23}$$

which is exactly what we already discussed above.

5.3.2 Additional within-group contacts

Suppose contacts basically follow the pattern just described, but that by an additional mechanism there are extra contacts with individuals having exactly the same h-state (it may be helpful, for the time being, to think of the h-state as being static). Then we can no longer derive an explicit expression for R_0, but we can deduce a rather simple nonlinear equation that R_0 has to satisfy. And from this equation we can derive a quantity, Q_0 say, with the property that $R_0 > 1$ if and only if $Q_0 > 1$ (here we assume that we have a subcritical situation when we restrict contacts to individuals having exactly the same h-state, since otherwise we will always have $R_0 > 1$). Again we list three variants of the assumption and the expression for the threshold quantity Q_0 that results:

$$k_{ij} = a_i b_j + c_j \delta_{ij} \qquad \Rightarrow Q_0 = \sum_{j=1}^{n} \frac{b_j a_j}{1 - c_j}, \tag{5.24}$$

$$k(\xi, \eta) \;=\; a(\xi)b(\eta) + c(\eta)\delta(\xi - \eta) \Rightarrow Q_0 = \int_\Omega \frac{b(\eta)a(\eta)}{1 - c(\eta)}\, d\eta, \qquad (5.25)$$

$$\Lambda(\eta)(\omega) \;=\; \alpha(\omega)b(\eta) + c(\eta)\delta_\eta(\omega) \;\Rightarrow\; Q_0 = \int_\Omega \frac{b(\eta)}{1 - c(\eta)}\alpha(d\eta). \qquad (5.26)$$

Here δ_{ij} is Kronecker's delta, i.e. $\delta_{ij} = 1$ if $i = j$ and zero otherwise. Furthermore, $\delta(\cdot - \eta)$ and δ_η both represent Dirac's delta 'function', i.e. the unit point measure concentrated in η. In fact the formulation of $k(\xi, \eta)$ has debatable mathematical underpinning, and the more precise formulation is exactly the one in terms of $\Lambda(\eta)(\omega)$.

Exercise 5.11 Verify at least one of the expressions for Q_0. Hint: For the special kernels write down the eigenvalue problems.

One can understand Q_0 in terms of 'superindividuals', which here correspond to subcritical epidemics within subpopulations consisting of individuals with one particular h-state. Indeed, if one individual is expected to infect c individuals, these collectively are expected to infect $c \cdot c = c^2$ individuals, et cetera. So the size of the 'clan' of one case is $1 + c + c^2 + c^3 + \cdots = (1 - c)^{-1}$ (see Exercise 1.39). At this clan level we are back to separable mixing and we can use the formula of the preceding subsection, which gives us the corresponding expression for Q_0. Note that we did implicitly assume that $\max c < 1$.

5.3.3 Finite-dimensional range

When the dimension of the range of the next-generation operator K is finite, but exceeds one, we may still reduce the determination of R_0 to the determination of the dominant eigenvalue of a matrix.

Exercise 5.12 Assume that $\Lambda(\eta)(\omega) = \sum_{i=1}^{n} \alpha_i(\omega)b_i(\eta)$. Show that R_0 is the dominant eigenvalue of the matrix L with entries

$$l_{ij} = \int_\Omega b_i(\eta)\alpha_j(d\eta).$$

Hint: Show first that the range of K is spanned by the α_i. Next compute how the coefficients with respect to this basis transform under K.

In Section 7.4 we shall consider an example having this form. But here we wish to understand the meaning of the assumption. So, what we are after is a version of a finite-dimensional range condition that allows an interpretation. Therefore assume that ξ has two components: a discrete one (which we indicate by i or j but also by m or f, for male and female respectively) and a continuous one (which we indicate by ζ and for which we have sexual activity as the motivating concrete example). We denote by $\widetilde{\Omega}$ the set over which the continuous variable ranges and let $\widetilde{\omega} \subset \widetilde{\Omega}$. For $\omega = (i, \widetilde{\omega})$ and $\eta = (j, \zeta)$ we use the notation

$$\Lambda(\eta)(\omega) = \Lambda_i(j, \zeta)(\widetilde{\omega}).$$

So $\Lambda_i(j, \zeta)(\widetilde{\omega})$ is the expected number of new cases with discrete component i and continuous component in $\widetilde{\omega}$, caused by one individual with h-state (j, ζ). We assume

that
$$\Lambda_i(j,\zeta)(\widetilde{\omega}) = \alpha_i(\widetilde{\omega})b_{ij}(\zeta),$$

which means that, conditional on the discrete component being i, the probability distribution of the continuous component is α_i, independently of the h-state of the individual that causes the infection. But dependence in the discrete component is still allowed, as described by b_{ij}. We will also refer to the finite-dimensional range condition as multigroup separable mixing.

Exercise 5.13 Deduce that, under this assumption, R_0 is the dominant eigenvalue of the matrix L with entries

$$l_{ij} = \int_{\Omega} b_{ij}(\zeta)\alpha_j(d\zeta).$$

As a concrete example, consider a population structured according to sex and sexual activity level. The population composition is described by the two probability distributions μ_f and μ_m. As before, we denote the mean by a bar and the variance by a double bar. Moreover, we introduce weighted distributions

$$\alpha_i(\widetilde{\omega}) = \frac{1}{\bar{\zeta}_i} \int_{\widetilde{\omega}} \zeta\mu_i(d\zeta), \quad i = f, m.$$

Assuming strict heterosexuality, we are led to define

$$b_{ij}(\zeta) = c_{ij}\zeta, \quad \text{with } c_{ij} = 0 \text{ if } i = j$$

(note that it is not necessarily the case that $c_{fm} = c_{mf}$, since the probability of transmission may be asymmetric).

The matrix L is then given by

$$\begin{pmatrix} 0 & \frac{c_{fm}}{\bar{\zeta}_m} \int_0^\infty \zeta^2\mu_m(d\zeta) \\ \frac{c_{mf}}{\bar{\zeta}_f} \int_0^\infty \zeta^2\mu_f(d\zeta) & 0 \end{pmatrix},$$

and we conclude that

$$R_0 = \sqrt{c_{fm}c_{mf}\left(\bar{\zeta}_m + \frac{\bar{\bar{\zeta}}_m}{\bar{\zeta}_m}\right)\left(\bar{\zeta}_f + \frac{\bar{\bar{\zeta}}_f}{\bar{\zeta}_f}\right)}.$$

We will study two examples of multigroup separable mixing in Section 5.5.

5.4 Submodels for the kernel

In this section, we address the dynamics of h-state change, the contact structure and the probability of transmission. We will then put the pieces together to give a more specific form of the kernel k.

A fruitful and often applicable way of modelling change in an individual's h-state is to regard h-state change as a Markov process. For this we introduce

$$P(\tau,\omega,\eta) = \text{probability that an individual originally}$$
$$\text{with h-state } \eta \text{ has an h-state with}$$
$$\text{value in } \omega \subset \Omega, \ \tau \text{ units of time later.}$$

(Note that the possibility $P(\tau, \Omega, \eta) < 1$ accounts for death.)

Examples:

1. *Static h-state*: $P(\tau, \omega, \eta) = \delta_\eta(\omega)$ (if death can occur one has to multiply with the survival probability as a function of τ and η).

2. *Age*: Let

$$P(\tau, \omega, \eta) = \frac{\mathcal{F}(\eta + \tau)}{\mathcal{F}(\eta)} \delta_{\eta+\tau}(\omega),$$

where $\mathcal{F}$ is the survival probability as a function of age. (Note that $\frac{\mathcal{F}(\eta+\tau)}{\mathcal{F}(\eta)}$ is the conditional probability to survive until at least age $\eta + \tau$, given that the individual was alive at age η.)

3. *Finite Markov chains*: Let v_η denote the unit vector whose ηth component equals 1 while all other components are zero. Then the probability vector $P = P(\tau, \eta)$ changes according to

$$\frac{dP}{d\tau} = \Sigma P - DP,$$

where Σ describes jumps and where the (diagonal) matrix M describes death. To the equation for change we have to add the initial condition that $P = v_\eta$ at $\tau = 0$. So formally,

$$P(\tau, \eta) = e^{(\Sigma - D)\tau} v_\eta \, ,$$

but of course the actual computation of $\exp[(\Sigma - D)\tau]$ from the 'data' $\Sigma - D$ is a laborious task. With forethought, we note already now that

$$\int_0^\infty e^{(\Sigma - D)\tau} d\tau = -(\Sigma - D)^{-1}.$$

The point is that the left-hand side makes clear that the ijth component of this matrix equals the expected time that an individual will spend in state i in the rest of its life after we have observed the individual in state j. The right-hand side tells us how to compute the matrix from the data $\Sigma - D$. The matrix $\Sigma - D$ will be called the Markov transition matrix (although actually it describes both transitions and death).

Concerning contacts, there is very little that can be said in any generality. So let us just observe that we need to specify

$$c(\xi, \zeta) \quad = \quad \text{probability per unit of time that an individual}$$
$$\text{with current h-state } \zeta \text{ has contact with an}$$
$$\text{individual with current h-state } \xi.$$

With regard to the probability of transmission, one has to keep in mind that whenever one is decomposing a quantity into multiplicative factors there is a certain degree of arbitrariness. If the probability of transmission is allowed to depend on the h-states of the individuals involved in a contact, it is logical to include that dependence in the ingredients that we discuss now. However, since we have already included a general function of two variables in the preceding paragraph, this gives a redundancy, which we prefer to avoid. Therefore we choose to describe the probability of transmission as a function h of the time τ elapsed since infection took place.

We now put the various pieces together. Being more specific entails a proliferation of modelling ingredients. Indeed, we have arrived at the identity

$$k(\xi, \eta) = \int_0^\infty h(\tau) \int_\Omega c(\xi, \zeta) P(\tau, d\zeta, \eta) \, d\tau,$$

which expresses k in terms of more easily mechanistically interpretable quantities at the expense of having to deal with three functions instead of one. In a bottom up approach one starts with a multitude of ingredients and aspires to combine the relevant information in a few computable numbers, such as R_0. Here we stárted from R_0, found that we need k to determine it, and have now found that we need to specify h, c and P to determine k and could go on to anatomise these. We will, however, do so only in more specific contexts and not in general. One example is treated in detail in Section 5.5 below.

Exercise 5.14 Assume the h-state space is a finite set. Let the h-state dynamics be described by the Markov transition matrix $\Sigma - D$. Let h be constant. Show that

$$k_{ij} = -h \sum_{l=1}^n c_{il} (\Sigma - D)_{lj}^{-1} .$$

Hint: Recall the remarks for a finite Markov chain above.

What remains to be done is to specify c. We do this in terms of the steady demographic state of susceptibles that the infective agent invades. This steady state can, for example, be obtained from a system of equations describing the dynamics of the susceptibles, distributed according to h-state, in the absence of infection. In applications, one could have the possibility to specify the demographic steady state from data (think of the age pyramid for example) or estimates (or even experimental design for certain problems related to animal or plant infections).

5.5 Extended example: two diseases

It has been observed in, for example, Africa that certain genital ulcer diseases, particularly chancroid and syphilis, can increase the risk of HIV infection. The damage that these ulcerative sexually transmitted diseases (we write USTDs for short) cause to the genital skin and membranes may facilitate both transmission and acquisition. AIDS has clearly been able to establish itself in the heterosexual population in Africa, as opposed to the situation in Europe and the USA. There transmission is highest in other subpopulations. Because of the higher prevalence of other STDs in Africa, one can pose a few obvious questions. To what degree can USTDs that are endemic in a population facilitate the spread of HIV into that population (for example the heterosexual population in Europe or in the USA)? What is the efficacy of control measures aimed at these USTDs in halting the spread of HIV in that population? At which aspects of the USTD should control measures then be aimed to be most effective? We will not attempt to answer these questions; we will, however, illustrate the building of submodels for the kernel—involving a major and a (relatively) minor disease—with which these questions could be studied. We do so in the form of a series of exercises. We give a few additional exercises where one can study the relative efficiency of various control measures aimed at the minor disease, using the results derived.

We call the major disease (in the sense of incurable) 'D' and the minor disease 'd'. We ask how the R_0 for invasion of D into a population where d is endemic depends on the parameters that govern the spread of d and D. Let us first look at (5.27) as a very simple model for D. For a change, we choose absolute numbers of individuals instead of densities as the dimension of our variables. We consider

$$\frac{dI}{dt} = \frac{\beta SI}{N} - (\mu + \sigma)I, \tag{5.27}$$

where S and I are population sizes of D-susceptibles and D-infecteds respectively, and σ describes the increased mortality rate due to disease D. For our purpose it is convenient to split β into two factors, $\beta = \rho p$, where ρ is the number of new contacts per individual per unit of time, assumed to be common to d and D transmission, and where p is the probability that D-transmission is successful upon contact.

Exercise 5.15 Let R_0^D be the basic reproduction ratio for the invasion of D into a homogeneous. Give an expression for R_0^D.

Assume that disease d is in an endemic steady state. We want to calculate R_0 for the disease D in a heterogeneous population, assuming that the susceptibility to D is, for individuals having d, v times as large as for individuals without d. What we have in mind is that meetings between individuals are totally random, but that the success ratio for infection transmission, given that contact takes place, is enlarged by a factor v. We denote by $w > 1$ the factor by which the success ratio is enlarged when a D-infectious individual is also suffering from d, and let p be the success ratio when both individuals involved in the contact are free from d. Consistency demands that $pvw \leq 1$. For our h-state space we take $\Omega = \{0, +\}$, where '0' means free of d, and '+' means having d. We assume that meetings occur independently of the h-state of the individuals involved and that the contact rate (i.e. the number of contacts per individual per unit of time) is given by ρ. Since we are interested in calculating R_0 for D-invasion, we start—as usual—with a population consisting of D-susceptibles only, and can therefore write $S \approx N$.

In Section 5.4 we noted the arbitrariness in assigning multiplicative factors to either infectivity or contacts. Write $b_{il} = hc_{il}$ for the combined factors describing the (constant) infectivity and contact pattern in the kernel as given in Section 5.4. Describe by N_0 and N_+ the steady (with respect to d) state population sizes of '0' and '+' individuals, in the absence of D.

Exercise 5.16 Check that the matrix $B = (b_{il})_{1 \leq i, l \leq 2}$ is given by

$$B = \frac{\rho p}{N} \begin{pmatrix} N_0 & N_0 w \\ N_+ v & N_+ vw \end{pmatrix}.$$

Before proceeding, we derive expressions for N_0 and N_+ in terms of N and parameters describing the spread of d. Let ζ denote the (constant) force of d-infection in the steady state and let γ be the probability per unit of time that d is cured by treatment (whereupon susceptibility to d returns). We do not concern ourselves with the question of how these parameters arise, we assume that they completely describe the endemic dynamics of d phenomenologically. Furthermore, let μ be the natural death rate.

Exercise 5.17 Give a system of two ODE for N_0 and N_+, based on the assumptions above (also recall Section 3.2), and check that the steady state of this system is characterised by

$$N_0 = \frac{\gamma + \mu}{\gamma + \mu + \zeta} N, \quad N_+ = \frac{\zeta}{\gamma + \mu + \zeta} N,$$

where $N = N_0 + N_+$.

Exercise 5.18 Give the Markov transition matrix $G := \Sigma - M$, as introduced in Section 5.4. Check that $-G^{-1}$ is given by

$$\frac{1}{(\mu + \sigma)(\mu + \sigma + \gamma + \zeta)} \begin{pmatrix} \mu + \sigma + \gamma & \gamma \\ \zeta & \mu + \sigma + \zeta \end{pmatrix}.$$

and that the next-generation matrix K is given by

$$K = -B\, G^{-1}.$$

Exercise 5.19 Does K have a one-dimensional range? How is the answer connected to a specific assumption we have made for the entries of the matrix B?

Exercise 5.20 Let B operate on a vector $x = (x_1, x_2)^\top$. Conclude that we can write

$$Bx = \frac{\rho p}{N}(x_1 + w x_2) \begin{pmatrix} N_0 \\ v N_+ \end{pmatrix},$$

and that therefore B has a one-dimensional range spanned by $(N_0, N_+ v)^\top$. Which vector therefore spans the range of K?

Exercise 5.21 Show that the eigenvalue problem $K\phi = \lambda\phi$ can be written as

$$-\frac{\rho p}{N}\left(\begin{pmatrix} 1 \\ w \end{pmatrix} \cdot G^{-1}\phi\right) \begin{pmatrix} N_0 \\ v N_+ \end{pmatrix} = \lambda\phi.$$

Check that an eigenvector of K is given by

$$\phi = \frac{\rho p}{N} \begin{pmatrix} N_0 \\ v N_+ \end{pmatrix}$$

and that the only non-zero eigenvalue of K is

$$-\frac{\rho p}{N} \begin{pmatrix} 1 \\ w \end{pmatrix} \cdot G^{-1} \begin{pmatrix} N_0 \\ v N_+ \end{pmatrix}.$$

Exercise 5.22 Check that R_0 is given by

$$R_0 = \rho p \frac{(\gamma + \mu)(\mu + \sigma + \gamma + \zeta w) + \zeta v w(\frac{\gamma}{w} + \mu + \sigma + \zeta)}{(\gamma + \mu + \zeta)(\mu + \sigma)(\mu + \sigma + \gamma + \zeta)}. \qquad (5.28)$$

Before doing the algebra, what do you expect the special case $w = v = 1$ to give?

In the following exercises we take a brief look at one way to extract information from the complex function of the parameters that (5.28) is. For this we note that R_0^D is a factor in the right-hand side and that we can therefore rewrite (5.28) as $R_0 = R_0^D F$, with

$$F := \frac{(\gamma + \mu)(\mu + \sigma + \gamma + \zeta w) + \zeta v w(\frac{\gamma}{w} + \mu + \sigma + \zeta)}{(\gamma + \mu + \zeta)(\mu + \sigma + \gamma + \zeta)}.$$

We now study only the multiplication factor F. The way in which F depends on its ingredients can give us a first idea of the relative influence that the various parameters determining d have on the ability of D to invade.

Exercise 5.23 i) First show that, since all parameters are positive, $F \geq 1$.

ii) Show that for large γ or small ζ, or for v and w close to one, $F \approx 1$. Is this obvious from the biological interpretation?

iii) Show that F is a strictly decreasing function of γ and a strictly increasing function of ζ.

iv) Show that for ζ relatively large (i.e. high infective pressure for disease d), we have $F \approx vw$.

One can draw graphs of F as a function of γ for various fixed values of ζ, or as a function of ζ for fixed γ (keeping the other parameters constant). One can then study questions such as: in attempts to lower F (and thereby to decrease R_0 for D), is it more efficient to aim control measures at increasing γ (improving medical care) or at decreasing ζ (e.g. by campaigning against unprotected sex)? Under what conditions is one preferable to the other? We will not go into that much detail.

The previous exercise hints that notably the product vw can be an important determinant of the size of F. This motivates us to look at the special case $v = w$ in more detail.

Exercise 5.24 i) Show that, for $v = w$, we can write

$$F = \left(1 - \frac{N_+}{N}\right) \frac{\mu + \sigma + \gamma + \zeta v}{\mu + \sigma + \gamma + \zeta} + \frac{N_+}{N} v^2 \frac{\frac{\gamma}{v} + \mu + \sigma + \zeta}{\mu + \sigma + \gamma + \zeta}.$$

ii) Make, in addition, the following assumptions:

$$\frac{\gamma}{\gamma + \mu} \approx 1, \qquad \frac{\sigma}{\gamma + \mu} \approx 0.$$

How could these be motivated biologically?

iii) Show that, with these assumptions, we have

$$F \approx 1 + \frac{N_+}{N}(v-1)\left(\frac{N_+}{N}(v-1)+2\right).$$

We conclude that, under the given assumptions, F approximately increases quadratically with both $v-1$ and the prevalence of d in the population.

In order to get even more familiar with the techniques from this chapter, we provide a further series of exercises for the d/D setting, with some added complications.

Exercise 5.25 Let us incorporate into our model the influence of disease D on the cure rate of disease d by introducing a factor z ($0 \le z \le 1$) that describes to what extent the probability per unit of time to recover from d is decreased by the presence of D. Retaining all assumptions in the previous calculations, how do the steady-state population sizes N_0 and N_+ change; how do B and the transition rate matrix G change? How does R_0 change? In the case that $v = w = 1$, will z matter?

In the remaining exercises we regard two extensions of these derivations to illustrate the idea of multigroup separable mixing (see Section 5.3.3). The first allows the individuals to be identified by sex, in addition to the marker 0 or +. Let the index f denote females and the index m males. We shall now allow some parameters to depend on sex, reflecting, for example, a higher probability of D-transmission from males to females than vice versa, or a higher contact rate for females if we concentrate on a subpopulation of prostitutes.

The four states $(f,0)$, $(f,+)$, $(m,0)$ and $(m,+)$, which we shall always consider in that order, naturally separate into two pairs $(f,0)$, $(f,+)$ and $(m,0)$, $(m,+)$ since the (f,m) distinction is static, while the $(0,+)$ distinction is dynamic. The dynamics are then described by

$$G = \left(\begin{array}{cc} G_f & 0 \\ 0 & G_m \end{array} \right),$$

with both G_f and G_m as the matrix in Exercise 5.18, where the parameters γ and ζ are allowed to be different for f and m. Consequently, we have

$$-G^{-1} = \left(\begin{array}{cc} -G_f^{-1} & 0 \\ 0 & -G_m^{-1} \end{array} \right).$$

The contact rates are ρ_f and ρ_m respectively, with consistency requiring that $\rho_f N_f = \rho_m N_m$ if all contacts are heterosexual, where N_f and N_m are the population sizes of females and males respectively. In self-explanatory notation, we have that $N_f = N_{f,0} + N_{f,+}$ and $N_m = N_{m,0} + N_{m,+}$ where $N_{f,0}$ and $N_{f,+}$ can be related to N_f and the parameters γ, μ and ζ with index f exactly as in Exercise 5.17, and similarly $N_{m,0}$ and $N_{m,+}$ can be related to N_m and the parameters γ, μ and ζ with index m.

Let p_{fm} be the D-transmission probability during a contact of a D-infectious male with a D-susceptible female, both of which are free from d. If the D-infectious individual is female while the D-susceptible individual is a male, the corresponding quantity is denoted by p_{mf}. The enhancement factors are v_f and v_m for d-infected female/male D-susceptibles and w_f and w_m for d-infected female/male D-infectious individuals.

Exercise 5.26 Convince yourself that the analogue of the matrix B from Exercise 5.16 is now given by

$$B = \begin{pmatrix} 0 & B_{fm} \\ B_{mf} & 0 \end{pmatrix},$$

where 0 is a 2×2 matrix containing only zeros, and where

$$B_{fm} = \begin{pmatrix} \rho_m \frac{N_{f,0}}{N_f} p_{fm} & \rho_m \frac{N_{f,0}}{N_f} p_{fm} w_m \\ \rho_m \frac{N_{f,+}}{N_f} p_{fm} v_f & \rho_m \frac{N_{f,+}}{N_f} p_{fm} v_f w_m \end{pmatrix}$$

and B_{mf} is the same, but with f and m interchanged throughout. Convince yourself that the range of B is spanned by two vectors

$$\frac{\rho_m}{N_f} p_{fm} \begin{pmatrix} N_{f,0} \\ N_{f,+} v_f \\ 0 \\ 0 \end{pmatrix}, \qquad \frac{\rho_f}{N_m} p_{mf} \begin{pmatrix} 0 \\ 0 \\ N_{m,0} \\ N_{m,+} v_m \end{pmatrix}.$$

Exercise 5.27 Use $K = -BG^{-1}$ and the results on the range of B of the preceding exercise to derive an explicit expression for R_0 in terms of $G_f^{-1}, G_m^{-1}, \rho_m, \rho_f,$ $p_{fm}, p_{mf}, v_f, v_m, w_f, w_m, \frac{N_{f,0}}{N_f}, \frac{N_{f,+}}{N_f}, \frac{N_{m,0}}{N_m}$ and $\frac{N_{m,+}}{N_m}$. Hint: Show first that R_0 is the dominant eigenvalue of the matrix

$$L = \begin{pmatrix} 0 & l_{fm} \\ l_{mf} & 0 \end{pmatrix},$$

with $l_{fm} = -\left(\frac{1}{w_m}\right) \cdot G_m^{-1} \psi_2$ and $l_{mf} = -\left(\frac{1}{w_f}\right) \cdot G_f^{-1} \psi_1$, where

$$\psi_1 = \frac{\rho_m}{N_f} p_{fm} \begin{pmatrix} N_{f,0} \\ N_{f,+} v_f \end{pmatrix}$$

and

$$\psi_2 = \frac{\rho_f}{N_m} p_{mf} \begin{pmatrix} N_{m,0} \\ N_{m,+} v_m \end{pmatrix}.$$

As a second extension, we regard (fixed) sexual activity level as the first component of the h-state. We let this trait take discrete values $i \in \{1, 2, ...\}$ (but this is not a necessity). Assume

$$\frac{dN(i,+)}{dt} = i\zeta N(i,0) - \gamma N(i,+) - \mu N(i,+)$$

and write $\phi = (\phi_0, \phi_1, ...)^\top$, with

$$\phi_i := \begin{pmatrix} \phi(i,0) \\ \phi(i,+) \end{pmatrix}.$$

The next-generation operator is now an infinite matrix acting on ϕ, where ϕ is an infinite array of two-vectors. Let the Markovian h-state dynamics for the ith two-vector be described by the matrix Σ_i.

Exercise 5.28 Write $N_i = N(i,0) + N(i,+)$. Give $N(i,0)$ and $N(i,+)$. Give Σ_i and G_i operating on the ith two-vector ϕ_i.

Exercise 5.29 Give a proportionate mixing expression for the meeting rate between $(i,\cdot)$-individuals and $(j,\cdot)$-individuals.

Exercise 5.30 Convince yourself that the ijth element of K is given by

$$K_{ij} = \frac{-\rho p i j}{\sum_k k N_k} \begin{pmatrix} N(i,0) & w N(i,0) \\ v N(i,+) & w v N(i,+) \end{pmatrix} G_j^{-1}.$$

Calculate $(K\phi)_i$. Show that the range of K is spanned by the infinite array of the two-vectors

$$i \begin{pmatrix} N(i,0) \\ v N(i,+) \end{pmatrix} \propto i \begin{pmatrix} \gamma + \mu \\ i v \zeta \end{pmatrix}.$$

Exercise 5.31 Show that

$$R_0 = \frac{-\rho p}{\sum_k k N_k} \begin{pmatrix} 1 \\ w \end{pmatrix} \cdot \sum_{j=0}^{\infty} j^2 G_j^{-1} \begin{pmatrix} N(j,0) \\ v N(j,+) \end{pmatrix}.$$

We leave it to those readers who are not yet exhausted to combine both extensions, that is, discuss heterosexual transmission taking sexual activity levels into account.

5.6 Pair formation models

In deterministic models in which, by assumption, all subpopulation sizes are infinite, two individuals never have contact twice.

Exercise 5.32 Digest this statement. Do you agree? Reflect on the following arguments: i) the expected number of contacts of some given individual in a given finite time interval is finite; ii) there are infinitely many candidates for a contact and they are 'chosen' with equal probability; iii) ergo, for any particular *couple* of individuals the probability of having contact in that time interval is zero.

Such an 'idealisation' seems questionable when, for instance, STDs are considered, but also when one thinks of families, schools, offices, ... As an alternative, we may conceive of a population as a network, with connections that form and break but exist for some extended period of time. There we think of transmission as being restricted to individuals that are connected. When every individual can be connected to at most one other individual, the resulting 'network' is simply a collection of singles and pairs, which makes the bookkeeping simple. It is this class of pair formation models that we consider in the present section.

What kind of difference do we expect? The key point is that, from the point of view of the infective agent, contacts between two infected individuals are wasted. Let us try to determine R_0 and look for such an effect.

As warm-up, we consider the pair formation and dissociation process by itself, in the absence of infection transmission. We use a Markov chain description as summarised in Figure 5.1.

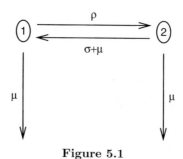

Figure 5.1

We assume an exponential lifetime distribution with parameter μ, a probability per unit of time ρ of acquiring a partner and a probability per unit of time σ that a partnership ends in a divorce. Since an individual can also return to the single state due to the death of the partner, the rate from 2 to 1 equals $\sigma + \mu$. The death of the individual that we consider takes it to a hypothetical state outside the system or, in other words, our probability distributions will be defective.

Exercise 5.33 i) Show that the transition matrix

$$G = \begin{pmatrix} -\rho - \mu & \sigma + \mu \\ \rho & -\sigma - 2\mu \end{pmatrix}$$

has determinant $D = \mu(\rho + \sigma + 2\mu)$ and that

$$-G^{-1} = \frac{1}{D} \begin{pmatrix} \sigma + 2\mu & \sigma + \mu \\ \rho & \rho + \mu \end{pmatrix}.$$

ii) Use these results to check that any living individual has an expected future lifespan $1/\mu$ and that, if it is single, it will spend a fraction $(\sigma + 2\mu)/(\rho + \sigma + 2\mu)$ of this time in that state, while, if it is in a pair, this fraction equals $(\sigma + \mu)/(\rho + \sigma + 2\mu)$.

iii) Show in two different ways that the expected number of partners of a single individual equals $\rho(\sigma + 2\mu)/D$: a) divide the expected time spent in a pair by the expected duration of a partnership; b) denote the quantity we wish to determine by Q and use first (and second) step analysis and the Markov property to derive the equation

$$Q = \frac{\rho}{\rho + \mu} \left(1 + \frac{\sigma + \mu}{\sigma + 2\mu} Q \right).$$

Next, let us distinguish between susceptibles and infectives. To make life simple at first, assume that a newly infected individual is infectious right away and that it has constant infectivity for the rest of its life. We also assume that the disease does not give rise to an extra death rate. What we want is to focus on a newly infected individual and to compute the expected number of new cases that it will make, i.e. R_0. Necessarily, a newly infected individual will be in a pair with another infective. Before it can infect anybody, it has to pass through the single state. After it has acquired a new partner, the infected individual that we consider may or may not infect the new partner. Note that we assume that the new partner is a susceptible. We are only

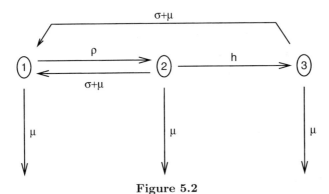

Figure 5.2

interested in the initial phase of the process when the infectives still form a negligible fraction of the population. We conclude that the states of the Markov chain need some modification; in particular the previous state 2 has to be split into two states, one for each possible state of the individual's partner. More precisely, we now consider Figure 5.2, where h is the probability per unit of time that the agent is transmitted from an infective to its susceptible partner.

We claim that

$$R_0 = \frac{\rho h(\sigma + \mu)}{\mu(\rho + \sigma + 2\mu)(h + \sigma + 2\mu)}.$$

Exercise 5.34 i) Show that the probability that a susceptible partner is infected equals $h/(h + \sigma + 2\mu)$.

ii) A newly infected individual has state 3. Show that the probability that it will enter state 1 is given by $(\sigma + \mu)/(\sigma + 2\mu)$.

iii) Argue that the expected number of partners of an infected single individual is, as calculated in Exercise 5.33, equal to $\rho(\sigma + 2\mu)/D$. Hint: does it matter that we consider an infective? Does it matter, as far as we restrict our attention to partnership status, whether the individual is in state 2 or in 3?

iv) Derive the expression for R_0.

Exercise 5.35 'We' give an alternative derivation of the expression for R_0.

i) Let P_{13} be the probability to ever arrive in 3 starting from 1. Use the Markov property to deduce that

$$P_{13} = \frac{\rho}{\rho + \mu} \left(\frac{h}{h + \sigma + 2\mu} + \frac{\sigma + \mu}{h + \sigma + 2\mu} P_{13} \right),$$

and solve for P_{13}.

ii) Let P_{33} be the probability to ever return to 3 when starting from 3. Show that

$$P_{33} = \frac{\sigma + \mu}{\sigma + 2\mu} P_{13}.$$

iii) Show that $R_0 = P_{33}(1 + R_0)$, and determine R_0.

iv) Observe that $R_0 > 1 \Leftrightarrow P_{33} > \frac{1}{2}$ and try to interpret the latter condition directly as a threshold condition.

Motivated by the preceding exercise, we make the following useful observation. A newly infected individual is in state 3 and can itself infect some other individual only when in state 2. From state 3 to state 2, it has to pass through state 1. So we can also base our bookkeeping on passage through 1. Of course, an infected individual in state 1 may re-enter state 1 after having infected someone else, i.e. from state 3. When we define

$$\widetilde{R}_0 = \text{expected number of individuals entering state 1 from state 3}$$
$$\text{due to one individual that just entered state 1 from state 3,}$$

then we also count such a re-entering individual. Note that clearly $\widetilde{R}_0$ has threshold value 1 but that it is, in general, different from R_0.

Exercise 5.36 Show that
$$\widetilde{R}_0 = 2\frac{\sigma + \mu}{\sigma + 2\mu}P_{13} = 2P_{33}.$$

This different way of doing the bookkeeping is of much help when dealing with different types of partnerships, as in Exercise 5.40 below. Before turning to further complications, however, let us look at simplifications and, more particularly, at the relation with R_0 for models that disregard pair formation.

Exercise 5.37 i) In the limit $\sigma, h \to \infty$ with $\frac{h}{\sigma} \to p$ we arrive at a model with contact rate ρ and success ratio $q = p/(p+1)$ for which $R_0 = \rho q/\mu$. Verify this statement.

ii) Take $h = p\sigma$ for fixed p. Show that R_0 is an increasing function of σ.

iii) Contemplate the difficulty of gauging epidemic models when comparing them.

Exercise 5.38 i) Suppose the disease causes an additive death rate κ. Modify the model and compute R_0.

ii) Formulate a model in which an individual enters a state of mourning after the death of its partner, from which it may jump with constant probability per unit of time to the single state.

Exercise 5.39 Suppose an infected individual has an exponentially distributed infectious period with parameter α, during which it has constant infectivity. Modify the original model and compute R_0.

Exercise 5.40 Suppose there are two types of pairs: serious ones (with divorce rate σ_1 and transmission probability/time h_1) and casual ones (with divorce rate σ_2 and transmission probability/time h_2). Assume that a single still acquires a partner with rate ρ but that it will then form a serious pair with probability f and a casual pair with probability $1 - f$. Elaborate the model formulation and compute R_0. Hint: Use the 'decoupling' argument about passage through the single state. This makes the problem one-dimensional, despite the fact that there are now two different states for newly infected individuals to be in.

Exercise 5.41 So far we have not taken the sex of an individual into account. If the sex ratio is $\frac{1}{2}$, does it matter? Can you think of situations in which it does matter?

5.7   Summary: a recipe for R_0

In this section we summarise the definition of R_0 in host populations consisting of various types of individuals, and give a recipe for its calculation. A key aspect of our characterisation is that we regard *generations* of only the infected individuals as they are distributed (at the moment of becoming infected) over all possible individual types (collected in the h-state space Ω). We then construct a *linear positive* operator that provides the next generation of infecteds (both the size and its distribution over Ω) in terms of the present generation. In the case that there are only finitely many types, the operator is (represented by) a matrix. Iteration of the next-generation operator describes the initial progression of infection within the heterogeneous population. Hence the value of the dominant eigenvalue/spectral radius of the operator relative to the threshold 1 determines whether generations of infecteds increase in size or decrease to zero. We therefore defined R_0 to be the dominant eigenvalue of the linear next-generation operator. The interpretation is the expected number of new cases produced by a *typical* infected individual during its entire infectious period, in a population consisting of susceptibles only. By 'typical' we mean an individual whose type is distributed over Ω according to the eigenvector corresponding to R_0. Once we suitably normalise this eigenvector, the vector can be interpreted as the probability distribution for the type at the moment of becoming infected (type at birth).

In general, the dominant eigenvalue (spectral radius) of a positive operator is not easy to compute. We have seen, however, that under additional assumptions (but for arbitrary Ω), R_0 can be expressed as a formula, rather than as the solution of an equation. The theory of linear positive operators, moreover, presents various approaches for the approximation of the spectral radius(see e.g. Krasnosel'skij et al. (1989)).

We now give a recipe, an algorithm, to compute R_0 from first principles. The basic steps are:

Identification Specify the relevant h-states and transmission routes and single out the states at birth from among them; we denote the set of states at birth by Ω_b.

Modelling Construct the elements $k(\xi, \eta)$ of the kernel, defined as the expected number of new cases of birth-state ξ that are produced by one infective with birth-state η.

Calculation Compute the dominant eigenvalue of the next-generation operator K defined by

$$(K\phi)(\xi) = \int_{\Omega_b} k(\xi, \eta)\phi(\eta)\, d\eta.$$

We make some additional comments about the individual steps before we give two exercises to illustrate the recipe.

Connected to the identification step is that one chooses the h-states in such a way that all relevant information can be incorporated. For example, adding transmission routes can increase the number of h-states that should be considered, in particular as states at birth. In general we can distinguish indirect transmission (via a vector, an intermediate host, contaminated food or environment), direct horizontal transmission between individuals, vertical transmission (i.e. from mother to unborn offspring) and *diagonal* transmission. (We propose the latter term for situations where horizontal transmission is also in a sense vertical since it is strictly related to birth (e.g. contact of individuals with infected afterbirth or aborted foetuses as in respectively scrapie and brucellosis).) It may happen (see Exercise 5.43) that an individual born by vertical transmission (so, 'born' in a double sense) behaves differently from a horizontally infected individual.

With regard to modelling the kernel k, one could differentiate between h-states that are characterised by one number and h-states that are higher-dimensional. Furthermore, one can differentiate between h-states that are static (for a given individual) and those that are dynamic. We have presented an approach for dynamic h-states in Section 5.4. As part of the construction of the kernel, one also has to specify the demographic steady state of susceptibles in the absence of infection.

With regard to the calculation of the dominant eigenvalue of K, we have seen various simplifying assumptions in Section 5.3 that lead to explicit expressions or to explicit criteria. We have also seen that one assumption (multigroup separable mixing) can be relevant for higher dimensional h-state space and that this assumption can lead to a reduction of the dimension of the space on which the operator acts.

We end this summary by showing how the recipe works in two examples. In Chapter 7 we will consider an additional example concerning BSE in cattle (Exercises 7.15 and 7.16).

Exercise 5.42 Imagine that plants are grown in a field (for flowers for example) where cuttings are taken at some per capita rate γ. The cuttings are planted in a nursery, where they will mature. Plants in the nursery are replanted in the field with rate ζ. The grower maintains a fixed population size of plants in the field (N_1) and plants in the nursery (N_2). Suppose that a fungus spreads in this host population and that this fungus can be directly transmitted between plants in the field (assuming mass action with transmission rate constant β_1), and between plants in the nursery (assuming mass action with transmission rate constant β_2). In addition, there is a probability p that a cutting taken from an infected plant is itself infected. Let μ_1 and μ_2 be the natural death rates of field and nursery plants, respectively. Finally, let ρ_1 and ρ_2 be the per capita infection-induced death rates of field and nursery plants respectively.

Calculate R_0.

Exercise 5.43 The bovine viral diarrhea virus (BVDV) in cattle has a complicated epidemiology. We base the example on a model by Cherry et al.,[3] and we deliberately give a detailed description to mimic a practical situation in which one wants to compute R_0.

Seven classes of individuals are considered (it is advised to draw a diagram of the classes and the connections between them): susceptibles (S, with death/birth rate μ), latently infected individuals (E, death rate μ), transiently infectious individuals (I, entered from E with rate α, death rate μ), recovered/immune

[3] B.R. Cherry, M.J. Reeves & G. Smith: Evaluation of bovine viral diarrhea virus control using a mathematical model of infection dynamics. *Prev. Vet. Med.*, **33** (1998), 91-108.

individuals (entered from the class I with (recovery) rate γ, split into three subclasses Z_1, Z_2, Z_3; see below) and persistently infected individuals (P). Persistently infected animals have reduced birth rate $\mu - a$ and increased death rate $\mu + b$ compared with transiently infected animals. We assume that calves born to persistently infected mothers are always persistently infected. The three recovered/immune classes have the following interpretation: individuals that are not pregnant (probability π_1) when they are infected move into class Z_1 upon recovery and have long-lasting immunity; animals pregnant for less than 150 days (probability π_2) when they become infected move into class Z_2 upon recovery; animals pregnant for more than 150 days (probability π_3) when they become infected move into class Z_3. From class Z_2 and Z_3 the individuals move into the class Z_1 of long-lasting immunity when they have given birth (with rates ϕ_2 and ϕ_3 respectively). The calves born from individuals in class Z_3 have active immunity to BVDV from birth. Calves from mothers in class Z_2, however, will either abort (probability $1 - \theta$) or are persistently infected while apparently healthy (probability θ). Both persistently and transiently infected animals can cause transient infections by horizontal transmission. We assume that this is described by mass action with transmission rate constants β_1 and β_2 for transient infections caused by transiently and persistently infected animals respectively). We will assume, as seems to be biologically reasonable, that $\beta_2 > \beta_1$, and that β_1 is small.

Let N be the total number of individuals in the herd, regulated by the farmer. First assume that $\beta_1 = 0$ (parts ii and iii), i.e. assume that transiently infected animals do not cause transient infections.

i) Convince yourself that the types at birth in this model are 'living separately' and 'living in the mother's womb'.

ii) Show that the next-generation matrix is given by

$$K = \begin{pmatrix} 0 & \frac{\beta_2 N}{\mu+b} \\ \frac{\alpha}{\mu+\alpha} \frac{\gamma\pi_2}{\mu+\gamma} \frac{\theta\phi_2}{\mu+\phi_2} & \frac{\mu-a}{\mu+b} \end{pmatrix},$$

and that consequently

$$R_0 = \frac{1}{2}B + \frac{1}{2}\sqrt{B^2 + 4A},$$

with

$$B = \frac{\mu - a}{\mu + b},$$

$$A = \frac{\alpha}{\mu + \alpha} \frac{\gamma\pi_2}{\mu + \gamma} \frac{\theta\phi_2}{\mu + \phi_2} \frac{\beta_2 N}{\mu + b}.$$

iii) One can show that $A + B$ has a threshold value that can be extracted (e.g. by using the Routh-Hurwitz criteria; see Edelstein-Keshet (1988)) from a description of the model as a system of eight nonlinear coupled ODE (see Cherry et al., loc. cit.). Note that $A + B$ does not equal R_0, but show that

$$R_0 > 1 \Leftrightarrow A + B > 1.$$

See also Section 6.2 for a more general comparison between R_0 and threshold quantities derived by local stability analysis.

iv) Assume that $\beta_1 > 0$. Give the next-generation matrix K for this case.

Exercise 5.44 Now do Exercise 3.27.

6

And everything else ...

This chapter is devoted to the initial real-time growth rate r, the probability of a minor outbreak, the final size, and the endemic level, with special attention for computational simplifications in the case of separable mixing.

6.1 Partially vaccinated populations (a first example)

To illustrate various aspects of definitions and calculations for heterogeneous populations, we shall treat in some detail a relatively simple example that is of some interest by itself, viz. the case of a population in which a fraction p is vaccinated. We assume that the description of Section 2.1 applies to the population before vaccination. In particular we have that

$$R_0 = c \int_0^\infty A(\tau) \, d\tau,$$

which we assume to be bigger than one, that r is the unique positive root of

$$1 = c \int_0^\infty e^{-r\tau} A(\tau) \, d\tau,$$

that the probability z of a minor outbreak is the unique root in $(0, 1)$ of the equation

$$z = e^{R_0(z-1)},$$

and that the fraction $s(\infty)$ that escapes from a major outbreak is likewise the unique root in $(0, 1)$ of the equation

$$s(\infty) = e^{R_0(s(\infty)-1)},$$

so that actually $z = s(\infty)$. We have in mind that this description concerns a local population in a metapopulation (i.e. a population consisting of separated local populations, by spatial or other characteristics, that are connected by migration of individuals; see Hanski and Gilpin (1997)) and we are also interested in transmission at the metapopulation level. A relevant quantity in this respect is

$$(1 - z)(1 - s(\infty)),$$

the probability of a major outbreak times the size of a major outbreak, since the R_0 at the metapopulation level will be proportional to this quantity. (Note that we use this quantity as a convenient approximation to the expected outbreak size, since it is easily computable.)

Now assume that, as a result of a vaccination campaign, in every local population a fraction p is vaccinated. From a medical or economic point of view, it is the suppression of clinical symptoms that determines the efficacy of a vaccine. From an epidemiological point of view, it is the reduction in susceptibility and infectivity that counts. We shall assume that vaccination reduces the susceptibility by a factor f and the infectivity by a factor ϕ, while leaving the time course of infectivity unaltered. By this we mean that a contact leads to transmission with probability

$fA(\tau)$ if the susceptible is vaccinated while the infective is not,

$\phi A(\tau)$ if the susceptible is not vaccinated but the infective is,

$f\phi A(\tau)$ if both are vaccinated,

where $0 \leq f, \phi \leq 1$. By allowing $f\phi$ to be positive, we acknowledge that the vaccine may not give full protection.

The question we now ask is: how are the analogues of R_0, r, z and $s(\infty)$ characterised for this new situation and what replaces $(1 - z)(1 - s(\infty))$? In particular, we want to know how the answer depends on p, f and ϕ, the three parameters characterising the vaccination coverage and the vaccine efficacy.

We now deal with a structured population, with two types of individuals: unvaccinated ones, which we denote by the index 1, and vaccinated ones, to be denoted by the index 2. Transmission is then described by the matrix

$$\begin{pmatrix} c(1-p)A(\tau) & c(1-p)\phi A(\tau) \\ cpfA(\tau) & cpf\phi A(\tau) \end{pmatrix}$$

during the early rise of an epidemic. In particular R_v, the reproduction ratio for the vaccinated population, is the dominant eigenvalue of the next-generation matrix

$$R_0 \begin{pmatrix} 1-p & (1-p)\phi \\ pf & pf\phi \end{pmatrix}.$$

Note that this matrix has a one-dimensional range spanned by the vector $(1 - p, pf)^\top$ which gives the proportions in which unvaccinated and vaccinated individuals will occur in early generations of the infected subpopulation.

Exercise 6.1 Why did we say 'early' generations?

Exercise 6.2 Show that the non-zero eigenvalue is

$$R_v = (1 - p + pf\phi)R_0.$$

Hint: For every matrix the trace is the sum of the eigenvalues, or, if this hint does not mean anything to you, use the fact that you know the eigenvector.

In the following we assume that $R_v > 1$, i.e. large outbreaks are still possible despite the vaccinations.

Exercise 6.3 Convince yourself that r_v, the real-time growth rate for the population with vaccination, is characterised by the condition that the matrix

$$\left(\begin{array}{cc} c(1-p)\int_0^\infty e^{-r_v\tau}A(\tau)\,d\tau & c(1-p)\phi\int_0^\infty e^{-r_v\tau}A(\tau)\,d\tau \\ cpf\int_0^\infty e^{-r_v\tau}A(\tau)\,d\tau & cpf\phi\int_0^\infty e^{-r_v\tau}A(\tau)\,d\tau \end{array} \right)$$

has dominant eigenvalue one. Next show that this amounts to the condition that

$$1 = R_v\frac{\int_0^\infty e^{-r_v\tau}A(\tau)\,d\tau}{\int_0^\infty A(\tau)\,d\tau}$$

by exploiting once more the one-dimensional range property.

Next we want to derive equations that characterise (determine) extinction probabilities. The plural is justified since we have to deal with the probability π_1 of extinction when starting with one unvaccinated infected individual, and the probability π_2 of extinction when starting with one vaccinated infected individual. An infected individual is making contacts according to a Poisson process with intensity c and with probability $1-p$ that the contacted individual will be unvaccinated, while with probability p it will be vaccinated. If the original infectious individual is itself unvaccinated, the transmission probabilities for these two cases are $A(\tau)$ and $\phi A(\tau)$ respectively, while if the original individual is actually vaccinated, these are $fA(\tau)$ and $f\phi A(\tau)$. With the same reasoning as in Section 1.2.2, we arrive at the system of equations

$$\begin{aligned} \pi_1 &= e^{R_0(1-p)(\pi_1-1)}e^{fR_0p(\pi_2-1)}, \\ \pi_2 &= e^{\phi R_0(1-p)(\pi_1-1)}e^{\phi f R_0p(\pi_2-1)}. \end{aligned} \tag{6.1}$$

Exercise 6.4 Give the derivation in some detail.

Exercise 6.5 Do you expect any special structure in this system? Do you see any special structure? Hint: The infectivity of the two types differs by a factor ϕ.

Exercise 6.6 Define $\xi := (1-p)(1-\pi_1) + fp(1-\pi_2)$. Show that ξ is determined by the equation

$$\xi = (1-p)(1 - e^{-R_0\xi}) + fp(1 - e^{-\phi R_0\xi}) \tag{6.2}$$

and that, once ξ is known, we have explicitly

$$\pi_1 = e^{-R_0\xi}, \quad \pi_2 = e^{-\phi R_0\xi}.$$

Thus we have reduced the system to a scalar problem. How would you interpret ξ? Hint: Victims are made in the proportions $1-p : fp$.

It remains to characterise the final sizes. Let σ_1 denote the fraction of the unvaccinated individuals that escapes a major outbreak and let σ_2 be the same quantity for the vaccinated individuals. Then consistency requires that

$$\begin{aligned} \sigma_1 &= e^{-R_0\{(1-p)(1-\sigma_1)+\phi p(1-\sigma_2)\}}, \\ \sigma_2 &= e^{-fR_0\{(1-p)(1-\sigma_1)+\phi p(1-\sigma_2)\}}. \end{aligned} \tag{6.3}$$

Exercise 6.7 Provide the arguments underlying these equations. Hint: Reread the derivation in Section 1.3.1 of the final-size equation in terms of a probabilistic consistency consideration.

Exercise 6.8 Define $\theta := (1-p)(1-\sigma_1) + \phi p(1-\sigma_2)$. Show that θ is determined by the equation

$$\theta = (1-p)(1 - e^{-R_0\theta}) + \phi p(1 - e^{-fR_0\theta}) \qquad (6.4)$$

and that, once θ is known, we have explicitly

$$\sigma_1 = e^{-R_0\theta}, \quad \sigma_2 = e^{-fR_0\theta}.$$

Note the analogy with equation (6.2) for ξ and, in particular, the role of reversal of ϕ and f in these equations. How would you interpret θ?

We now claim that the relevant quantity to consider when thinking about transmission at the metapopulation level is $\xi\theta$. Indeed, if a local population is challenged from outside with a standard 'amount' of infectious material, the probability that an unvaccinated or a vaccinated individual is infected is proportional to respectively $1 - p$ and fp, with the same constant of proportionality. So, the probability that a major outbreak occurs is proportional to ξ. If a major outbreak occurs, it will hit a fraction $1 - \sigma_1$ of the fraction $1 - p$ of unvaccinated individuals and a fraction $1 - \sigma_2$ of the fraction p of vaccinated individuals. Since we are interested in the output of infectious material towards other local populations that may result from this outbreak, we give the vaccinated individuals a weighting factor ϕ to account for the reduction in infectivity. So, θ is indeed a measure for the output of infectious material that results from a major outbreak.

Thus we have completed the task we set out to perform. As a brief excursion into applied issues, we address the question whether it is more efficient to reduce ϕ or to reduce f, when there is a choice (admittedly this will not happen very often; yet it may be that several vaccines exist and that one has to make a comparison). More precisely, we ask which combination of f and ϕ, satisfying the constraint $f\phi = k$, a constant less than one, minimises the product $\xi\theta$. We shall put $\phi = \frac{k}{f}$ and use f as a free parameter in the interval $[k, 1]$.

Exercise 6.9 Analyse how the product $\xi\theta$ depends on f when $\phi f = k$. Interpret the results. Hint: Exploit the symmetry related to interchanging f and ϕ. Moreover, it is helpful to concentrate on the extreme situations $R_0 \to \infty$ and $R_v = (1-p+pk)R_0$ (see Exercise 6.2) only slightly bigger than one. *Warning:* This is a rather technical exercise.

We conclude that the extreme cases $f = k$ and $f = 1$ are equally good (or bad?), but that they show a trade-off between the likeliness of a major outbreak and the size of a major outbreak. Moreover, we see that what is considered optimal depends on quantitative aspects, no generally valid criterion emerges.

6.2 The intrinsic growth rate r

In Section 1.2.3, our starting point was (a special case of) the equation

$$i(t) = c \int_0^\infty A(\tau) i(t - \tau) \, d\tau \tag{6.5}$$

for the incidence i in the initial phase of an epidemic. A natural question is therefore:

What is the analogue of this equation when we account for heterogeneity in the host population?

The details of the answer depend on the choices we make (in particular concerning the factoring of the expected number of transmissions, like the product $cA(\tau)$ above; recall Section 5.4), but the overall structure is determined by the bookkeeping, that we hope by now is becoming familiar. Our choice is to write

$$i(t, \xi) = N(\xi) \int_0^\infty \int_\Omega A(\tau, \xi, \eta) i(t - \tau, \eta) d\eta d\tau \tag{6.6}$$

where $N(\xi)A(\tau, \xi, \eta)$ is the rate at which an individual that was itself infected while having h-state η is expected to generate new cases with h-state ξ at time τ after its own infection, and where $i(t, \cdot)$ is the h-state specific incidence at time t. The h-state specific population density $N(\xi)$ describes the virgin steady state. The word 'choice' in the above refers to the fact that we have written N as a factor, even though we confine ourselves to stating the interpretation of the product $N(\xi)A(\tau, \xi, \eta)$. The point is that the precise interpretation of $A(\tau, \xi, \eta)$ depends on features of the contact process, as considered in Sections 1.3.2 and 1.3.3. These features, in fact, only matter if we compare different host populations or consider what happens after the initial phase, when susceptible individuals are depleted. We will refer to (6.6) as the equation for the (linearised) real-time evolution (or real-time growth).

In Section 1.2.3, we proceed by substituting the Ansatz $i(t) = ke^{rt}$ into (6.5), which led to the characteristic equation

$$1 = c \int_0^\infty A(\tau) e^{-r\tau} \, d\tau. \tag{6.7}$$

If the host population is heterogeneous, the corresponding Ansatz is

$$i(t, \xi) = e^{rt} \Psi(\xi), \tag{6.8}$$

or, in words, that i factorises into the product of a function of t (which then actually has to be an exponential since its translates should be multiples of the function itself) and a function of h-state ξ. We find that r and Ψ should be such that

$$K_r \Psi = \Psi \tag{6.9}$$

has a positive solution Ψ, where

$$(K_r \phi)(\xi) = N(\xi) \int_\Omega \int_0^\infty A(\tau, \xi, \eta) e^{-r\tau} \, d\tau \, \phi(\eta) \, d\eta. \tag{6.10}$$

In other words, K_r should have eigenvalue 1. Since the kernel is positive and since Ψ should be positive, we can say slightly more if we assume irreducibility: K_r should have dominant eigenvalue one.

Exercise 6.10 Verify that R_0 is the dominant eigenvalue/spectral radius of K_0. Hint: Convince yourself that, in the notation of Section 5.2, $k(\xi, \eta) = N(\xi) \int_0^\infty A(\tau, \xi, \eta) \, d\tau$.

We expect, of course, that

$$\text{sign}(R_0 - 1) = \text{sign } r. \tag{6.11}$$

The key element for a proof is a lemma that states that the spectral radius of K_r is a continuous and strictly decreasing function of r. A precise formulation requires precise assumptions on Ω, A and N. We refrain from a general elaboration here and refer, on the one hand, to the literature[1] and on the other hand to the end of this section for a special case.

The next question then is:

Does $r < 0$ imply that all solutions of (6.6) decay to zero exponentially?

In fact, we have not even addressed this question for the homogeneous case of Section 1.2.3 (but see Exercise 3.9). There are two aspects that need attention:

- For what $\lambda \in \mathbb{C}$ does (6.6) have a non-trivial solution of the form $i(t, \xi) = e^{\lambda t} \Psi(\xi)$?
- When all such λ have negative real part, can we conclude that all solutions decay to zero exponentially?

For the second aspect we refer to the literature.[2] The key observation related to the first aspect is that $|e^{\lambda t}| \leq e^{\text{Re}\lambda t}$ which has as a corollary that the spectral radius of K_λ is less than or equal to the spectral radius of $K_{\text{Re}\lambda}$. So all such λ lie to the left of r in the complex plane. Or, in other words, the sign of r determines whether or not there are such λ in the right-half plane.

The observation that r is the rightmost of all such λ is also the key to the conclusion that, when $r > 0$, every positive solution of (6.6) exhibits exponential growth with exponent r. In summary, the intrinsic growth rate r is the unique real root of the 'equation'

$$\text{spectral radius of } K_r = 1. \tag{6.12}$$

Just as in the case of R_0, one can derive computational simplifications of this rather implicit characterisation, by adopting assumptions that give K_r a special structure. We do not elaborate on this theme, but simply refer back to Section 5.3 for inspiration.

Exercise 6.11 Assume that $A(\tau, \xi, \eta) = a(\xi)b(\tau, \eta)$. Derive an equation for r.

---

[1] See e.g. J.A.P. Heesterbeek: R_0, PhD-thesis, University of Leiden, 1992 (which needs some corrections, W. Desch & R. Grimmer personal communication); H. Inaba: Threshold and stability results for an age-structured epidemic model. *J. Math. Biol.*, **28** (1990), 411-434; H.J.A.M. Heijmans: The dynamical behaviour of the age-size distribution of a cell population. In: J.A.J. Metz & O. Diekmann (1986), pp. 185-202.

[2] O. Diekmann, S.A. van Gils, S.M. Verduyn Lunel & H.-O. Walther: *Delay Equations: Functional-, Complex-, and Nonlinear Analysis*. Springer-Verlag, Berlin, 1995; G. Gripenberg, S-O. Londen & O. Staffans: *Volterra Integral and Functional Equations*. Cambridge University Press, Cambridge, 1990.

Why does R_0 have such a predominant position in epidemic theory, when one could just as well characterise success of invasion in terms of the sign of r, where, furthermore, r is a quantity that relates more clearly to something that can be observed? The reason is that the characterisation of R_0 is more *explicit*, which often helps to derive *explicit* formulas, or approximations, and to compute numerical values.

Many modellers are used to writing down systems of ordinary differential equations and to derive a threshold condition by determining the stability of the infection-free (or virgin) steady state (e.g. by verifying the Routh-Hurwitz criteria; see Edelstein-Keshet (1988) for these criteria). These modellers are not used to introducing infectivity kernels A and calculating R_0 from those. Sometimes (or should we say many times?) they even distrust whether indeed the threshold value one of R_0, as defined in the preceding chapter, corresponds exactly to their stability condition. Often, the appearance of the two conditions is so different that the correspondence cannot be decided by just staring at them (see Exercise 5.42-iii for an example of correspondence). Even more often, the Routh-Hurwitz criteria remain implicit and do not yield an explicit formula straight away. The remainder of this section is intended for modellers who recognise themselves in the above description.

Earlier (in particular in Sections 2.1 and 5.4) we have considered the situation in which infected individuals could be in a finite (say n) number of different states, with state transitions occurring according to a rate matrix Σ, and death occurring according to a diagonal rate matrix D. Here we introduce a third matrix T, to describe reproduction (in an epidemiological sense, so the ijth element of T is the rate at which an infected individual with current h-state j produces secondary cases with state at birth i). We shall show that the zero steady state of the linear system

$$\frac{dx}{dt} = (T + \Sigma - D)x \tag{6.13}$$

is asymptotically stable if and only if $R_0 < 1$, where R_0 is the dominant eigenvalue of the next-generation operator

$$K = -T(\Sigma - D)^{-1}. \tag{6.14}$$

As already explained in Section 5.4, the fact that K is the next-generation matrix follows once we realise that the matrix exponential

$$e^{\tau(\Sigma-D)}$$

applied to the ith unit vector e_i yields the probability distribution over the various h-states at time τ after being infected in state i, and that consequently

$$K = \int_0^\infty T e^{\tau(\Sigma-D)} \, d\tau = -T(\Sigma - D)^{-1}.$$

Here we assume that Σ and D are such that the integral converges, which, actually, is equivalent to $-(\Sigma - D)^{-1}$ being positive.

Lemma 6.12 *Let H be a real matrix with positive off-diagonal elements (i.e. $h_{ij} \geq 0$ for $i \neq j$). Then $e^{\tau H}$ is a positive matrix. Moreover, for the spectral bound $s(H)$ defined by*

$$s(H) = \sup\{\mathrm{Re}\lambda | \lambda \text{ is an eigenvalue of } H\} \tag{6.15}$$

we have the equivalence

$$s(H) < 0 \Leftrightarrow \left(\det H \neq 0 \text{ and } -H^{-1} \geq 0 \right). \tag{6.16}$$

Proof For suitably large $\theta > 0$ we have that $H + \theta I \geq 0$, and hence it follows from the Taylor series definition of the matrix exponential (see e.g. Hirsch & Smale (1974)) that

$$e^{\tau(H+\theta I)} \geq 0.$$

Hence

$$e^{\tau H} = e^{-\tau\theta} e^{\tau(H+\theta I)} \geq 0.$$

Now assume that $s(H) < 0$. Then $\det H \neq 0$, since $\det H = 0$ would imply that $\lambda = 0$ is an eigenvalue of H and hence that $s(H) \geq 0$, a contradiction. It follows from $s(H) < 0$ that the integral

$$\int_0^\infty e^{\tau H} \, d\tau$$

converges. Clearly, $\int_0^\infty e^{\tau H} \, d\tau \geq 0$ since $e^{\tau H} \geq 0$. Moreover, using the Taylor series once again, we see that

$$H \int_0^t e^{\tau H} \, d\tau = \int_0^t e^{\tau H} \, d\tau \, H = e^{tH} - I$$

(where I is the identity matrix), which, by taking the limit $t \to \infty$, implies that

$$\int_0^\infty e^{\tau H} \, d\tau = -H^{-1}.$$

We conclude that $-H^{-1} \geq 0$.

To prove the converse, assume that H^{-1} exists and is negative. Let ϕ be such that $H\phi = r\phi$ with $\phi \geq 0$ and r real. Applying H^{-1} to this identity, we see that necessarily r must be negative. The fact that such a combination of ϕ and r must exist and that $r = s(H)$ follows from the Perron-Frobenius Theorem 5.3 applied to $H + \theta I$. This concludes the proof of Lemma 6.12.

Theorem 6.13 *Let T be a positive matrix, Σ a positive off-diagonal matrix and D a positive diagonal matrix. Assume that the spectral bound $s(\Sigma - D)$ is negative. Let r denote the spectral bound $s(T + \Sigma - D)$ and let R_0 denote the dominant eigenvalue of the positive matrix $K = -T(\Sigma - D)^{-1}$. Then*

$$r < 0 \Leftrightarrow R_0 < 1. \tag{6.17}$$

Proof First note that $r = 0 \Leftrightarrow R_0 = 1$ since $(T + \Sigma - D)\phi = 0 \Leftrightarrow -T(\Sigma - D)^{-1}\psi = \psi$ with

$$\psi = (\Sigma - D)\phi, \quad \phi = (\Sigma - D)^{-1}\psi.$$

Now assume that $r < 0$. Then, by Lemma 6.12, $(T + \Sigma - D)^{-1} \leq 0$, which implies that

$$(\Sigma - D)^{-1}(-T(\Sigma - D)^{-1} - I)^{-1} \geq 0. \tag{6.18}$$

Let ψ be a positive eigenvector corresponding to the dominant eigenvalue R_0 of K. For $R_0 > 1$ we can rewrite the relation $K\psi = R_0\psi$ as

$$(-T(\Sigma - D)^{-1} - I)\psi = (R_0 - 1)\psi \geq 0.$$

Applying the negative matrix $(\Sigma - D)^{-1}$ to this inequality leads to a contradiction with the inequality (6.18). So, when $r < 0$, neither $R_0 = 1$ nor $R_0 > 1$ is possible. Hence $r < 0 \Rightarrow R_0 < 1$.

Next assume that $R_0 < 1$. Then, applying Lemma 6.12 to $K - I$, we find that $(-T(\Sigma - D)^{-1} - I)^{-1} \leq 0$ and hence $(\Sigma - D)^{-1}(-T(\Sigma - D)^{-1} - I)^{-1} \geq 0$, and consequently

$$-(T + \Sigma - D)^{-1} \geq 0. \tag{6.19}$$

Let ϕ be a positive eigenvector of $T + \Sigma - D$ corresponding to the eigenvalue r, i.e. let $(T + \Sigma - D)\phi = r\phi$. If we apply the inequality (6.19) to this identity we obtain, for $r > 0$, the contradiction that $-\phi \geq 0$. So when $R_0 < 1$, neither $r = 0$ nor $r > 0$ is possible. Hence $R_0 < 1 \Rightarrow r < 0$. This concludes the proof of Theorem 6.13.

An early version of this theorem is presented in a paper by Nold (1980).[3] The assumption on $s(\Sigma - D)$ is certainly satisfied when all columns of Σ add to zero and the diagonal elements of D are all strictly positive, which is the usual situation in epidemic models.

The special situation of just one possible state at birth corresponds to

$$Tx = (h \cdot x)\, b,$$

with b a given positive vector (preferably normalised such that its elements sum to one). In this situation we obtain

$$R_0 = h \cdot (-(\Sigma - D)^{-1}b)$$

as the explicit formula.

We need to point out that linearisation near the infection-free steady state of a nonlinear ODE model leads to a system of higher dimension than (6.13), since the (various kinds of) susceptibles and removed individuals also occur in the bookkeeping. However, the invariance of the infection-free situation guarantees that this system decouples. And provided one has stability within the invariant infection-free subspace, the stability is completely governed by (6.13). We therefore emphasise the crucial conclusion:

> Within the context of ODE models, a demographically stable infection-free steady state is asymptotically stable if and only if $R_0 < 1$.

As indicated above (mainly by pointing out references) this conclusion extends to infinite-dimensional models, but the precise formulation of the assumptions becomes a lot more subtle and cumbersome in that case.

Finally, we mention that also in more general h-state spaces and in the case of age structure, one can find a relation between the formulation in terms of an infectivity

---

[3] A. Nold: Heterogeneity in disease-transmission modeling. *Math. Biosci.*, **52** (1980), 227-240

function A and a formulation in terms of a system of (ordinary or partial) differential equations. The function A, however, needs to satisfy a restrictive condition (see Exercises 6.14 and 6.15).

Exercise 6.14 Assume

$$A(\tau, \eta, \xi) = \beta(\xi, \eta)e^{-\alpha(\eta)\tau}.$$

Let $I(t, \xi)$ be the total size of the infective subpopulation with h-state ξ at time t. Give an expression for I and show by differentiation and manipulation that (6.6), in the nonlinear version with $N(\xi)$ replaced by $S(t, \xi)$, simplifies to

$$\frac{dI}{dt}(t, \xi) = S(t, \xi) \int_{\Omega} \beta(\xi, \eta)I(t, \eta)\, d\eta - \alpha(\xi)I(t, \xi).$$

Exercise 6.15 Let $\Omega = \mathbb{R}_{\geq 0}$, and let the h-state be 'age'. Again regard equation (6.6) in the nonlinear version, but use a and a' as variables. Assume

$$A(\tau, a, \eta) = \beta(a, \eta + \tau)e^{-\int_{\eta}^{\eta+\tau} \alpha(\sigma)\, d\sigma}$$

(note that an individual that was infected with h-state η necessarily has h-state $\eta + \tau$ at infection-age τ, given that it is still alive). Define again the total size of the infective subpopulation of age a at time t,

$$I(t, a) = \int_0^a i(t - \tau, a - \tau)e^{-\int_{a-\tau}^a \alpha(\sigma)\, d\sigma}\, d\tau,$$

and carry out the same manipulation as in the previous exercise to show that I satisfies

$$\frac{\partial I}{\partial t} + \frac{\partial I}{\partial a} = S(t, a) \int_0^{\infty} \beta(a, a')I(t, a')\, da' - \alpha(a)I(t, a),$$

with $a' = \eta + \tau$.

6.3   Some generalities

In this section we review aspects of the time course, final size, endemic level and probability of extinction of the formulation of general models for the spread of infectious agents in heterogeneous host populations. We do so in the form of exercises. Ideally, these should be elaborated in detail. We hope that, the hurried reader— who has no time for that many exercises—will still get some information and gain some understanding by just reading them and checking the main ideas. In contrast to Section 6.1, we have absorbed the contact intensity c into the function A, to simplify the notation.

Exercise 6.16 Spell out the assumptions that underlie the equation

$$\frac{\partial S}{\partial t}(t, x) = S(t, x) \int_{\Omega} \int_0^{\infty} A(\tau, x, \eta)\frac{\partial S}{\partial t}(t - \tau, \eta)\, d\tau\, d\eta. \tag{6.20}$$

Do they apply to plants in a spatial domain Ω? Do they allow for dynamic h-states? Is inflow of new susceptibles or loss of immunity incorporated?

Exercise 6.17 Derive from (6.20) by integration the final-size equation

$$\ln \frac{S(\infty, x)}{S(-\infty, x)} = \int_{\Omega} \left(\int_0^{\infty} A(\tau, x, \eta) \, d\tau \right) \{S(\infty, \eta) - S(-\infty, \eta)\} \, d\eta. \qquad (6.21)$$

For later use, it is convenient to derive, as a first step, the equation

$$\ln \frac{S(t, x)}{S(-\infty, x)} = \int_{\Omega} \left(\int_0^{\infty} A(\tau, x, \eta) \, d\tau \right) \{S(t - \tau, \eta) - S(-\infty, \eta)\} \, d\eta. \qquad (6.22)$$

Note that $S(-\infty, x) = N(x)$, i.e. the population composition before the infectious agent entered the population.

Exercise 6.18 Derive the alternative form of (6.21)

$$\frac{S(\infty, x)}{S(-\infty, x)} = e^{-\int_{\Omega}(\int_0^{\infty} A(\tau, x, \eta) \, d\tau) S(-\infty, \eta)\{1 - \frac{S(\infty, \eta)}{S(-\infty, \eta)}\} \, d\eta} \qquad (6.23)$$

by a probabilistic consistency consideration as in Sections 1.3.1 and 6.1.

Exercise 6.19 How would you define R_0? Do you agree that (6.21) should have a nontrivial solution for $R_0 > 1$? (the trivial solution being $S(\infty, \eta) = S(-\infty, \eta)$). How would you prove that this is indeed the case? Hint: Use monotone iteration.

Exercise 6.20 When the aim is to characterise the endemic level in a population with demographic turnover, we should replace (6.6) by

$$\frac{\partial S}{\partial t}(t, x) = B(x) - \mu(x) S(t, x) - i(t, x), \qquad (6.24)$$

$$i(t, x) = S(t, x) \int_{\Omega} \int_0^{\infty} A(\tau, x, \eta) i(t - \tau, \eta) \, d\tau \, d\eta \qquad (6.25)$$

where the h-state x is static.

i) Check that you know what B, μ and i are supposed to describe.

ii) Calculate explicitly what the infection-free steady state is.

iii) Do you recall how R_0 is defined in this setting?

iv) Derive equations for $\overline{S}$ and $\overline{i}$ characterising the endemic steady state.

v) For the mathematically inclined: Do you see any relation between $R_0 > 1$ and the existence of a non-trivial steady state? Hint: Do not care too much about technical details at this point. The problem is actually not at all easy and involves the notion of ejective fixed points for nonlinear maps defined on cones.[4]

We turn to the probability that only a minor outbreak occurs when nevertheless $R_0 > 1$. We start by giving a description in terms of

$$p_k(x) := \text{ the probability that an individual of type } x \text{ begets } k \text{ offspring}$$

---

[4] See e.g. R.D. Nussbaum: *The Fixed Point Index and Some Applications.* Seminaire de Mathématiques Superieures, les Presses de l'Université de Montreal, 1985.

and

$$m(x,\omega) \quad := \quad \text{the probability that the type at birth of a child of an}$$
$$\text{individual of type } x \text{ belongs to the subset } \omega \text{ of } \Omega,$$

while postponing a discussion of how $p_k(x)$ and $m(x,\omega)$ are related to $S(-\infty, x)$ and $A(\tau, x, \eta)$. The basic assumption is that offspring are produced independently, i.e. given that k offspring are produced, the type at birth of each of these is drawn from the probability distribution $m(x, \cdot)$.

Let

$$\pi(x) := \text{probability of extinction when starting with one individual of type } x.$$

Exercise 6.21 Convince yourself that consistency requires that

$$\pi(x) = \sum_{k=0}^{\infty} p_k(x) \left(\int_{\Omega} \pi(\eta) m(x, d\eta) \right)^k \tag{6.26}$$

Exercise 6.22 Let $f(y, x)$ be defined as

$$f(y, x) := S(-\infty, y) \left(\int_0^{\infty} A(\tau, y, x) \, d\tau \right).$$

Check that the specifications

$$p_k(x) = \frac{1}{k!} \left(\int_{\Omega} f(y, x) \, dy \right)^k e^{-\int_{\Omega} f(y,x) \, dy} \tag{6.27}$$

and

$$m(x, \omega) = \frac{\int_{\omega} f(y, x) \, dy}{\int_{\Omega} f(y, x) \, dy} \tag{6.28}$$

are consistent with earlier assumptions about the transmission process. Use these specifications to rewrite (6.26) as

$$\pi(x) = \sum_{k=0}^{\infty} \frac{1}{k!} \left(\int_{\Omega} \pi(\eta) f(\eta, x) \, d\eta \right)^k e^{-\int_{\Omega} f(y,x) \, dy}. \tag{6.29}$$

6.4   Separable mixing

The general theme that emerges from the preceding section is that, on a formal level, the addition of structure complicates the bookkeeping, but that it does not fundamentally alter the kind of relations that exist between various quantities. When it comes to doing calculations, however, the difference between the structured and the unstructured cases is enormous. In Section 5.3.1 we showed how the assumption of separable mixing facilitates the computation of R_0 (see also Exercise 6.11). In mathematical terms, the point is that whenever operators have a one-dimensional range we can work with scalar quantities. In terms of the interpretation, we can say that whenever the h-state at the moment of becoming infected is following an a priori given distribution (in particular independently of the h-state of the infecting individual), all individuals are identical in a stochastic sense and therefore we know how to take averages. The aim of this section is to demonstrate that this principle is not restricted to R_0, but extends to other aspects of the spread of infectious agents. The general conclusion is that the assumption of separable mixing allows us to work with scalar quantities. Note that Section 6.1 has already offered a simple and relatively concrete example.

Throughout this section we assume that

$$A(\tau, x, \eta) = a(x)b(\tau, \eta). \tag{6.30}$$

Exercise 6.23 Formulate in words what (6.30) means. What will be the distribution of the h-state of newly infected individuals?

Exercise 6.24 Using (6.30) and the Ansatz

$$\frac{\partial S}{\partial t}(t, x) = -\frac{dw}{dt}(t)a(x)S(t, x), \tag{6.31}$$

for some function w to be determined, rewrite equation (6.20) and integrate from $-\infty$ to t to obtain

$$w(t) = \int_0^\infty \int_\Omega b(\tau, \eta)S(-\infty, \eta)\left(1 - e^{-a(\eta)w(t-\tau)}\right) d\eta \, d\tau. \tag{6.32}$$

Hint: Note that (6.31) implies that $S(t, x) = S(-\infty, x)e^{-a(x)w(t)}$ (provided we require $w(-\infty) = 0$).

So once we determine w from the scalar integral equation (6.32), we find S by substitution:

$$S(t, x) = S(-\infty, x)e^{-a(x)w(t)}. \tag{6.33}$$

To obtain the final-size equation, we simply take the limit $t \to \infty$ in (6.32) to deduce

$$w(\infty) = \int_\Omega \int_0^\infty b(\tau, \eta) \, d\tau \, S(-\infty, \eta)\left(1 - e^{-a(\eta)w(\infty)}\right) d\eta, \tag{6.34}$$

which is a nonlinear scalar equation for the unknown $w(\infty)$.

Exercise 6.25 Compute R_0 and show that (6.34) has a non-trivial solution for $R_0 > 1$.

Exercise 6.26 Derive (6.34) from (6.23) using (6.30) and the appropriate Ansatz.

Exercise 6.27 Introduce the assumption (6.30) in the setting of Exercise 6.20, while making the Ansatz

$$i(t, x) = S(t, x)a(x)v(t). \tag{6.35}$$

You should obtain the system of equations

$$\frac{\partial S}{\partial t}(t,x) = B(x) - \mu(x)S(t,x) - S(t,x)a(x)v(t), \qquad (6.36)$$

$$v(t) = \int_0^\infty \int_\Omega b(\tau,\eta)S(t-\tau,\eta)a(\eta)\, d\eta\, v(t-\tau)\, d\tau. \qquad (6.37)$$

From these derive the scalar equation

$$1 = \int_0^\infty \int_\Omega b(\tau,\eta)d\tau \frac{a(\eta)B(\eta)}{\mu(\eta) + \bar{v}a(\eta)}d\eta \qquad (6.38)$$

for the steady state value of the quasi-incidence v. Check that a unique positive solution exists provided that $R_0 > 1$.

Exercise 6.28 In the setting of Exercise 6.21, assume that

$$m(x,\omega) = v(\omega) \qquad (6.39)$$

i.e. assume independence of the distribution of the h-state at birth from the h-state of the mother. Define

$$z = \int_\Omega \pi(\eta)v(d\eta), \qquad (6.40)$$

and derive from (6.26) the scalar equation

$$z = \sum_{k=0}^\infty z^k \int_\Omega p_k(\eta)v(d\eta). \qquad (6.41)$$

Finally, check that once z is known we can recover π via the formula

$$\pi(x) = \sum_{k=0}^\infty z^k p_k(x). \qquad (6.42)$$

Exercise 6.29 Exploiting the mild degree of arbitrariness in the multiplicative decomposition (6.30), normalise a such that

$$\int_\Omega a(\eta)S(-\infty,\eta)\, d\eta = 1. \qquad (6.43)$$

Show that the formulae of the preceding exercise apply with

$$v(\omega) = \int_\Omega a(\eta)S(-\infty,\eta)\, d\eta \qquad (6.44)$$

$$p_k(x) = \frac{1}{k!}\left(\int_0^\infty b(\tau,x)\, d\tau\right)^k e^{-\int_0^\infty b(\tau,x)\, d\tau}. \qquad (6.45)$$

7

Age structure

7.1 Demography

Especially in the context of infectious diseases among humans, 'age' is often used to characterise individuals. Partly this reflects our system of public health administration (and, perhaps, our preoccupation with age). Indeed, we can exploit that data on the distribution of the random variable 'age at (first) infection' contain information about the prevailing force of infection in an endemic situation.

There is, however, also a more 'mechanistic' reason to incorporate age structure: patterns of human social behaviour and sexual activity correlate with age. In addition, the effect that the infective agent has on the host sometimes depends heavily on the age of the host (e.g. in polio) or it may depend on another aspect of the host, such as pregnancy, which correlates with age (e.g. in rubella).

Age is a dynamic variable, but its dynamics is very simple: $\frac{da}{dt} = 1$ by definition. In this short chapter we elaborate some of the material of Chapters 5 and 6 for this special case. We shall also deal with endemic steady states and the inverse problem of estimating R_0 from data about the average age at infection, seropositivity as a function of age, etc. Finally, we discuss vaccination schedules as one of the major applied issues of age-structured epidemic models. First of all we give a very brief introduction to the key notion of mathematical demography: the stable age distribution.

The cohort *survival function* $\mathcal{F}_d(a)$ (d for death, a for age) describes the probability that an arbitrary newborn individual will survive at least until age a. The age-specific *force of mortality* $\mu(a)$, i.e. the per capita probability per unit of time of dying, is related to $\mathcal{F}_d(a)$ by

$$\mu(a) = -\frac{d}{da} \ln \mathcal{F}_d(a) = -\frac{\mathcal{F}'_d(a)}{\mathcal{F}_d(a)} \Leftrightarrow \mathcal{F}_d(a) = e^{-\int_0^a \mu(\alpha)\, d\alpha}.$$

In a density-independent situation (in other words, when a *linear* model applies) the total population size will eventually grow exponentially with a certain rate, which is traditionally denoted by r (but note that this now refers to the growth rate of the *host* population and not, as before, to the subpopulation of infected hosts). Moreover, the distribution with respect to age will stabilise to a fixed shape, the normalised *stable age distribution* given explicitly in terms of $\mathcal{F}_d(a)$ and r by

$$N(a) = Ce^{-ra}\mathcal{F}_d(a), \tag{7.1}$$

with $C = (\int_0^\infty e^{-ra} \mathcal{F}_d(a)\, da)^{-1}$. The factor e^{-ra} reflects that the relative contribution of an individual has to be discounted as the individual ages, since meanwhile the total population is changing (cf. the explanation of $R_{0,\text{relative}}$ in Section 3.3, below Exercise 3.21).

Exercise 7.1 i) Show that the mean age at death of a cohort (i.e. a group of individuals born at more or less the same time, say within a certain year) is given by

$$\int_0^\infty a\mu(a)\mathcal{F}_d(a)\, da.$$

ii) Show that the mean age of those dying at more or less the same time, say within a certain year, is given by

$$\frac{\int_0^\infty a\mu(a)e^{-ra}\mathcal{F}_d(a)\, da}{\int_0^\infty \mu(a)e^{-ra}\mathcal{F}_d(a)\, da}.$$

iii) Reflect upon the difference.

iv) Elaborate for $\mu(a) \equiv \mu$. Note that the quantity calculated in i) is often called the *life expectancy* of a newborn individual.

Exercise 7.2 Analyse the meaning of the following statement: 'Fast-growing populations have a steep age pyramid'.

Remark: One frequently sees a formulation of age-dependent population growth in terms of a partial differential equation (first derived by McKendrick[1]) to describe ageing and dying,

$$\frac{\partial n}{\partial t} + \frac{\partial n}{\partial a} = -\mu n,$$

and a boundary condition $n(t,0) = \int_0^\infty \beta(\alpha)n(t,\alpha)\, d\alpha$ to describe the 'inflow' of newborns. Here $\beta(a)$ is the age-specific fecundity, i.e. the probability per unit of time of giving birth. Let $B(t) := n(t,0)$ be the total birth rate at time t. The equation above can be reformulated as a so-called renewal equation

$$B(t) = \int_0^\infty \beta(a)\mathcal{F}_d(a)B(t-a)\, da.$$

The population growth rate r is then found from the ingredients β and μ as the unique real root of the so-called Euler-Lotka equation

$$1 = \int_0^\infty e^{-ra}\beta(a)\mathcal{F}_d(a)\, da$$

by substituting the Ansatz $B(t) = ce^{rt}$ in the renewal equation.

Exercise 7.3 How would you define in words the basic reproduction ratio R_0 in this demographic context? And can you give a formula for it?

---

[1] A.G. McKendrick: Applications of mathematics to medical problems. *Proc. Edin. Math. Soc.*, **44** (1926), 98-130.

7.2   Contacts

How does the expected number of contacts per unit of time that an individual of age a has with individuals of age α depend on the population size and composition? To make such a question meaningful, we have to be more specific about the precise interpretation of contact. For example, do we refer to sexual contacts or to the inhaling of aerosols just breathed out by someone else? But even after a further specification, the question is a very hard one! Therefore most modellers adopt a very pragmatic approach by simply assuming something that keeps the equations relatively tractable and does not require the estimation of very many parameters. Our later assumptions will be very much in that spirit. At this point we try to keep the model relatively general and flexible by introducing a *contact coefficient* $c(a, \alpha)$ having, by definition, the meaning that an individual of age α has per unit of time $c(a, \alpha)N(a)$ contacts with individuals of age a. An individual therefore has $\int_0^\infty c(a, \alpha)N(a)\, da$ contacts in total. All the difficulties mentioned in Chapter 1 concerning the dependence of c on population size are intensified here, now that we also have to think about dependence on population composition. Since we have nothing to add to what has already been said in Chapter 1, we shall think here of a mild disease in one particular population and not worry about scaling of c.

An obvious question presents itself: are contacts necessarily symmetric or, more precisely, should c and N satisfy the relation

$$c(a, \alpha)N(a)N(\alpha) = c(\alpha, a)N(\alpha)N(a),$$

which amounts to $c(a, \alpha) = c(\alpha, a)$?

Exercise 7.4 i) Contemplate the kind of contacts for which the answer is 'yes' and those for which it is 'no'. Hint: Also think of parents caring for children.

ii) Contemplate the indeterminacy in decomposing 'probability of transmission' and 'probability of transmission, given a contact', and how this bears upon the symmetry of c. In other words, reflect upon the inherent vagueness of 'contacts'.

7.3   The next-generation operator

The dynamics of the h-state is very simple in the case of age: an individual of age a has τ units of time later age $a+\tau$. The probability that it has not died in this time interval of length τ equals $\mathcal{F}_d(a+\tau)/\mathcal{F}_d(a)$ (here we assume that the death rate is the same for infected and uninfected individuals; so the formulae below have to be adapted when infection entails a serious risk of death for the host). We need one more ingredient: the probability of transmission given a contact. In principle this could depend on the ages of both individuals involved. Due to Exercise 7.4-ii above, we are aware of an element of freedom in our description. Since c is already a general function of two variables, it would be redundant to introduce yet another such function. Let our last ingredient be

$$h(\tau, \alpha) \quad := \quad \text{probability of transmission of the infective agent,}$$
$$\text{given a contact between a susceptible (of arbitrary age)}$$
$$\text{and an individual that was itself infected while having age } \alpha.$$

You may wonder why we allowed for this dependence on α. What we have in mind is that—as a rule and up to a certain point—older individuals are bigger and so may distribute (much) larger quantities of the infective agent around them. Or, working in the other direction, the immune system of very young individuals may be less effective in dealing with the agent.

The next-generation operator is given by

$$(K\phi)(a) = \int_0^\infty k(a,\alpha)\phi(\alpha)\,d\alpha.$$

Exercise 7.5 i) Repeat in words the rationale underlying the following formula for the kernel k of the next-generation operator (cf. Chapter 5)

$$k(a,\alpha) = \int_0^\infty h(\tau,\alpha)c(a,\alpha+\tau)N(a)\frac{\mathcal{F}_d(\alpha+\tau)}{\mathcal{F}_d(\alpha)}\,d\tau.$$

ii) Show that under the assumption that, for certain functions f and g,

$$c(a,\alpha) = f(a)g(\alpha),$$

R_0 is given by

$$R_0 = \int_0^\infty \psi(\alpha)f(\alpha)N(\alpha)\,d\alpha, \tag{7.2}$$

with

$$\psi(\alpha) = \int_0^\infty h(\tau,\alpha)g(\alpha+\tau)\frac{\mathcal{F}_d(\alpha+\tau)}{\mathcal{F}_d(\alpha)}\,d\tau. \tag{7.3}$$

Reformulate the assumption on c in words, perhaps starting with the special case $f = g$.

iii) Assume that, for certain functions f_k and g_k with $k = 1, ..., n$ we have

$$c(a,\alpha) = \sum_{k=1}^n f_k(a)g_k(\alpha).$$

Specify an $n \times n$ matrix M such that R_0 is the dominant eigenvalue of M.

Many childhood diseases have an infectious period that is two orders of magnitude shorter than the average human lifespan. In that case we might approximate $\mathcal{F}_d(\alpha+\tau)/\mathcal{F}_d(\alpha)$ by 1 and $c(a,\alpha+\tau)$ by $c(a,\alpha)$ and next put $H(\alpha) = \int_0^\infty h(\tau,\alpha)\,d\tau$. We shall call this the 'short-disease approximation'.

Exercise 7.6 Simplify the expressions obtained in Exercise 7.5-ii and 7.5-iii on the basis of the short-disease approximation.

Exercise 7.7 Suppose animals enter a farm at a constant rate v and stay there for a fixed period M during which they 'mix' uniformly with the other animals at the farm. Define age = 0 at the moment of entering the farm.

i) Assume that death does not occur on the farm, but interpret removal from the farm as a type of death at age M (which indeed it most likely is). Specify $\mathcal{F}_d(a)$.

ii) What can you say about $c(a, \alpha)$ in this situation?

iii) Assume that infected individuals become immune at an age-independent rate, while having constant (in particular age-independent) infectivity during the infectious period. Derive an expression for R_0.

Exercise 7.8 Suppose that, as in Section 6.3.2, we assume that $c(a, \alpha)$ is split into a part that has separable mixing and a part giving additional contacts with individuals of the same age. Mathematically this would pose no problems. Is it a meaningful assumption from a biological point of view?

7.4 Interval decomposition

We now turn to *the* example of a finite-dimensional range situation (see Chapter 5). The idea is to discretise age by forming age intervals, which together cover precisely all feasible ages, and to discretise c accordingly. This means that a function of two continuous variables is replaced by a matrix. Mathematically we formulate this by introducing intervals I_i, $i = 1, 2, ..., n$, which are non-overlapping (i.e. $I_i \cap I_j = \emptyset$ for $i \neq j$) and together cover the positive axis (i.e. $\mathbb{R}_+ = \cup_{i=1}^n I_i$), and by requiring that

$$c(a, \alpha) = c_{ij} \text{ for } a \in I_i \text{ and } \alpha \in I_j$$

and certain given numbers c_{ij}, $1 \leq i, j \leq n$. The intervals can conveniently be adapted to the school system, the public health administrative system, et cetera. That the numbers c_{ij} are 'given' is a euphemism, and we shall return to this issue below.

At first sight we do not recognise c as being of the form $\sum_{k=1}^n f_k(a)g_k(\alpha)$. Yet it is! To demonstrate that, it is helpful to introduce the characteristic function χ_I of a set I defined by

$$\chi_I(a) = \begin{cases} 1 & \text{if } a \in I, \\ 0 & \text{otherwise.} \end{cases}$$

Indeed, with this piece of notation at hand, we can write

$$c(a, \alpha) = \sum_{k,l=1}^n c_{kl} \chi_{I_k}(a) \chi_{I_l}(\alpha)$$

$$= \sum_{k=1}^n \chi_{I_k}(a) \sum_{l=1}^n c_{kl} \chi_{I_l}(\alpha),$$

which is of the indicated form since we can take $f_k = \chi_{I_k}$ and $g_k = \sum_{l=1}^n c_{kl} \chi_{I_l}$.

Exercise 7.9 Elaborate for this special case the expression of Exercise 7.5-iii in the short-disease approximation of Exercise 7.6.

7.5 Inferring parameters from endemic data

Let us *assume* that a steady (= stationary = time-independent) endemic situation occurs. By analogy with the situation described in Section 7.1, we can introduce the cohort 'remain uninfected' function $\mathcal{F}_i(a)$ (*i* for infection, *a* for age) describing the probability that an arbitrary new-born individual will, given that it survives until at least age a, be uninfected (in the sense of 'never before being infected') at age a. Accordingly, we can introduce the age-specific *force of infection* $\lambda(a)$, i.e. the per susceptible probability per unit of time of being infected, by

$$\lambda(a) = -\frac{d}{da} \ln \mathcal{F}_i(a) \Leftrightarrow \mathcal{F}_i(a) = e^{-\int_0^a \lambda(\alpha)\,d\alpha}. \tag{7.4}$$

In the demographic context, the survival function is a purely descriptive statistical object. In the context of infectious diseases, a postulated mechanism is responsible for the observed 'survival' function $\mathcal{F}_i(a)$, and hence there should be a consistency condition that serves as an equation from which, at least in principle, $\mathcal{F}_i$ (or, equivalently, λ) can be determined. Indeed, the interpretation of all the ingredients involved requires that

$$\lambda(a) = \int_0^\infty \int_0^\infty h(\tau, \alpha)c(a, \alpha + \tau)\lambda(\alpha)S(\alpha)\frac{\mathcal{F}_d(\alpha + \tau)}{\mathcal{F}_d(\alpha)}\,d\tau\,d\alpha, \tag{7.5}$$

where

$$S(a) = N(a)\mathcal{F}_i(a), \tag{7.6}$$

i.e. S is the relative density of susceptibles. This certainly requires an explanation.

The basic idea is that we are dealing with an innocent disease that does not affect population numbers. Moreover, we adhere to the interpretation of c as discussed in Section 7.2 (this is a subtle issue deserving careful attention, in particular when comparing host populations of different sizes).

Let P denote the total population size. Then the age-specific incidence is given by $P\lambda(a)S(a)$, and, according to the assumptions preceding Exercise 7.5, the number of new cases per unit of time resulting from this steady incidence equals

$$S(a) \int_0^\infty \int_0^\infty h(\tau, \alpha)c(a, \alpha + \tau)P\lambda(\alpha)S(\alpha)\frac{\mathcal{F}_d(\alpha + \tau)}{\mathcal{F}_d(\alpha)}\,d\tau\,d\alpha$$

where the first factor is due to the fact that a contact with an individual of age a is with probability $\mathcal{F}_i(a)$ with a susceptible (read it as $N(a)\mathcal{F}_i(a)$ or $N(a)\frac{S(a)}{N(a)}$, whichever you prefer). Consistency then requires that this quantity is equal to $P\lambda(a)S(a)$. If we factor out $PS(a)$, we arrive at (7.5).

Since $\mathcal{F}_i(a)$ depends on λ, the consistency condition is a nonlinear integral equation for λ. A rigorous mathematical analysis of this equation[2] leads in particular to the conclusion that for $R_0 > 1$ there is indeed a non-negative solution. See also (the elaboration of) Exercise 6.22.

Exercise 7.10 (cf. Exercise 3.6) Show that the *mean age at infection* $\bar{a}$ is given by

$$\bar{a} = \frac{\int_0^\infty a\lambda(a)\mathcal{F}_i(a)\mathcal{F}_d(a)\,da}{\int_0^\infty \lambda(a)\mathcal{F}_i(a)\mathcal{F}_d(a)\,da}. \tag{7.7}$$

---

[2] H. Inaba: Threshold and stability results for an age-structured epidemic model. *J. Math. Biol.*, **28** (1990), 411-434; D. Greenhalgh: Threshold and stability results for an epidemic model with an age-structured meeting rate. *IMA J. Math. Appl. Med. Biol.*, **5** (1988), 81-100.

Then consider the special case in which every individual lives exactly M units of time. Use partial integration to rewrite $\bar{a}$ in the form

$$\bar{a} = \frac{\int_0^M \mathcal{F}_i(a)\, da}{\int_0^M \lambda(a)\mathcal{F}_i(a)\, da}.$$

Exercise 7.11 Show that linearisation at $\lambda \equiv 0$ leads us back from the right-hand side of (7.5) to the next-generation operator. Hint: Interpret λN as ϕ.

Exercise 7.12 i) Assume $c(a, \alpha) = \sum_{k=1}^n f_k(a) g_k(\alpha)$. Reduce the equation for λ to a finite-dimensional problem. Pay particular attention to the case $n = 1$.

ii) Elaborate for the special case of interval decomposition and the short-disease approximation.

iii) Suppose that the c_{ij} are unknown, but that we can measure in some way the force of infection in the various age intervals. Suppose that we want to determine the c_{ij} from these data. How many equations do we have? And how many unknowns? So, can we realise our objective?

iv) In the literature one finds reference to the WAIFW matrix ('Who Acquires Infection From Whom'; see e.g. Anderson & May, (1991)). Can you guess what matrix is meant? Hint: Do not take the name too literally.

v) Design additional assumptions on the c_{ij} that reduce the number of unknowns to the number of equations. Hint: Consult Anderson and May (1991) and Greenhalgh and Dietz (1994).[3]

Exercise 7.13 i) Consider the special case $n = 1$ and $f(a) \equiv 1$ (in other words, suppose that data about $\mathcal{F}_i(a)$ suggest that $\mathcal{F}_i(a) = e^{-Qa}$, for some Q, is not too bad an approximation). How would you estimate the force of infection for Q?

ii) Additionally, assume that g is constant and that h is independent of α, and, finally that $\mathcal{F}_d(a)$ is an exponential function, say $e^{-\mu a}$. Show that

$$R_0 = 1 + \frac{Q}{\mu + r}.$$

Show that, under the same conditions, the mean age of infection

$$\bar{a} = \frac{1}{\mu + Q},$$

so that for $r \approx 0$ we have

$$R_0 = \frac{L}{\bar{a}},$$

where L is the life expectancy.

---

[3] D. Greenhalgh & K. Dietz: Some bounds on estimates for reproductive ratios derived from the age-specific force of infection. *Math. Biosc.*, **124** (1994), 9-57.

iii) Assume again that f is identically 1 and that $\mathcal{F}_i(a) = e^{-Qa}$. Assume in addition that ψ (defined by (7.3)) is approximately constant. Express R_0 in terms of the observable quantities $N(a)$ and Q as

$$R_0 = \frac{\int_0^\infty N(a)\, da}{\int_0^\infty N(a)e^{-Qa}\, da}.$$

iv) Now repeat ii) for the situation where individuals live until a fixed age L and then die, while retaining the other assumptions. Show that in that case $\bar{a} = 1/Q$ and that

$$R_0 \approx \frac{L}{\bar{a}},$$

when $r \approx 0$ and $QL \gg 1$.

Exercise 7.14 How should one compare the risk of dying at an early age due to inoculation with the risk of dying at a later age due to infection? The current exercise relates to a historic treatment of such questions. Assume that individuals survive a given disease with probability θ. Assume further that the disease has negligible duration compared with the expected life-time of individuals.

i) Express the probability to be alive at age a in terms of $\mathcal{F}_d, \mathcal{F}_i$ and θ.

ii) Next, in order to decompose mortality into causes related to the disease and other causes, compute the probability $P(a)$ to be alive at age a, given that one does not die from other causes.

iii) Argue that elimination of the infection leads to a survival function that is obtained from $\mathcal{F}_d(a)$ by multiplication with $P^{-1}(a)$.

iv) Show that, for a constant force of infection, P^{-1} describes a logistic curve, and compute the limit of $P^{-1}(a)$ for a tending to infinity. This result was obtained by Daniel Bernoulli in 1760 concerning smallpox, in what is probably the first paper (published much later in 1766) dealing with mathematical epidemiology.[4]

In the following two exercises we regard the calculation of R_0 and r for BSE in an age-structured cattle population.

Exercise 7.15 We focus on the supposed two, main, ways of transmission for BSE: through consumption of contaminated food and through vertical transmission from mother to foetus. On farms with dairy cows at the start of the epidemic, animals that died were brought to a rendering factory and turned into meat and bonemeal that was subsequently fed back to cattle. We wish to express R_0 in terms of the (mostly age-dependent) functions and parameters that could describe the transmission cycle.

Let $b(a)$ be the per capita birthrate, $\mu(a)$ the 'natural' per capita death rate (by culling), and $\nu(\tau)$ the infection-age-dependent culling rate for symptomatic animals. Regard $\mathcal{F}_d(a)$ and $\mathcal{F}_i(\tau) = \exp(-\int_0^\tau \nu(\sigma)\, d\sigma)$ as the age-dependent

---

[4] D. Bernoulli: Essai d'une nouvelle analyse de la mortalité causée par la petite vérole et des avantages de l'inoculation pour la prévenir. *Mém. Math. Phys. Acad. Roy. Sci., Paris*, 1766, 1-45.

demographic survival function and the infection survival function respectively. We assume, for convenience only, that all infected animals are rendered (completely) and that the rendering process has no influence on the 'agent'. Let $\beta(a)$ be the susceptibility as a function of age. Finally, let $\gamma(\tau)$ describe the infectiousness of an infected individual, irrespective of the route by which it became infected, and let m be the probability of maternal transmission. The demographic steady state $\overline{S}(a) = \mathcal{F}_d(a)/\int_0^\infty \mathcal{F}_d(\alpha)d\alpha$ represents the stable age distribution of the cattle population (held artificially constant by the farmers). Assume that the infected cattle feed is distributed randomly over the cattle herd.

i) Argue that the age of animals infected by feed will be distributed according to the density function
$$\frac{\beta(a)\mathcal{F}_d(a)}{\int_0^\infty \beta(a)\mathcal{F}_d(a)\, da}.$$

ii) What are the types at birth in this system?

iii) Then R_0 is the dominant eigenvalue of
$$K = \left(\begin{array}{cc} k_{11} & k_{12} \\ k_{21} & k_{22} \end{array} \right),$$

with k_{11} the expected number of new feed-infected individuals from a feed-infected individual and with similar interpretations for the other three elements. So
$$R_0 = \frac{1}{2}(k_{11} + k_{22}) + \frac{1}{2}\sqrt{(k_{11} + k_{22})^2 - 4(k_{11}k_{22} - k_{12}k_{21})}.$$
Express the elements of K in terms of the ingredients of the model.

iv) Although direct horizontal transmission between animals is unlikely, one could investigate the influence on R_0 of infection by other than maternal or feed sources. How does the above calculation change if we include a parameter w to summarise direct mass-action transmission?

Exercise 7.16 The real-time evolution of the BSE infection in an age-structured cattle population is described by
$$i(t, a) = \overline{S}(a)\beta(a) \int_0^\infty \int_0^\infty A(\alpha, \tau)i(t - \tau, \alpha)\, d\tau\, d\alpha$$

if we restrict ourselves to the dominant mode of transmission via feed and neglect all other routes.

i) Give an expression for the kernel $A(\alpha, \tau)$ using the ingredients in Exercise 7.15.

ii) Show that
$$i(t, a) = f(a)e^{rt}$$
is a solution of this integral equation if f is an eigenfunction of an operator M corresponding to eigenvalue 1, where M is defined as
$$(Mf)(a) = \overline{S}(a)\beta(a) \int_0^\infty \int_0^\infty A(\alpha, \tau)f(\alpha)e^{-r\tau}\, d\tau\, d\alpha.$$

Show that f is indeed the function we are looking for if and only if r satisfies

$$\int_0^\infty \int_0^\infty A(\alpha,\tau)\overline{S}(a)\beta(a)e^{-r\tau}\,d\tau\,d\alpha = 1. \tag{7.8}$$

iii) Argue that a unique value $r^* > 0$ exists if $R_0 > 1$ and that r^* can be found numerically by computing the unique zero of a function $g(r) - 1$, where g is defined by the left-hand side of (7.8), and hence is monotonically decreasing.

7.6   Vaccination

As an applied problem concerning age-structured models, we regard vaccination as a control strategy. We do so in two different situations: i) we investigate under what conditions a given agent can successfully invade a partially vaccinated population (or, more precisely, how large a fraction of the population do we have to keep vaccinated in order to prevent the agent from establishing); ii) the more realistic situation that a vaccination campaign is started when the agent has become endemic, and we ask the question whether the vaccination effort could be sufficient to eliminate the agent.

In the first situation the idea is to calculate a reproduction ratio R_v for invasion of a population of susceptibles in a demographic steady state where a vaccination schedule v is in operation. Schedule v can prevent the agent from establishing if $R_v < 1$ can be accomplished. Let

$$\mathcal{F}_v(a)$$

denote a vaccination 'survival function', i.e. the conditional probability that an individual of age a that is alive is still susceptible (and has not been made immune by vaccination). To arrive at explicit and manageable criteria, we adopt the one-dimensional range condition of Exercise 7.5-ii.

Exercise 7.17 Argue that R_v is given by

$$R_v = \int_0^\infty \int_0^\infty h(\tau,\alpha)g(\alpha+\tau)\frac{\mathcal{F}_d(\alpha+\tau)}{\mathcal{F}_d(\alpha)}f(\alpha)N(\alpha)\mathcal{F}_v(\alpha)\,d\tau\,d\alpha.$$

Exercise 7.18 In the easiest case, let a fraction q of the population be vaccinated at birth. Show that in order to prevent establishment we must have

$$q > 1 - \frac{1}{R_0},$$

where R_0 is given by (7.2).

Exercise 7.19 Now suppose we vaccinate a fraction q of the individuals of age a_v. Show that the appropriate inequality for the minimal proportion to be vaccinated in order to prevent establishment is

$$q > \frac{R_0 - 1}{C\int_{a_v}^\infty \int_0^\infty h(\tau,\alpha)g(\alpha+\tau)\mathcal{F}_d(\alpha+\tau)f(\alpha)e^{-r\alpha}\,d\tau\,d\alpha}.$$

In the second situation, when the agent is endemic in the population, developing a procedure based on R_0 no longer makes sense. We do still assume that a steady state has arisen. We have to discount the available susceptibles not only by the probability to be unvaccinated, but also by the probability to have escaped infection so far. Therefore, instead of (7.6), we obtain for the steady-state density of susceptibles in the endemic situation

$$S(a) = N(a)\mathcal{F}_v(a)\mathcal{F}_i(a),$$

and furthermore the force of infection λ satisfies the nonlinear integral equation (7.5).

Exercise 7.20 Assume separable mixing $c(a, \alpha) = f(a)g(\alpha)$.

i) Show that necessarily $\lambda(a) = Qf(a)$ for a constant Q that has to satisfy

$$1 = \int_0^\infty \psi(\alpha)N(\alpha)\mathcal{F}_v(\alpha)e^{-Q\int_0^\alpha f(a)\,da}f(\alpha)\,d\tau\,d\alpha, \qquad (7.9)$$

where $\psi(\alpha)$ is given by (7.3). The same expression, apart from notation, was derived by Dietz and Schenzle (1985).[5]

ii) In the short-disease approximation (Section 7.3), show that (7.9) can be approximated by

$$1 = \int_0^\infty f(\theta)g(\theta)H(\theta)N(\theta)\mathcal{F}_v(\theta)e^{-Q\int_0^\theta f(a)\,da}\,d\theta.$$

If one further assumes the same age dependence in activity level ($f = g$), one can use data about the endemic state to estimate f, Q and the demographic ingredients and subsequently calculate whether or not a given vaccination schedule suffices to eliminate the agent, i.e. causes $R_v < 1$. We refer once more to Dietz and Schenzle (1985) for more information.

Finally, we make a 'remark' about the mean age at infection in relation to control measures such as vaccination.

Exercise 7.21 Consider the formula (7.7) for the mean age at infection, and specifically the case of a constant force of infection (3.10). What is the influence on $\bar{a}$ of vaccinating a fraction of the population? This is a general phenomenon and can lead to dangerous situations, since seriousness of complications arising from childhood infections is often positively correlated with age. See Anderson and May (1991) and the references given there.

---

[5] K. Dietz & D. Schenzle: Proportionate mixing models for age-dependent infection transmission. *J. Math. Biol.*, **22** (1985), 117-120.

8

Spatial spread

8.1 Posing the problem

As an example, think of a fungal pathogen affecting an agricultural crop. A farmer having ascertained that his field is affected wants to know: How fast is the infection spreading? What fraction of the yield do I stand to lose if I do not spray with fungicides? The trade-off here could be that spraying is expensive and bad for the environment. Suppose that harvest is three months away. Do I take the loss of plants or do I invest in fungicide and accept the concomitant pollution?

So, the key question is: How fast is the infection spreading?

A student of the preceding chapters might be inclined to answer the farmer by first drawing Figure 8.1 and next, in an attempt to be pragmatic rather than scrupulous, saying that the relevant part of the curve is to a good approximation described by Ce^{rt}, with C determined from the current situation and r from (one hopes) known data about spore production and dispersal. The aim of this chapter is to provide the student with better ingredients for an answer. In particular, we will explain that it is much more likely that the fraction of the crop affected will grow as a quadratic function of time.

The main point is that the infestation is localised in patches, called foci, which expand more or less radially. The population growth parameter r, however, describes population growth that is uniform (in space), as can be concluded from the eigenfunction, which is constant (in fact also when we consider R_0). This will be elaborated below.

Thus we arrive at the following set of questions: Do models predict radial expansion of epidemic fronts? If so, how can we determine the *speed* of the front from the model ingredients? What are the conditions that promote wave-like expansion?

8.2 Warming up: the linear diffusion equation

Let u denote the density of a pest species. Let this species inhabit a very large domain without structure (no hedges, roads, canals, rivers, mountains, ...). We take the plane $\mathbb{R}^2$ as an idealised representation of this domain and let $x \in \mathbb{R}^2$ denote a spatial position. Note that $\mathbb{R}^2$ is homogeneous (i.e. translation-invariant: points are interchangeable) and isotropic (the same structural properties in every direction).

We assume that the species grows at a net per capita rate κ. In addition, it disperses. If movement is completely random, we may use the diffusion equation to

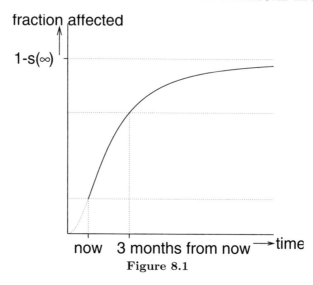

Figure 8.1

describe dispersal (the underlying assumption being that the flux is proportional to the gradient). Hence we postulate that

$$u_t = D\triangle u + \kappa u, \tag{8.1}$$

where $u_t = \frac{\partial u}{\partial t} = \frac{\partial}{\partial t} u(t, x)$ is the partial derivative of u with respect to time, where D is a species-dependent constant (called the diffusion constant) representing the mean square distance covered per unit of time, and where

$$\triangle u = \frac{\partial^2 u}{\partial x_1^2} + \frac{\partial^2 u}{\partial x_2^2}$$

is shorthand for the sum of the second partial derivatives of u with respect to the two coordinate directions.

The chief advantage of (8.1), and the reason to consider it here, is that we can solve this equation explicitly. The explicit expression allows us to pinpoint certain quantitative and qualitative properties, in particular the phenomenon of an *asymptotic speed of propagation* c_0, which equals the *minimal speed* for which *plane travelling-wave solutions* exist. Armed with this knowledge, we then study, in subsequent sections, the robustness of the conclusions: Do the qualitative phenomena extend to the nonlinear realm? And to models involving kernels $A(\tau, x, \xi)$?

The solution of (8.1),

$$u(t, x) = \frac{1}{4\pi Dt} e^{-\frac{|x|^2}{4Dt} + \kappa t}, \tag{8.2}$$

describes the effect of a localised (at $x = 0$) disturbance (at $t = 0$) of the unstable steady state $u \equiv 0$ (read as: u is identically zero).

Exercise 8.1 Verify that u given in (8.2) satisfies the equation (8.1).

Exercise 8.2 Verify that $\int_{\mathbb{R}^2} u(t, x)\, dx = e^{\kappa t}$ for $t > 0$.

Exercise 8.3 Show that uniformly for $|x| \geq \varepsilon$ we have

$$\lim_{t \downarrow 0} u(t, x) = 0$$

and conclude that u is indeed concentrated in $x = 0$ at $t = 0$. (In the language of distributions and measures, we have $\lim_{t \downarrow 0} u(t, x) = \delta(x)$, with δ the Dirac 'function' (measure/distribution) concentrated in $x = 0$.)

What can be said about the behaviour of $u(t, x)$ for large t? Of course the factor $\frac{1}{t}$ goes to zero, but $e^{\kappa t}$ goes to infinity much faster. So, if, on the one hand, we fix x and let t tend to infinity, we find that u grows exponentially with rate κ. If, on the other hand, we fix t and let $|x| \to \infty$ (that is, we observe far ahead in space) we find that u is negligibly small. Thus it appears that the limits $t \to \infty$ and $|x| \to \infty$ cannot be interchanged: the order matters. In such situations of non-uniform convergence, one expects to see a *transition layer* (in which transition from one extreme, zero, to the other extreme, infinity, is made) once we approach infinity in (t, x) space in a taylor-made fashion. We can immediately infer from the explicit expression what this 'taylor-made fashion' is: we have to avoid the exponent in (8.2) going to either $+\infty$ or $-\infty$ (the factor $1/t$ necessitates that we be a little more precise about the first of these possibilities, viz. we have to avoid approaching $+\infty$ too quickly; it is indeed this factor $1/t$ that makes the precise characterisation of the transition layer rather subtle). We refrain from a more precise study, and restrict ourselves to the observation that, for any $\varepsilon > 0$, for $t \to \infty$

$$u(t, x) \to \begin{cases} 0 & \text{if } |x|^2 > (4D\kappa + \varepsilon)t^2, \\ \infty & \text{if } |x|^2 < (4D\kappa - \varepsilon)t^2. \end{cases}$$

In suggestive words, we could say that we distinguish the 'not yet' region, being the exterior of a disc, the radius of which grows like $2t\sqrt{D\kappa + \frac{\varepsilon}{4}}$, and the 'already over' region, being the interior of a disc, the radius of which grows like $2t\sqrt{D\kappa - \frac{\varepsilon}{4}}$. (Here ε is a positive number that can be taken arbitrarily small. It relieves us from going into the details of the subtle limiting behaviour that occurs when x grows like $2t\sqrt{D\kappa} + O(\ln t)$.) Yet another way of expressing this result is to state that

$$c_0 := 2\sqrt{D\kappa}$$

is the *asymptotic speed of propagation* of the disturbance. We conclude that the linear diffusion equation (8.1) displays radial expansion of a disturbance with a well-defined speed that can be easily computed from the parameters.

Our next aim is to characterise c_0 in a completely different way, viz. as the minimal speed of plane travelling waves. The point is that this characterisation carries over much more easily to other situations, in which explicit calculations are often impossible.

A plane wave travelling in the direction specified by a given unit vector ν is described by a solution of the form

$$u(t, x) = w(x \cdot \nu - ct), \tag{8.3}$$

(here $x \cdot \nu$ is the inner product of the vectors x and ν, i.e. $x \cdot \nu = x_1\nu_1 + x_2\nu_2$).

Exercise 8.4 Show that, in order for (8.3) to define a solution of (8.1), w should satisfy

$$Dw'' + cw' + \kappa w = 0. \tag{8.4}$$

Next argue that this requires w to be of the form $w(\xi) = \exp(\lambda \xi)$, with λ satisfying the so-called characteristic equation

$$D\lambda^2 + c\lambda + \kappa = 0. \tag{8.5}$$

Conclude that

$$\lambda = \lambda_\pm = \frac{-c \pm \sqrt{c^2 - 4D\kappa}}{2D}. \tag{8.6}$$

We want *positive* solutions, because of the interpretation. Oscillating solutions are characterised by complex $\dot\lambda$. So we should require $\lambda_\pm$ to be real; that is, we should have $c^2 - 4D\kappa \geq 0$, i.e. $c \geq 2\sqrt{D\kappa} = c_0$. We conclude that plane travelling wave solutions exist for all speeds c that exceed a threshold c_0 and that the minimal plane-wave speed c_0 coincides with the asymptotic speed of propagation.

Exercise 8.5 For $c = c_0$ we have $\lambda_\pm = -\frac{c_0}{2D}$ and

$$w(\xi) = e^{\lambda_\pm(x \cdot \nu - c_0 t)} \sim e^{-\lambda_\pm c_0 t} = e^{\frac{c_0^2}{2D}t} = e^{2\kappa t}$$

for large t. What do you conclude from this?

8.3 Verbal reflections suggesting robustness

Consider a steady state (zero/infection-free) that is unstable and such that any physically feasible (in particular *positive*) perturbation triggers a transition towards another steady state ('infinity' in the case of the linear diffusion equation, i.e. 'after the epidemic'). To this local dynamics, add a spatial component and coupling, which means that perturbations at some point generate perturbations at nearby points. How fast do perturbations spread?

Imagine space to be homogeneous and isotropic. Travelling plane waves are uniform in all directions but one. So they manifest how disturbances travel in one direction (although this direction is arbitrary because of the isotropy). And the speed tells us how fast the spread will be.

Probably to the irritation of Fisher, plane waves do not come with a unique speed, but rather with a continuum of possible speeds, bounded only at one side. Why should the minimal speed be the truly relevant one?

To fix a unique solution, the partial differential equation (PDE) (8.1) has to be supplemented with an initial condition. At least as a thought experiment, we can therefore manipulate the solution. The following explanation is due to J.A.J. Metz. Imagine a series of fireworks placed in a row, with fuses of varying lengths. By lighting the fuses, one can create a travelling wave of explosions. By choosing the lengths of the fuses appropriately, one can achieve any speed one wants. If, however, the fireworks also have the tendency to kindle their nearest neighbours, this process of self-kindling will dominate as soon as one tries to achieve, by manipulation of the fuses, a speed that is too low. Therefore, the minimal plane-wave speed corresponds to the inherent speed of the self-infection mechanism!

Travelling plane-wave solutions are examples of similarity solutions, i.e. solutions depending only on a certain combination of the independent variables. They show up whenever the dynamics are equivariant under a group of transformations (in this case translations and rotations). They are often the quintessence of intermediate asymptotic behaviour, when the transients reflecting the initial conditions have died out, but where the final state has not yet been achieved everywhere and boundary conditions (every real domain is finite!) do not yet impinge upon the natural dynamics.[1]

We conclude that whenever

- local dynamics consists of a transition from an unstable steady state to a stable one,
- perturbations spread, i.e. there is some form of coupling of local dynamics,
- space is homogeneous and isotropic,

we are bound to find that

- travelling plane-wave solutions exist for all speeds c exceeding a minimal speed c_0,
- the minimal wave speed c_0 is the asymptotic speed of propagation associated with the self-triggering mechanism.

Many of these conclusions remain valid if there is homogeneity but no isotropy.[2]. Of course, the minimal plane-wave speed will then depend on the direction ν. When defining what asymptotic propagation means, one should then blow up not a disc but rather another (convex) set defined on the basis of the function $c_0(\nu)$.

Exercise 8.6 Suppose a species spreads by windborne propagules. Assume there is a prevailing wind direction. Let the unit vector σ point to this wind direction and let θ be the wind velocity. Then (8.1) should be replaced by

$$u_t = D\triangle u - \theta\sigma \cdot \nabla u + \kappa u,$$

where ∇u is the gradient (i.e. the vector of partial derivatives $\nabla u = (\partial u/\partial x_1, \partial u/\partial x_2)^{\top}$) and so $\sigma \cdot \nabla u = \sigma_1 \frac{\partial u}{\partial x_1} + \sigma_2 \frac{\partial u}{\partial x_2}$ is the directional derivative in the direction σ. In case you wonder about the minus sign, put $D = 0$ and $\kappa = 0$ and check that $u(t, x) = \phi(\sigma \cdot x - \theta t)$ satisfies the equation for any function ϕ, and that such solutions correspond to plane waves travelling in the direction σ. Look at Figure 8.2, turn it into an animation of a travelling wave and verify that the direction of propagation is to the right.

Now let us look for wave solutions

$$u(t, x) = w(\sigma \cdot x - ct)$$

travelling with speed c in the direction σ. Show that such solutions exist for all $c \geq c_0 + \theta$ with $c_0 = 2\sqrt{D\kappa}$ as before. Next look for travelling waves in the opposite direction. What do you conclude?

---

[1] G.I. Barenblatt, *Similarity, Self-similarity and Intermediate Asymptotics*. Plenum, New York, 1979.

[2] H.F. Weinberger: Long-time behaviour of a class of biological models. *SIAM J. Math. Anal.* **13** (1982), 353-396; F. van den Bosch, O. Diekmann & J.A.J. Metz: The velocity of spatial population expansion. *J. Math. Biol.* **28** (1990), 529-565.

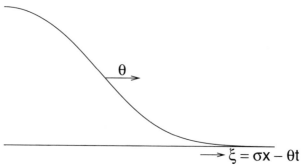

Figure 8.2

This seems an appropriate place for the following side-remark. Within the general framework, it is possible that a population (pest or not) grows while being blown off to ever more removed regions (i.e. to infinity, in mathematical jargon). The spectral radius criterion would say $R_0 > 1$, so the population grows. A local observer, however, would only notice a very temporary growth, followed by local extinction. Within a more measure-theory-oriented approach, growth is defined in terms of a local indicator, the *Perron root*. In the situation just described, it follows that the Perron root is less than one and this is interpreted as 'the population does not grow'. We refer to Jagers (1995) for a precise definition of the Perron root and note that Shurenkov (1992) has shown that the spectral radius and the Perron root coincide whenever the state space is compact.[3]

8.4   Linear structured population models

Let $B(\tau, x, \xi)$ be the expected rate at which an individual born at position ξ in space produces offspring at position x at age τ. Then straightforward bookkeeping considerations suggest that (8.1) be replaced by the time-translation invariant integral equation

$$u(t, x) = \int_0^\infty \int_\Omega B(\tau, x, \xi) u(t - \tau, \xi) \, d\xi \, d\tau, \tag{8.7}$$

where Ω represents the area in which the species lives (readers are invited to read the interpretation of the right-hand side of (8.7) aloud to themselves, to check their understanding of it). When $\Omega = \mathbb{R}^2$ and we assume homogeneity, B should be a function of the *relative position* $x - \xi$ rather than of x and ξ separately. When, in addition, we assume isotropy, we can consider B as a function of the *distance* $|x - \xi|$ only. Here we restrict our attention to that situation, that is, we consider (with slight notational abuse of the symbol B)

$$u(t, x) = \int_0^\infty \int_{\mathbb{R}^2} B(\tau, |x - \xi|) u(t - \tau, \xi) \, d\xi \, d\tau. \tag{8.8}$$

---

[3] P. Jagers: The deterministic evolution of general branching populations. In: O. Arino, D. Axelrod & M. Kimmel (eds.), *Mathematical Population Dynamics*. Wuerz, Winnipeg, 1995; V.M. Shurenkov: On the relationship between spectral radii and Perron roots. Preprint 1992-17, Department of Mathematics, Chalmers University Göteborg, 1992.

Tempted by our analysis of the linear diffusion equation and the robustness considerations of the preceding section, we look for travelling plane-wave solutions, i.e. we put

$$u(t, x) = w(x \cdot \nu - ct) \tag{8.9}$$

and deduce that w should satisfy

$$w(\theta) = \int_{-\infty}^{\infty} V_c(\zeta) w(\theta - \zeta) \, d\zeta, \tag{8.10}$$

where, by definition,

$$V_c(\zeta) = \int_0^{\infty} \int_{-\infty}^{\infty} B(\tau, \sqrt{(\zeta + c\tau)^2 + \sigma^2}) \, d\sigma \, d\tau, \tag{8.11}$$

which, nota bene, does not depend on ν.

Exercise 8.7 Derive (8.10) in detail.

Exercise 8.8 Derive the characteristic equation

$$1 = \int_{-\infty}^{\infty} e^{-\lambda\zeta} V_c(\zeta) \, d\zeta \tag{8.12}$$

by inserting the trial solution $w(\theta) = e^{\lambda\theta}$ into (8.10).

Exercise 8.9 We give the right-hand side of (8.12) a name, i.e. we write $1 = L_c(\lambda)$, where

$$L_c(\lambda) := \int_{-\infty}^{\infty} e^{-\lambda\zeta} V_c(\zeta) \, d\zeta. \tag{8.13}$$

Show that

$$L_c(\lambda) = \int_{-\infty}^{\infty} e^{-\lambda\alpha} \int_0^{\infty} \int_{-\infty}^{\infty} e^{\lambda c\tau} B(\tau, \sqrt{\alpha^2 + \sigma^2}) \, d\sigma \, d\tau \, d\alpha, \tag{8.14}$$

and show from this that

$$L_c(0) = \int_0^{\infty} \int_{\mathbb{R}^2} B(\tau, |\eta|) \, d\eta \, d\tau = R_0, \tag{8.15}$$

$$\frac{dL_c}{d\lambda}(0) = c \int_0^{\infty} \int_{\mathbb{R}^2} \tau B(\tau, |\eta|) \, d\eta \, d\tau > 0 \tag{8.16}$$

for $c > 0$, and that

$$\frac{d^2 L_c}{d\lambda^2}(\lambda) > 0 \tag{8.17}$$

for all λ and all $c \geq 0$, and finally that for every $\lambda < 0$, $L_c(\lambda)$ is a monotonically decreasing function of c with limit zero for $c \to \infty$.

Conclude from all this that, whenever $R_0 > 1$, the set $\{c : \text{there exists } \lambda < 0 \text{ such that } L_c(\lambda) < 1\}$ consists of a half-line (c_0, ∞).

Exercise 8.10 Establish that c_0 can be characterised, together with the corresponding value of λ, say λ_0, as the solution of the pair of equations (8.12) and

$$\frac{dL_c}{d\lambda}(\lambda) = 0. \tag{8.18}$$

We conclude that, starting from the modelling ingredient $B(\tau, |\eta|)$, one can
constructively define a minimal plane-wave speed c_0 by the pair of equations (8.12),
(8.18). It remains to ascertain that c_0 thus defined is also the asymptotic speed of
propagation of disturbances. We postpone remarks on this issue to the next section,
where we deal with the nonlinear problem.

Exercise 8.11 When the species considered is actually an infectious agent exploiting
a host population, and if we assume mass-action contacts, we have

$$B(\tau, x, \xi) = S_0(x)A(\tau, x, \xi), \tag{8.19}$$

where A is our familiar epidemic model ingredient and $S_0(x)$ is the host density
as a function of position. Do you agree? If so, check that travelling front solutions
require a uniform host density S_0 (as may be an appropriate assumption within
fields of agricultural crops, or, if we think of fields as host individuals, for fields
within a region).

We refer to Thieme (1979)[4] for estimates of the speed of propagation using only
lower bounds for S_0 and to Shigesada and Kawasaki (1997) for numerical studies
of the speed of propagation when high- and low-density host population patches
alternate.

8.5   The nonlinear situation

Nonlinearity leads to boundedness, but, under suitable assumptions, nothing much
changes otherwise.

Models in population genetics, combustion and population dynamics lead to
nonlinear diffusion equations

$$u_t = D\triangle u + f(u) \tag{8.20}$$

with a nonlinear function f having properties as displayed graphically in Figure 8.3.

So, forgetting for a moment about space, zero is an unstable steady state and
any positive perturbation ultimately leads to the stable steady state p. Hence the
considerations of Section 8.3 apply. But how should one compute c_0?

If we linearise (8.20) at $u \equiv 0$, we obtain equation (8.1) with $\kappa = f'(0)$, to which
we can associate a speed c_0. Is this the right one?

To show that (8.20) has travelling plane-wave solutions for every $c \geq c_0$ is a matter
of phase-plane analysis, for which we refer to the literature.[5] To show that $u(t, x)$ tends
to zero outside a ball that expands with speed larger than c_0, while tending to p inside

---

[4] H.R. Thieme: Density-dependent regulation of spatially distributed populations and their
asymptotic speed of spread. *J. Math. Biol.*, **8** (1979), 173-187.
[5] K.P. Hadeler & F. Rothe: Travelling fronts in nonlinear diffusion equations. *J. Math. Biol.*, **2**
(1975), 251-263; A.I. Volpert, V.A. Volpert & V.A. Volpert: *Travelling Wave Solutions of Parabolic*

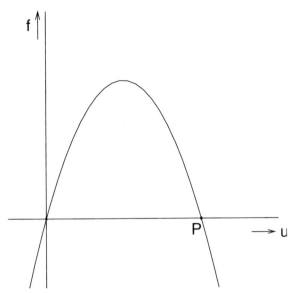

Figure 8.3

a ball that expands with speed smaller than c_0, is not easy but also not impossible. The intricate proof, for which we refer to Aronson & Weinberger and Diekmann & Temme (loc. cit.), involves comparison arguments based on the maximum principle. Both results involve the condition

$$f(u) \leq f'(0)u, \tag{8.21}$$

which reflects that the living conditions are optimal at very low densities. When the condition is not satisfied, for example due to an Allee effect, one may have so-called 'pulled' waves, the speed of which is not determined by the low-density situation. Also for this case many results are known (see the references already quoted).

We conclude that, provided the nonlinearity f satisfies certain interpretable and reasonable conditions, there is a well-defined asymptotic speed of propagation that can be calculated from the appropriate linearisation.

In the epidemic context, the starting point is equation (6.14) with $S(-\infty, x) = S_0$, independent of x. Introducing

$$u(t, x) = -\ln \frac{S(t, x)}{S_0} \tag{8.22}$$

and manipulating as in Section 6.3, we arrive at the nonlinear integral equation

$$u(t, x) = \int_0^\infty \int_{\mathbb{R}^2} S_0 A(\tau, x, \xi) g(u(t - \tau, \xi)) \, d\xi \, d\tau \tag{8.23}$$

Systems. AMS Translations of Mathematical Monographs, Vol. 140, 1994; D.G. Aronson & H.F. Weinberger: Nonlinear diffusion in population genetics, combustion, and nerve pulse propagation. In: *Partial Differential Equations and Related Topics.* J.A. Goldstein (ed.), Lecture Notes in Mathematics, Vol. 446, Springer-Verlag, Berlin, 1975, pp 5-49; O. Diekmann & N.M. Temme (eds.): *Nonlinear Diffusion Problems.* Mathematisch Centrum, Amsterdam, 1976.

with

$$g(u) = 1 - e^{-u}. \tag{8.24}$$

We observe two things:

- the linearisation at $u \equiv 0$ is of the form (8.7);
- $g(u) \leq g'(0)u$ (the 'virgin' situation is optimal for the infective agent).

Motivated by these observations, we expect that one can prove for u satisfying (8.23) that c_0 is the asymptotic speed of propagation, where

- c_0 is calculated from S_0 and A (which is, of course, assumed to depend on $|x - \xi|$ only) as in the preceding section;
- u tends to the familiar final size within any ball that expands with speed less than c_0.

That these expectations are warranted was shown in detail by Thieme and Diekmann around 1978, using comparison methods in the spirit of Aronson & Weinberger; more complicated situations were dealt with extensively somewhat later in a series of papers by J. Radcliffe and L. Rass (see the forthcoming book by Radcliffe & Rass (2000)). Pioneering work had been done much earlier by Fisher, Skellam, Kolmogorov-Petrovski-Piscounov, Kendall and later Mollison (see the references in Metz & Van den Bosch (1995)[6]).

We conclude that for our familiar epidemic model, one can compute the asymptotic speed of propagation from the model ingredients S_0 and A, viz. by solving (8.12) and (8.18).

8.6   How to make the theory operational?

In our top-down approach we are still far from the bottom. As a next step, one should introduce parametrised families of kernels A on the basis of a mixture of mechanistic and pragmatic considerations. In addition, it is useful to derive approximation formulae for c_0 involving moments (both of time-type and of space-type) of the kernel A. Using experimental data to estimate parameters (preferably in an independent manner) and to verify predictions, one can then assess the theory.[7]

In the ideal version of this chapter, as we planned it, this section would consider a case study (e.g. rabies) in detail. Time constraints prevented the realisation of this ideal.

8.7   Summary: The speed of propagation

Within the context of idealised models, we have unambiguously defined the (asymptotic) speed c_0 of the spatial propagation of an infection, and we have

---

[6] J.A.J. Metz & F. Van den Bosch: Velocities of epidemic spread. In: Mollison (1995), pp. 150-186.

[7] See e.g. F. Van den Bosch, J.A.J. Metz & O. Diekmann: The velocity of spatial population expansion. *J. Math. Biol.*, **28** (1990), 529-565; M.A. Lewis & S. Pacala: Modelling and analysis of stochastic invasion processes. *J. Math. Biol.*, in press, 1999; U. Dieckmann, R. Law & J.A.J. Metz (eds.): *The Geometry of Ecological Interactions*. Cambridge University Press, Cambridge, to appear 2000; Shigesada & Kawasaki (1997); McGlade (1999).

characterised c_0 in terms of the basic model ingredients in such a way that the computation of c_0 from the ingredients is rather simple. Thus we added one more indicator of the infectiousness of an agent to the list (consisting so far of R_0, r, the probability of a major outbreak, the size of a major outbreak, and the endemic level). For many ecological or agricultural systems, this is actually the most relevant indicator!

9

Macroparasites

9.1 Introduction

As we have seen in Chapter 4, the defining mathematical distinction between microparasites and macroparasites is that for macroparasites, as a rule, re-infection through the environment is essential to get an increase in individual infectious load and consequent infectious output. In this chapter, we give a brief introduction to the consequences that this distinction has for formulating epidemic models for macroparasites. For the largest part, we concentrate on the definition and calculation of R_0.

Typically, macroparasites are multicellular organisms (e.g. helminths and other worm-like parasites) where definite stages in a life cycle can be distinguished. Several of these stages live outside living hosts. We will regard mostly two stages, adults living within a host and larvae (hatched from eggs produced by the adults and shed by the hosts) living in the environment of the host, since many features can already be illustrated in this minimal setting. Larvae are then infective to hosts and uptake can be by, for example, ingestion or skin penetration.

Exercise 9.1 Reflect on whether, as in the microparasite case, infection-age of an infected individual could be adopted as a basis to model infectious output.

The mathematical distinction between micro- and macroparasites has consequences for the way R_0 is defined. What matters for R_0 is the infectious output of infected individuals and the contact patterns of these individuals with susceptibles. In the microparasite case we have seen that we can describe the infection within an individual as an autonomous process, disregarding further influence of the individual's environment. The definition of R_0 for microparasites as the expected number of new host individuals infected per infected host is a direct consequence of these considerations, and it makes good sense to follow individual hosts, possibly of various types. In the macroparasite case, however, the influence of the environment is essential in order to describe the infection pressure acting on hosts. In addition, the obligatory environmental stages bring with them the consequence that direct contacts between hosts do not generally provide a good description of the spread of the infectious output to other hosts. Thinking about R_0 as the number of hosts infected per host does not relate as closely to the actual biological processes as in the case of microparasites. It makes little sense to follow individual hosts, since what happens to the parasite in the environment is crucial. It makes more sense to follow parasites in the traditional demographic spirit: one chooses a reference point in the life cycle (usually 'being newborn', but here we prefer 'becoming/being adult'), and calculates for one individual at the reference point, the expected number of offspring reaching the reference point. The life cycle then has as a consequence that R_0 could be described as the expected number of new adult parasites (i.e. stages that end up in the major host) produced per adult parasite.

Exercise 9.2 i) R_0 for microparasites refers to the situation that the susceptible hosts are in a demographic steady state and that each contact an infective makes is with a susceptible. What do you think the corresponding assumption will be in the case of macroparasites?

ii) It was pointed out in 1959 by Macdonald and later by Nåsell that for parasites with obligatory sexual reproduction in the main host, the assumption needed to characterise R_0 presents a conceptual difficulty. Can you describe it? Hint: How many parasites will likely be present in an infected host in the early phase of an epidemic?

iii) We expand on this difficulty in the context of a model. The following system of differential equations has been considered as a model for schistosomiasis:[1]

$$\frac{dI}{dt} = -\delta I + C(m)(N - I),$$

$$\frac{dm}{dt} = -\mu m + A\frac{I}{N}.$$

Here I is the density of infected snails (the intermediate host; for details of the life cycle see e.g. Anderson & May (1991)) and m is the mean number of adult worms in infected humans. So μ corresponds to the per capita death rate of worms inside the human host, δ to the per capita death rate of infected snails and N to the total density of snails (assumed to be constant). A is proportional to the rate at which an infected snail produces the free-living stage that can penetrate the human skin, and C is proportional to the rate of production of eggs by the female adult worms (which includes the hatching of eggs to produce the free-living stage that can infect snails in water). The dependence of C on the mean number of *paired* worms per human is here incorporated as a dependence on m. The point is, however, that the fact that it takes two to reproduce should be reflected in properties of C. In so-called hybrid models[2] (i.e. mixing stochastic and deterministic components) one takes for C an increasing function that is zero at zero and, most importantly, has a vanishing derivative at zero. This latter assumption reflects that, for small values of m, parasites will not usually be able to find a mate within the human host.

The aim of this exercise is to show, basically via phase-plane analysis of the system given above, that the number of non-trivial steady states is either zero

---

[1] W.M. Hirsch, H. Hanish & J.P. Gabriel: Differential equation models of some parasitic infections: methods for the study of asymptotic behaviour. *Comm. Pure Appl. Math.*, **38** (1985), 733-753.
[2] See Nåsell (1985); J.P. Gabriel, H. Hamisch & W.M. Hirsch: Worm's sexuality and special function theory. In: Gabriel, Lefèvre & Picard (1990), pp. 137-144.

or two, and that in the latter case one needs to introduce a substantial parasite load to escape from the domain of attraction of the infection-free steady state (and to converge to the stable endemic level).

So, assume that the isocline

$$I = \frac{C(m)N}{\delta + C(m)N}$$

has a sigmoid shape. First interpret the steady-state condition graphically. Then address the following questions: How does the number of steady states depend on the parameter A? What is R_0? Is there a critical value for A? If so, characterise it and describe the dynamical behaviour for A above the critical value.

In Section 9.3 we show how to calculate R_0 for macroparasites in unstructured populations and populations with finite discrete structure. In Section 9.4 we briefly indicate the corresponding theory to characterise invasion into periodic environments (for both macro- and microparasites). In Section 9.5 we touch upon a subtle pathology that is a direct consequence of the macroparasite definition.

We start, however, by describing a typical way in which spread of macroparasites is modelled, where we draw on work by Kostizin, Anderson and May, Hadeler and Dietz, Kretzschmar and Adler, and others.

9.2   Counting parasite load

The derivation of the bookkeeping equations based on infection-degree is similar to a derivation in Exercise 1.33, and serves as another example of a stochastic epidemic model. Let $p_i(t)$ be the number of hosts carrying i adult parasites at time t, with i a natural number. The term 'susceptibles' could now be reserved for those hosts carrying no parasites, p_0, but actually the term does not make much sense anymore, since every host is susceptible to additional infection. In fact, we will use the customary term 'naive' to denote host individuals that never have experienced parasites. Note, however, that these individuals are not necessarily the same group as those without parasites. If one takes acquired immunity upon re-infection into account, it might be that individuals who have just been cleared of parasites, and so would enter the category p_0, are different from naives, since the ones that had parasites before can have a different immune status because of this previous presence. We refrain, however, from using a special name for these hosts.

Let $L(t)$ denote the density of larvae in the environment at time t. We distinguish between the following events/mechanisms:

- adult parasites die in the host with per capita probability per unit of time μ;
- larvae die in the environment with per capita rate ν;
- the host experiences a force of infection β, in the sense that the probability per unit of time that a larva enters is β; ultimately, we have to close the feedback loop and relate β to L, for instance as $\beta = \theta L$, for some constant θ;
- upon entering a host, a larva transforms instantaneously into an adult;
- an adult parasite produces larvae with probability per unit of time λ.

We assume that parasites act independently from each other (yet there is interaction through the death of the host). We disregard host demography and therefore view transmission dynamics as a pure immigration-death process for the parasites. To derive differential equations for the dynamics of the p_i, we consider the events that can lead to a change over an interval of length Δt in a host carrying i adult parasites to start with.

Exercise 9.3 What do we have to assume about Δt and what consequences does this have for the events themselves and for the corresponding probabilities. Hint: The events described above follow exponential probability distributions.

Exercise 9.4 Describe the two key events within a host and give their probabilities in the interval of length Δt.

Exercise 9.5 Use Exercise 9.4 to show that the differential equation for p_i, $i \geq 1$, is given by

$$\frac{dp_i}{dt}(t) = -(\beta + i\mu)p_i(t) + (i+1)\mu p_{i+1}(t) + \beta p_{i-1}(t) \tag{9.1}$$

and that the differential equation for p_0 is given by

$$\frac{dp_0}{dt}(t) = -\beta p_0(t) + \mu p_1(t). \tag{9.2}$$

For the larvae, we need some additional assumptions. We need to specify how larvae come into contact with hosts. Assume that this is governed by mass action, with proportionality constant θ, say. Let N be the (constant) number of hosts and let $P(t)$ be the total number of adult parasites at time t in the N hosts. In the present formulation both p_i and L are numbers, but all relations continue to hold when both are interpreted as densities, i.e. numbers per unit area.

Exercise 9.6 Convince yourself that an equation incorporating these assumptions and governing the changes in the larval population is

$$\frac{dL}{dt} = \lambda P - \theta N L - \nu L, \tag{9.3}$$

where

$$P(t) = \sum_{i=1}^{\infty} i p_i(t).$$

Moreover, argue that $\beta = \theta L$.

Exercise 9.7 Suppose that next we do take host demography into account and that we postulate a per capita birth rate b of hosts, a natural host death rate d and an additional mortality rate caused by the parasites. For the latter assume that each parasite the host harbours increases the death rate by κ. All newborns are free from parasites. Show that (9.1) and (9.2) become

$$\begin{aligned} \frac{dp_i}{dt}(t) &= -(\beta + d + i(\mu + \kappa))p_i(t) \\ &\quad + (i+1)\mu p_{i+1}(t) + \beta p_{i-1}(t), \quad i > 0, \end{aligned} \tag{9.4}$$

$$\frac{dp_0}{dt}(t) = b\sum_{i=0}^{\infty} p_i(t) - (d+\beta)p_0(t) + \mu p_1(t). \tag{9.5}$$

This model was first formulated by Kostizin (1934).[3]

For the analysis of the infinite system of ordinary differential equations there are two common routes, both making use of the special structure of the system. The first involves probability generating functions, which we have encountered already in Section 1.2.2, reducing the study of the infinite system of ODE to the study of a first-order partial differential equation, which can be solved by the method of characteristics. The second, more common, route involves a brute force approximation that leads to a simplified model of two ODE for the average number of parasites per host and the density of larvae. We derive both types of equations below, but refrain from a detailed analysis.

Imagine a single host, randomly chosen at time t from the (large) population. Then $p_i(t)/N(t)$ is the probability that this host has i parasites.

Exercise 9.8 Convince yourself that a law-of-large-numbers argument reconciles this interpretation with the original definition of the $p_i(t)$.

We summarise some aspects of the detailed information on parasite distribution $p_i(t)$ into the variable $P(t)$, the number of adult parasites present in the population at time t, and $N(t)$, the total number of hosts at time t:

$$P(t) := \sum_{i=0}^{\infty} i p_i(t), \quad N(t) := \sum_{i=0}^{\infty} p_i(t).$$

In our present set-up, host demography is ignored and therefore N is constant. Some calculation shows that, starting from (9.1) and (9.2), one obtains the following differential equation for P:

$$\frac{dP}{dt} = -\mu P + \beta N. \tag{9.6}$$

Let us first give, based on the interpretation, an expression for R_0 for the system (9.3), (9.6). Consider one adult. Its expected lifespan is $\frac{1}{\mu}$, and during that time it is expected to produce larvae at a rate λ. Larvae have an average lifespan given by $\frac{1}{\nu+\theta N}$, and a larva has probability per unit of time θN of making it into an adult. Therefore

$$R_0 = \frac{\lambda}{\mu} \frac{\theta N}{\nu + \theta N}.$$

Let $g(t, z)$ be the probability generating function of the $p_i(t)$. That is,

$$g(t, z) := \sum_{i=0}^{\infty} p_i(t) z^i.$$

Exercise 9.9 Show that from the system (9.1), (9.2) one can derive the linear first-order partial differential equation

$$\frac{\partial g}{\partial t}(t, z) = \beta(z - 1)g(t, z) - \mu(z - 1)\frac{\partial g}{\partial z}(t, z). \tag{9.7}$$

---

[3] V.A. Kostizin: *Symbiose, parasitisme et evolution.* Hermann, Paris, 1934.

Exercise 9.10 Check equation (9.6). Hint: Differentiate the PDE (9.7) with respect to z and put z equal to 1.

Exercise 9.11 As a follow-up to Exercise 9.7, show that the differential equations for P and N are given by

$$\frac{dN}{dt} = (b-d)N - \kappa P, \tag{9.8}$$

$$\frac{dP}{dt} = -(\mu+d)P + \beta N - \kappa \sum_{i=0}^{\infty} i^2 p_i, \tag{9.9}$$

if we take (the particular assumptions on) host demography into account. Hint: Either proceed by direct calculation using the definitions of P and N and equation (9.4) and (9.5), or proceed as in Exercise 9.10, starting from the PDE for the generating function related to (9.4) and (9.5):

$$\frac{\partial g}{\partial t}(t,z) = bN(t) - dg(t,z) + \beta(z-1)g(t,z) + (\mu - (\kappa+\mu)z)\frac{\partial g}{\partial z}(t,z).$$

Readers wishing even more practice can derive this PDE for the system (9.4), (9.5).

Exercise 9.12 The PDE (9.7) can be solved by the method of characteristics. Let the initial distribution be given by $g(0,z) = g_0(z)$. Show that the solution of (9.7) is

$$g(t,z) = g_0(e^{-\mu t}(z-1)+1)\exp\left(\frac{\beta}{\mu}(z-1)(1-e^{-\mu t})\right).$$

Deduce that for $t \to \infty$ the distribution becomes Poisson with parameter β/μ. Does this result surprise you?

To link the adults with the dynamics of the larvae in the environment, we note once again that our mass-action assumption of uptake of larvae by hosts implies that $\beta = \theta L(t)$ in (9.6) (and (9.9)). In the absence of host demography, one can then summarise the model into the two-dimensional system (9.3), (9.6).

Exercise 9.13 We explore the system from Exercises 9.7 and 9.11 a little further. We see that when we include host demography three differential equations arise (for P, L and N), but that this system is not closed. Anderson & May (1978)[4] initiated the following approximation to close the three-dimensional system. If we have information (e.g. from field studies) on parasite distributions $p_i(t)$, we can try to use this to postulate a formula for the $p_i(t)$ and use the formula to try and express $\sum i^2 p_i$ in terms of P and N, despite the fact that we know that this is incompatible with the assumptions underlying the bookkeeping in the first place. So this is a mixed strategy, in which we combine empirical information and theoretical considerations which are not (necessarily) consistent. In the Anderson

---

[4] R.M. Anderson & R.M. May: Regulation and stability of host-parasite population interactions. *J. of Animal Ecol.*, **47** (1978), 219-267.

& May approach the idea was that aggregation would have a stabilising influence on the dynamics of the system. We will come back to this at the end of this section.

Typically, data show parasite distributions to be highly aggregated, i.e. a small fraction of the hosts carries a large fraction of the parasite population. First express the parasite distribution as a fraction by defining $r_i(t) := \frac{p_i(t)}{N(t)}$. Let the generating function of the r_i-distribution be $g(t, z) = \sum_{i=0}^{\infty} r_i(t) z^i$. Note that $g'(t, 1)$, defined as $g'(t, 1) = \frac{\partial g}{\partial z}(t, z)|_{z=1}$ is equal to the mean of the distribution.

i) Check that for the parasite distribution we have that the mean parasite load is $g'(t, 1) = \frac{P}{N}$.

ii) Show that $\sum_{i=0}^{\infty} i^2 p_i = N \sum_{i=0}^{\infty} i^2 r_i = N(g''(t, 1) + g'(t, 1))$. Note that the variance of the parasite distribution is given by $g''(t, 1) + g'(t, 1) - (g'(t, 1))^2$ and that we can rewrite

$$\frac{1}{N} \sum_{i=0}^{\infty} i^2 p_i = \text{mean} \left(\frac{\text{variance}}{\text{mean}} + \text{mean} \right).$$

iii) Traditionally,[5] a negative binomial distribution is adopted for the r_i, with parameters $m = P/N$, i.e. the mean, with $0 \leq m \leq 1$, and $k > 0$, which is a measure of aggregation (a smaller k means that parasites are more aggregated). The generating function is given by

$$g(t, z) = \left(1 + \frac{m}{k}(1 - z) \right)^{-k}.$$

Use this expression to show that

$$\sum_{i=0}^{\infty} i^2 p_i = N \left(\frac{k+1}{k} \frac{P^2}{N^2} + \frac{P}{N} \right)$$

which enables us to close the system of differential equations for P and L by changing (9.9) into

$$\frac{dP}{dt} = P \left(\frac{\theta N L}{P} - (\mu + d + \kappa) - \kappa \frac{k+1}{k} \frac{P}{N} \right). \qquad (9.10)$$

Exercise 9.14 i) If we could give a mechanistic argument to express the variable L in terms of P and N, we could reduce the system even further to a more tractable two-dimensional system. Suppose that the larvae in the environment have a (much) shorter lifespan than the adult parasites living in the host (i.e. during the average lifetime of an adult, many larvae will have come and gone). This difference in time scale could then be exploited by taking the extreme view that,

---

[5] R.M. Anderson & R.M. May: Helminth infections of humans: mathematical models, population dynamics and control. *Adv. in Parasitol.*, **24** (1985), 1-101.

relative to the time scale of adult dynamics, the larvae are in fact in a (quasi) steady state (cf. Section 10.2). Show that, with this assumption, we obtain

$$\beta N = \frac{\lambda \theta P N}{\nu + \theta N}.$$

ii) To summarise, the closing 'trick' of the previous exercise and the time-scale argument of the present exercise reduce the originally infinite-dimensional system to

$$\frac{dN}{dt} = (b - d)N - \kappa P \tag{9.11}$$

$$\frac{dP}{dt} = P\left(\frac{\lambda \theta N}{\nu + \theta N} - (\mu + d + \kappa) - \kappa \frac{k + 1}{k} \frac{P}{N}\right). \tag{9.12}$$

Exercise 9.15 i) Use the same heuristic argument as that following equation (9.6) to show that for the system (9.11), (9.12) we have

$$R_0 = \frac{\lambda}{\mu + d + \kappa} \frac{\theta N}{\nu + \theta N}.$$

When θN becomes large relative to ν, we obtain

$$R_0 = \frac{\lambda}{\mu + d + \kappa}.$$

ii) Note that we obtain the same expression if we apply the heuristic argument to the three-dimensional system (9.3), (9.8), (9.10). The time-scale argument for L therefore has no influence on R_0. Is this to be expected?

Exercise 9.16 To study the supposed effect of aggregation on stability (see Exercise 9.13), 'we' do some more analysis on the system (9.11), (9.12) along the lines of Section 3.3. Four cases of different long-term behaviour of the system (9.11), (9.12) can be identified (Anderson & May (1978), see ref. in Exercise 9.13) after the parasite is introduced into a naive host population. Assume $b > d$. When no parasites are present, the host population size grows exponentially with rate $b-d$. Four regions of parameter space can be identified, using the rate of production of larvae λ as a bifurcation parameter.

i) Show that for $0 < \lambda < \lambda_0 := \mu + d + \kappa$ the parasite cannot invade successfully and that the host population will asymptotically (i.e. when $t \to \infty$) grow exponentially.

ii) Show that for $\lambda_0 < \lambda < \lambda_1 := b + \mu + \kappa$ the invasion is successful and both the parasite and host populations will grow. Show that the mean parasite load $\frac{P}{N}$, however, goes to zero, and that N will asymptotically grow exponentially with the same rate as before.

Hint: First derive a differential equation for $\frac{P}{N}$.

iii) Show that for $\lambda_1 < \lambda < \lambda_2 := \mu + d + \kappa + (b - d)\frac{k+1}{k}$ both populations will asymptotically grow exponentially with the same reduced rate which is smaller than $b - d$, but positive. The mean parasite load $\frac{P}{N}$ approaches a constant for $t \to \infty$.

iv) Show that for $\lambda > \lambda_2$ there exists a steady state $(\overline{N}, \overline{P})$ for the system (9.11), (9.12). One can show that this steady state is locally asymptotically stable. In this situation the parasite is able to regulate the host population by bringing the exponential growth to a standstill.

Kretzschmar and Adler[6] have studied the dynamics of the full infinite dimensional model with host demography and have compared this with the two-dimensional approximation by Anderson and May given above. In the infinite-dimensional model the thresholds λ_0 and λ_1, and the concomitant behaviour given in Exercise 9.16, are as in the approximation, but the threshold λ_2 is different. The infinite-dimensional model can be shown[7] to have a fixed distribution of parasites, which is approached asymptotically, and which has a variance-to-mean ratio that is larger than one. In other words, a distribution is approached that is overdispersed. This would justify the Anderson and May approximation (although there one assumes the same distribution at all times, rather than a changing distribution that only asymptotically reaches a fixed overdispersed shape). In the approximation, λ_2 depends on the aggregation parameter k of the assumed negative binomial distribution. The problem with this is that k is not a biological parameter—it is at best an unknown function of the biological parameters in the system. As a consequence, it can be difficult to draw detailed biological conclusions. Kretzschmar and Adler (loc.cit.) show that, rather than by aggregation, the system is stabilised by the ratio of the variance to the mean. The infinite-dimensional model and the two-dimensional approximation discussed behave in much the same way.

If the two models discussed above are used as bases to test the influence of additional mechanisms on stability, we find that differences in the predicted dynamics start to occur between the infinite-dimensional and two-dimensional models. For one such additional mechanism, where the parasites are now able not only to influence the death rate of hosts, but also to reduce the host's fertility, it has been shown that periodic solutions are possible.[8]

Exercise 9.17 Let ζ be the per (parasite) capita reduction of host fertility. Two possibilities to model the influence of all the parasites within the host in reducing the host's fertility are to add all contributions (i.e. to take $i\zeta$ for a host carrying i parasites), or to multiply them (i.e. to take ζ^i). What is the preferred choice and why?

---

[6] F.R. Adler & M. Kretzschmar: Aggregation and stability in parasite-host models. *Parasitology*, **104** (1992), 199-205; M. Kretzschmar & F.R. Adler: Aggregated distributions in models for patchy populations. *Theor. Pop. Biol.*, **43** (1993), 1-30.

[7] M. Kretzschmar: Comparison of an infinite dimensional model for parasitic diseases with a related 2-dimensional system. *J. Math. Anal. Appl.*, **176** (1993), 235-260.

[8] O. Diekmann & M. Kretzschmar: Patterns in the effects of infectious diseases on population growth. *J. Math. Biol.*, **29** (1991), 539-570.

9.3   The calculation of R_0 for life cycles

In the previous section we have already derived an expression for R_0 by a heuristic argument for parasites with two stages in the life cycle. This argument can be formalised for parasites with n stages, possibly living in different hosts and environments, and for the case that there is heterogeneity in the parasite, the hosts and the environments. The idea is to consider two rates d_i and m_i in the absence of density-dependent effects. Let the stage-i individuals of the parasite give rise to stage-$(i+1)$ individuals at a rate m_i (notably, maturation, of one larval stage into the next, or actual production of new parasites). Without loss of generality, we can assume that stage n, that determines inflow into stage 1 to close the cycle, is a reproduction stage leading to the birth of new parasites. Let $d_i > 0$ be the rate of leaving stage i (by death or by transition to stage $i+1$).

Exercise 9.18 i) Give d_1, d_2, m_1, and m_2 in the case of the system (9.3), (9.8), (9.10).

ii) We can generalise the heuristic argument by considering that there are n links in the life cycle and that each link is characterised by a multiplication factor for its contribution to the propagation of the parasite species. Note that m_i in a sense gives the reproduction rate of stage-$(i+1)$ individuals and that sojourn time in stage i is exponentially distributed with parameter d_i. Argue that

$$R_0 = \prod_{i=1}^{n} \frac{m_i}{d_i}.$$

Exercise 9.19 We can formalise the above using the ideas introduced in Section 5.4 to calculate R_0. For this we view the progress through the life cycle as a continuous-time Markov chain on $\{1, ..., n\}$. Suppose we start with one newborn parasite in stage 1; how many stage 1 parasites is it expected to produce? Reproduction giving rise to new parasites of stage 1 is described by the vector $(0, ..., 0, m_n)^\top$ by assumption. Give the transition matrix G and realise that

$$R_0 = -(0 \cdots 0\, m_n)G^{-1}\begin{pmatrix} 1 \\ 0 \\ \vdots \\ 0 \end{pmatrix} = \prod_{i=1}^{n} \frac{m_i}{d_i}.$$

Hint: Let $\mathcal{P}(a) = (\mathcal{P}_1(a), ..., \mathcal{P}_n(a))^\top$ be the vector giving the probabilities that an individual born in stage 1 is in stage i at age a. Calculate the time this individual is expected to spend in stage n and multiply by the reproduction in that stage.

Since R_0 describes the success of invasion into the parasite-free steady state, we can use the same reasoning as in Section 6.2 to characterise it. We give this approach because it can be generalised in a natural way to characterise invasion into a periodic environment (Section 9.4). We write x_i for the mean parasite load for stage-i parasite individuals. By this we mean the average number of parasites per host, for those stages in the life cycle living in hosts, and the density of parasites, for those stages living in another environment (such as in the water or on a pasture). Then, linearised around the steady state $(N, x_1, ..., x_n) = (\overline{N}, 0, ..., 0)$, we can describe the dynamics by

$$\frac{dx}{dt}(t) = Ax(t) = (M - D)x(t),$$

where

$$M = \begin{pmatrix} 0 & 0 & \cdots & 0 & m_n \\ m_1 & 0 & \cdots & 0 & 0 \\ 0 & m_2 & \ddots & 0 & 0 \\ \vdots & \ddots & \ddots & & \vdots \\ 0 & \cdots & 0 & m_{n-1} & 0 \end{pmatrix}, \quad D = \text{diag}(d_1, ..., d_n).$$

The same arguments as in Section 6.2 then show that the spectral radius $\rho(K)$ of $K = MD^{-1}$ determines the stability of the parasite-free steady state.

Exercise 9.20 i) Note that K^n is a diagonal matrix in which all diagonal elements are the same. Show that $R_0 := \rho(K)^n = \rho(K^n)$ has the required properties (threshold 1 and right interpretation) and that

$$\rho(K^n) = \prod_{i=1}^{n} \frac{m_i}{d_i}.$$

ii) Argue that taking $R_0 = \rho(K)$ would be biologically inconsistent. Hint: Remember Exercise 9.15-ii.

If there is heterogeneity in the system, we can calculate R_0 in a similar fashion as in both Exercises 9.19 and 9.20. We restrict ourselves to finite discrete and static structures. First note that we may identify any difference in type between hosts or environments with a (virtual) difference in types for the particular parasite stage that inhabits these hosts or environments. We can therefore restrict ourselves to the case where we recognise k_i types $(i, 1), ..., (i, k_i)$ of parasites of stage i, for $i \in \{1, ..., n\}$. The cycle length remains n, but now a cycle can consist, at least in theory, of any combination of one type from each of the stages, in the order $(1, i_1) \to (2, i_2) \to \cdots \to (n, i_n) \to (1, j_1)$, with $i_r \in \{1, ..., k_r\}, r \in \{1, ..., (k_1 + \cdots + k_n)n\}$ and $j_1 \in \{1, ..., k_1\}$. (For completeness we note that we can fit the arguments below into standard Markov-chain theory by introducing a map π from the space of all types to $\{1, ..., n\}$ that orders the types lexicographically and then to define rates m_{rs} as the rate of giving rise to type $\pi^{-1}(r)$ individuals by a type $\pi^{-1}(s)$ individual, and d_r as the rate of leaving its current stage for an individual of type $\pi^{-1}(r)$.)

Exercise 9.21 Generalise the heuristic argument of Exercise 9.18-ii by combining diagonal sojourn matrices D_i with stage-transition matrices M_i (so now we exclude type change within one and the same stage) and form products to arrive at one cycle matrix and argue that R_0 should be determined by the product of the n factor matrices $M_i D_i^{-1}$.

For the matrix approach we can then generalise the matrices M and D to

$$
M = \begin{pmatrix}
0 & 0 & \cdots & 0 & M_{1n} \\
M_{21} & 0 & \cdots & 0 & 0 \\
0 & M_{32} & \ddots & 0 & 0 \\
\vdots & \ddots & \ddots & \ddots & \vdots \\
0 & \cdots & 0 & M_{n,n-1} & 0
\end{pmatrix}, \quad D = \mathrm{diag}(D_1, ..., D_n),
$$

where the entries of M and D are all matrices of appropriate dimensions. Again define $K = MD^{-1}$ and let $K_{rs} = M_{rs}D_r^{-1}$, for $r, s \in \{1, ..., n\}$; then

$$
R_0 = \rho\left(\left(\prod_{j=2}^{n-1} K_{j,j-1}\right) K_{1n} \right).
$$

9.4   Seasonality and R_0

The existence of environmental stages in the life cycles of macroparasites entails that temporal variation in the environment that influences, for example, developmental rates and death rates could have a marked influence on the transmission dynamics and the ability to invade a naive population. Also, in microparasitic infections, a fluctuating environment can influence the dynamics. There the influence is not on the parasite itself, the development of which we regarded as an autonomous process within the host, but more on the host. We have already encountered examples such as seasonality in contact rate caused by the school system (in connection to, e.g., measles). Another example is the class of vector-transmitted infections with insect-vectors, where the emergence of the insects can be strongly influenced by seasonal variations (e.g. in temperature, humidity, etc.).

In general, few methods are available to incorporate temporal variation. If the variation is not periodic, success of invasion could be studied by calculating dominant Lyapunov exponents. If variation is periodic (e.g. seasonal), we can make use of the Floquet theory for stability of the trivial steady state of systems of ordinary differential equations with periodically varying rate constants.[9] The case of a macroparasite with n stages in its life cycle and a microparasite with n discrete static classes of individual (host) types, can be dealt with in the same formalism. Let $x(t) = (x_1(t), ..., x_n(t))^\top$ denote the vector describing the average parasite load of stage-i parasites, or the vector describing the density of infectives of type i, $i = \{1, ..., n\}$. We assume that all temporal variation included in the system has a common minimal period and we scale this period to 1. In the invasion phase, we regard the linearised real-time evolution as in the previous section:

$$
\frac{dx}{dt}(t) = A(t)x(t) = (M(t) - D(t))x(t), \quad x(0) = x_0, \tag{9.13}
$$

where now $A(t+1) = A(t)$ for all t, and where 'linearised' refers to the absence of density dependent effects in the case of macroparasites and to a demographic steady

---

[9] J.A.P. Heesterbeek & M.G. Roberts: Threshold quantities for helminth infections. *J. Math. Biol.*, **33** (1995), 415-434.

state for the susceptibles in the case of microparasites. For microparasites compare
Section 6.2, with $A(t) = T(t) + \Sigma(t) - D(t)$.

Now let $\Phi(t)$ be the so-called standard fundamental solution of (9.13), i.e.

$$\frac{d\Phi}{dt}(t) = A(t)\Phi(t), \quad \Phi(0) = I,$$

where I is the identity matrix (see e.g. Hale (1969)). This matrix can be obtained by
solving (9.13) consecutively with the n unit vectors of $\mathbb{R}^n$ as initial conditions and
taking the resulting vectors as columns. Then $\Phi(t) \geq 0$ is a positive matrix for all
$t \geq 0$ (because $\mathbb{R}^n_+$ has to be positively invariant; the appropriate assumptions that
guarantee this are $T(t) \geq 0$, $\Sigma(t)$ has positive off-diagonal elements, $D(t)$ is diagonal
and positive). The solution to (9.13) is given by

$$x(t) = \Phi(t)x_0.$$

The stability of the steady state can be determined from the constant matrix E defined
by

$$E = \Phi(1),$$

which is such that $\Phi(t + 1) = \Phi(t)E$. Floquet theory (see e.g. Hale (1969), Hirsch &
Smale (1974)) asserts that E exists and is uniquely determined by $A(t)$, and standard
theory gives that the stability of the trivial steady state of (9.13) is determined by the
value of the dominant eigenvalue $\lambda_d(E)$ of E (dominant Floquet multiplier) relative to
the threshold 1. Essentially, $\lambda_d(E)$ describes the asymptotic behaviour of the discrete
sequence $x(0) \rightarrow x(1) \rightarrow x(2) \rightarrow \cdots$.

Mathematically speaking, $\lambda_d(E)$ is the most obvious choice as threshold quantity for
periodic systems. It is certainly possible to calculate it numerically. The matrix E can
be obtained by successively solving the system (9.13) over a single period only, with the
unit vectors of $\mathbb{R}^n$ as initial conditions. What $\lambda_d(E)$ is lacking is an interpretation at
the individual level such as the one that R_0 possesses (no wonder, since the whole idea
of generations vanishes if the absolute moment of birth matters). There is no obvious
relation between $\lambda_d(E)$ and the matrix $A(t)$ (see Hale (1969), p. 121, for an example
of a periodic matrix $A(t)$ whose eigenvalues have negative real parts for every fixed
value of t, but still unbounded solutions exist and the dominant Floquet multiplier is
larger than 1). It is therefore not hopeful to try and distill a threshold condition from
a matrix obtained by somehow taking a weighted average of $A(t)$ over time.

9.5   A 'pathological' model

In the previous sections we aimed at characterising R_0 for macroparasites in
homogeneous, heterogeneous (with respect to population structure) and finally time-
periodic circumstances. The theory presented, however, was based on a description of
the parasite population based on mean parasite loads, and the distribution of parasites
was neglected. How could one create a theory of R_0 for the setting in terms of $p_i(t)$ as
introduced in Section 9.2? In this short section we look at an atypical model for the
spread of macroparasites that shows that the differences between microparasites and
macroparasites have a subtle consequence for R_0 (in addition to the subtlety already
mentioned in Exercise 9.2-ii). The consequence is that there can be a difference in

threshold between the situation where we characterise invasion success of a parasite by an increase in its numbers and the situation where we, in addition, require that also its abundance increases in the sense that the population of infected hosts has to increase. The model originates in work of Barbour.[10]

In the models treated in Section 9.2 only a single larva could enter the host at any given time (cf. Exercise 9.2). In parasites where clumps of eggs are shed by hosts into the environment (e.g a common water supply or a pasture) larvae will typically be picked up by the main hosts in clumps as well. Now suppose we model this in the following way, which is akin to microparasites. We pretend that hosts have a certain rate of contact to other hosts and that the number of parasites entering a host depends on the parasite load of the 'infecting' host and that this number is drawn from an 'infection distribution' with mean η (which takes into account the fact that a large proportion of free-living parasites do not make it back into a new host). This is comparable to hosts infecting a common pool and susceptibles drawing from that pool in the following way: first they choose a random infected host and then they draw from the infection distribution a number of parasites that enter simultaneously upon infection (depending on the parasite load of the selected infecting host, but without the infecting host experiencing a drop in its own parasite load). More concretely, let c be the 'contact' rate of hosts and let q_{ij} be the probability that the 'contact' of the target host is with an infected host carrying j parasites and that the target host becomes infected with i parasites. We assume that $\sum_{i \geq 0} q_{ij} = 1$ for each j and $\sum_{k \geq 1} k q_{kj} = j\eta$.

Let $r_i(t)$ be the fraction of hosts carrying i parasites at time t. We neglect host demography (which is a crucial assumption) and take μ to be the death rate of adult parasites in the host, as before. Since we are only interested in R_0, we neglect re-infection and assume that all target hosts are naive. Hosts can become infected with arbitrary numbers of parasites (with certain probabilities). Once an individual is infected, we have only death of parasites to deal with, and passing on of infection to others. The dynamics of the r_i are governed by (cf. Section 9.2)

$$\frac{dr_i}{dt}(t) = -i\mu r_i(t) + (i+1)\mu r_{i+1}(t) + c \sum_{j=1}^{\infty} r_j(t) q_{ij}, \qquad i \geq 1. \tag{9.14}$$

Exercise 9.22 We can derive (9.14) also from the equation for the incidence, involving the infection-age-dependent kernel A, as introduced for direct host-to-host transmission in Chapter 5. Let $x(i, j, \tau)$ be the probability of having i parasites left, τ time units after being infected with j parasites. Let $z(t, k)$ be the incidence at time t (as a fraction of the population) of infected hosts who are 'born' with k parasites. Convince yourself (cf. Section 5.4) that z satisfies

$$z(t, k) = \int_0^{\infty} \sum_{j \geq 1} \left(c \sum_{i=1}^{j} x(i, j, \tau) q_{ki} \right) z(t - \tau, j) \, d\tau$$

---

[10] A.D. Barbour: Threshold phenomena in epidemic theory. In: *Probability, Statistics and Optimization*. F.P. Kelly (ed.), Wiley, New York, 1994, pp 101-116; A.D. Barbour, J.A.P. Heesterbeek & C.J. Luchsinger: Thresholds and initial growth rates in a model of parasitic infection. *Ann. Appl. Prob.*, **6** (1996), 1045-1074.

$$= \int_0^\infty \sum_{j \geq 1} A(\tau, k, j) z(t - \tau, j) \, d\tau.$$

and that the fraction of the population carrying i parasites, $r_i(t)$, can be calculated as

$$r_i(t) = \int_0^\infty \sum_{j \geq i} x(i, j, \tau) z(t - \tau, j) \, d\tau.$$

The next-generation operator is given by

$$(K\phi)(k) = \sum_{j \geq 1} \int_0^\infty A(\tau, k, j) \, d\tau \, \phi(j).$$

Now assume that parasites die independently from each other in the host with per capita rate μ and find the binomial distribution

$$x(i, j, \tau) = \binom{j}{i} e^{-i\mu\tau} (1 - e^{-\mu\tau})^{j-i}.$$

Show that with this choice we obtain (9.14).

Exercise 9.23 i) Let us calculate R_0 for the model described in this section. First give a heuristic argument that shows that

$$R_0 = \frac{c\eta}{\mu}.$$

ii) Let a matrix U be defined by the right-hand side of the system (9.14) acting on the infinite-dimensional vector r with elements r_i. Then (9.14) can be rewritten as

$$\frac{dr}{dt} = Ur,$$

and the vector $\phi(j) = j$ is an eigenvector of $U^\top$ with eigenvalue $c\eta - \mu$.

iii) For the formulation of Exercise 9.22, first show that

$$A_{kj} := \int_0^\infty A(\tau, k, j) \, d\tau = \frac{c}{\mu} \sum_{i=1}^j \frac{q_{ki}}{i},$$

and next that

$$(K\phi)(k) = \frac{c}{\mu} \sum_{j \geq 1} \sum_{i=1}^j \frac{q_{ki}}{i} \phi(j).$$

Hint: The expected sojourn time in the class of hosts carrying i parasites is $(i\mu)^{-1}$, if parasites die independently from each other. This can also be calculated by working out the expectation of $x(i, j, \tau)$ with respect to τ.

iv) Since K is a strictly positive matrix, there is only one positive eigenvalue with a positive eigenvector, and this eigenvalue is defined to be R_0. Since the spectrum of a matrix is the same as the spectrum of its transpose, we can also, at least formally, look for eigenvalues and eigenvectors of $K^\top$ (or alternatively look for a left-eigenvector of K). Show that the vector $\phi(j) = j$ is an eigenvector of the transposed matrix $K^\top$ corresponding to the eigenvalue $\frac{c\eta}{\mu}$.

Analysis of the system (9.14) shows the following threshold behaviour. If a parasite enters a naive population, $R_0 = \frac{c\eta}{\mu}$ characterises the asymptotic behaviour of $\sum_{i \geq 1} i r_i(t)$; i.e. the value relative to the threshold 1 determines whether or not the parasite population will grow or decline. For an epidemic, however, we think more in terms of a rise or decline in the number of infected hosts after invasion, and this turns out to be more delicate. Concretely, one can show (using tools that are beyond this text) that for $\eta \leq e$, both parasite and infected host behaviour is governed by R_0. For $\eta > e$, however, the behaviour of the hosts is characterised by the quantity $\frac{ce \ln \eta}{\mu} \leq R_0$ relative to the threshold 1. In particular, in the case that $\eta > e$, an epidemic among the hosts only occurs if $\frac{ce \ln \eta}{\mu} > 1$. What apparently can happen is that for $R_0 > 1 > \frac{ce \ln \eta}{\mu}$ the invasion is successful in that parasites increase, but they become aggregated in too few hosts to be able to let the infected hosts increase epidemically.

It is an open problem to see what the effect of introducing host demography and, more specifically, mortality caused by the parasites would be on the above phenomenon. Intuitively one would expect an influence if, for example, hosts carrying more parasites have a higher death rate.

10

What is contact?

If the title of a chapter is a question, one expects perhaps that the chapter contains the answer. Not so here.

10.1   Introduction

In this section we reflect on the various aspects of contacts and the contact process. The interpretation of the word contact varies for different infectious agents, and only refers to happenings—to resurrect a term invented by Ross (1911)—where infection transmission could occur.

We first note that a contact can refer to two separate types of happenings, i.e. both the actual event of a transmission opportunity and the pairing of two individuals during which several such opportunities can arise. In general, pairings last longer than transmission events. Depending on the question studied and the time scales involved, a situation may call for separate modelling of the two types of contact.

The two most important aspects of contacts for infection transmission are 1) the number of contacts per unit of time and 2) the number of *different* individuals with whom these contacts occur. Without going into too much detail, we reflect on various possibilities concerning both aspects. Aspect 1 is concerned not only with variation in the number of transmission opportunities during a pairing, but also with the duration of the pairing (Section 10.2). We regard three cases, ordered by increasing time scale of contact duration: instantaneous contacts, pairings that are short-lived compared with the other time scales in the system, and finally pairings that are not short-lived on the infection time scale. Aspect 2 concerns spatial or social networks with variation in the set of potential contactees. This could be the entire population, a dynamic subset of the population or a fixed subset of the population. Moreover, the (sub)sets can contain finitely or (for mathematical reasons) infinitely many individuals. In the latter case one may assume that, when contacts are random, the probability to contact the same individual twice is negligible. We give some of the flavour of these aspects in Section 10.4 on graphs and networks. In Section 10.3 we briefly address the interesting new aspects that arise in heterogeneous populations where different types of individuals are recognised. We deal with consistency conditions, with gauging the effects of subdividing the population into subsets of mixing individuals, and with core groups.

10.2   Contact duration

Assume that there are infinitely many potential contactees. We only regard happenings involving two individuals.

Assume contact duration is very short. In other words, regard contacts as effectively instantaneous, while the contact intensity (number of contacts per unit of time) is proportional to the density of the two types of individuals involved. One then speaks of mass-action kinetics. As a metaphor one can take gaseous molecules colliding elastically in a reaction vessel (a context where the term 'law of mass action' originated). Pairings have infinitesimally short duration and all transmission opportunities during a pairing are compressed into this one instant. The contact rate between susceptibles and infectives is proportional to the product SI of the respective densities. If S and I represent *numbers* of individuals, it is proportional to SI/N, where N is the total population size. Although it is an extremely simple description of the contact process, there are situations in which mass action can be a good option, for example if contact duration is very short compared with, say, the average length of the infectious period.

Exercise 10.1 Regard a single infected individual. How does the number of contacts per unit of time for this individual rise under mass action as a function of the population density N? Is this a realistic description of the contact process? Can you give examples of happenings where it is certainly not realistic? What would the dependence of the contact rate on N look like for these examples?

We now study more closely the case where contacts have non-negligible duration. This implies that an individual is—purely due to time limitation—not able to engage in more and more contacts per day, say, even if contact opportunities continue to increase. In the following series of exercises we give a way to model these saturation effects mechanistically. For this, let N be the population density and suppose individuals have some form of pairwise contacts of exponentially distributed duration. We introduce the fraction $C(N)$ of the population that is engaged in a contact at any given time. We have that $NC(N)$ is the density of pairs in the population at any given time (actually, the density is $\frac{1}{2}NC(N)$, since pairs consist of two individuals; we will come back to this below). Since only a fraction SI/N^2 of the random pairings is between a susceptible and an infective, the density of pairings consisting of an infective and a susceptible, at any given time, is given by

$$NC(N)\frac{S}{N}\frac{I}{N} = C(N)\frac{SI}{N}. \tag{10.1}$$

Exercise 10.2 i) Verify that the alternative (which is to consider the fraction of susceptibles that are engaged in contact at any given time) leads to the same expression.

ii) Give an argument to show that $C(N)$ can also be interpreted as the fraction of the time that an individual will spend on contacts, given that the population density equals N.

Exercise 10.3 Give some properties that the function $C(\cdot)$ will reasonably have. When does (10.1) revert to mass action? Can you give an example of a situation where $C(\cdot)$ would be approximately constant over its entire range?

There are of course many functional forms that have the properties in Exercise 10.3. Often, models are studied without further specification and by only using these properties to derive results.[1] If one studies the contact process per se, it is interesting to have an explicit expression for $C(N)$. Based on Holling's time budget argument ('Holling's disc equation') to derive the functional response (i.e. the number of prey eaten per predator per unit of time, as a function of prey density), Dietz[2] suggested the following form:

$$C(N) = \frac{aN}{1 + bN} \qquad (10.2)$$

(in chemostat models this is referred to as the Monod equation and in chemical reaction kinetics as the Michaelis-Menten function). We give both a heuristic and a formal derivation of this relation.

Exercise 10.4 The idea of functional response in predator-prey interaction is that the number of prey caught per predator per unit of time as a function of prey density is limited by the time spent handling and eating prey that has been caught. Holling assumed that the number of prey caught per predator is proportional to prey density and search duration. Let T_h be the expected 'handling time' of a single prey that has been caught, and let T be the total time available to a predator for food gathering and eating. Let N be the prey density and let a be the search efficiency (the product of the area covered per unit of time while searching and the probability that a prey in this area is actually detected). If Z is the number of prey caught by a predator in time T, express the actual search time as a function of T, T_h and Z. Express Z as a function of N, a and the search time and derive the formula for the functional response.

Exercise 10.5 A slightly more formal (but not essentially different) derivation makes the time-scale argument explicit. The idea is to distinguish two types of predator ('searching', with index '0', and 'busy handling prey', with index '1') and to assume that the switches in type are fast compared with changes in the density N of prey and in the density of predators due to births and deaths (note that, among other things, this requires that prey greatly outnumber predators). One then regards one predator on the time scale of type change. If e is the rate of return from the handling of prey (i.e. $1/e = T_h$), and if we assume that predators meet prey according to the law of mass action, we can write

$$\frac{dp_0}{dt} = -aNp_0 + ep_1,$$

where p_i is the probability that the predator is in state i, $i = 0, 1$. The non-zero steady state $\bar{p}_0$ is usually referred to as a quasi- (or pseudo-) steady state. Can you explain why? Express $\bar{p}_0$ as a function of N and derive the expression for the functional response.

---

[1] See e.g. H. Thieme: Epidemic and demographic interaction in the spread of potentially fatal diseases in growing populations. *Math. Biosc.*, **111** (1992), 99-130; J. Zhou, H.W. Hethcote: Population size dependent incidence in models for diseases without immunity. *J. Math. Biol.*, **32** (1994), 809-834.
[2] K. Dietz: Overall population patterns in the transmission cycle of infectious disease agents. In: Anderson & May (1982).

Exercise 10.6 Suppose we now want to justify Dietz's formula (and to give an interpretation of its parameters) in the case of contacts between individuals. Where do we run into trouble? Hint: Does time limitation matter for the prey in the Holling argument?

Instead of trying to rationalise an ad hoc function such as (10.2), we can try and derive an expression from mechanistic assumptions and then show that it has the desired properties. In the exercises below, we give an argument for the unstructured case, which generalises to structured populations.[3]

Exercise 10.7 Consider a population consisting of singles, with density $X(t)$ (i.e. those individuals that are at time t not engaged in a pairwise contact), and pairs, with density $P(t)$. Consistency requires that $N(t) = X(t) + 2P(t)$. Give a rational underpinning of the following system:

$$\frac{dX}{dt} = -\rho X^2 + 2\sigma P,$$

$$\frac{dP}{dt} = \frac{1}{2}\rho X^2 - \sigma P.$$

Denote by S_1 the density of susceptible singles, by I_1 the density of infective singles, by S_2 the density of pairs consisting of two susceptibles, by M the density of pairs consisting of a susceptible and an infective (so M means 'mixed'), and, finally, by I_2 the density of pairs consisting of two infectives.

Assume (for the time being) that infectives stay infectious for ever and that

$$\frac{dS_1}{dt} = -\rho S_1(S_1 + I_1) + 2\sigma S_2 + \sigma M,$$

$$\frac{dI_1}{dt} = -\rho I_1(S_1 + I_1) + 2\sigma I_2 + \sigma M,$$

$$\frac{dS_2}{dt} = \frac{1}{2}\rho S_1^2 - \sigma S_2, \tag{10.3}$$

$$\frac{dM}{dt} = \rho S_1 I_1 - \sigma M - \beta M,$$

$$\frac{dI_2}{dt} = \frac{1}{2}I_1^2 - \sigma I_2 + \beta M,$$

where ρ is the reaction rate constant for pair formation, σ is the probability per unit of time that a pair dissolves into two singles, and β is the probability per unit of time that transmission occurs in a mixed pair. A key point here is that the pair formation/dissolution process proceeds independently of the susceptible/infective distinction. The processes are illustrated in Figure 10.1 (note the symmetry).

Exercise 10.8 Define $X = S_1 + I_1$ and $P = S_2 + M + I_2$. Show that

$$\frac{dX}{dt} = -\rho X^2 + 2\sigma P, \tag{10.4}$$

$$\frac{dP}{dt} = \frac{1}{2}\rho X^2 - \sigma P.$$

---

[3] J.A.P. Heesterbeek & J.A.J. Metz: The saturating contact rate in marriage and epidemic models. *J. Math. Biol.*, **31** (1993), 529-539.

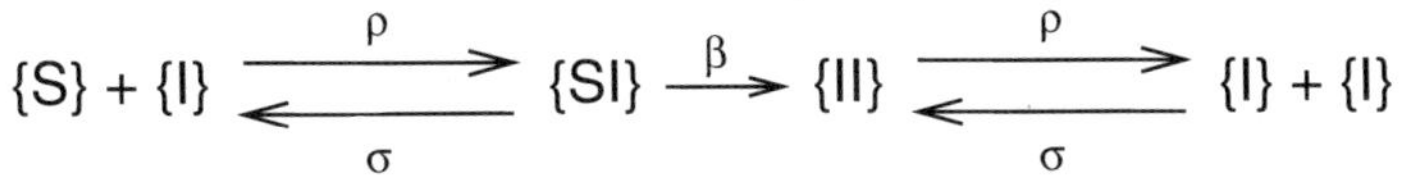

$$\{S\} + \{I\} \underset{\sigma}{\overset{\rho}{\rightleftharpoons}} \{SI\} \overset{\beta}{\longrightarrow} \{II\} \underset{\sigma}{\overset{\rho}{\rightleftharpoons}} \{I\} + \{I\}$$

Figure 10.1

Exercise 10.9 Verify that the steady state of the system (10.4) is given by

$$\overline{X} = \frac{\sqrt{1 + 4\nu N} - 1}{2\nu}, \qquad \overline{P} = \frac{1 + 2\nu N - \sqrt{1 + 4\nu N}}{4\nu}, \qquad (10.5)$$

where N equals the total population size and where $\nu := \frac{\rho}{\sigma}$. What can you say about the stability and the domain of attraction?

Hint: Use that $X + 2P = N$.

Exercise 10.10 Introduce the density of susceptibles $S = S_1 + 2S_2 + M$ and the density of infectives $I = I_1 + 2I_2 + M$. Show that

$$\frac{dS}{dt} = -\beta M, \qquad (10.6)$$

$$\frac{dI}{dt} = \beta M.$$

Explain how this result could have been obtained directly from the interpretation.

We are now ready to introduce the time-scale argument. The idea is that β is very small relative to both ρ and σ. Consequently, S and I will change slowly relative to the speed at which the pair formation/dissolution process equilibrates. To incorporate and elaborate this technically, we eliminate the βM terms from the systems (10.3) and (10.6), and compute the steady-state values of S_1, I_1, S_2, M and I_2 for the resulting system. These are quasi-steady-state values, since in fact S and I will be slowly changing. To compute how S and I will change on the long time scale, we simply substitute the quasi-steady-state value of M into the system (10.6).

Exercise 10.11 Verify that the quasi-steady state of the system (10.3) is given explicitly by

$$\overline{S}_1 = \overline{X}\frac{S}{N}, \quad \overline{I}_1 = \overline{X}\frac{I}{N}, \quad \overline{S}_2 = \overline{P}\left(\frac{S}{N}\right)^2, \qquad (10.7)$$

$$\overline{M} = 2\overline{P}\frac{S}{N}\frac{I}{N}, \quad \overline{I}_2 = \overline{P}\left(\frac{I}{N}\right)^2,$$

and explain the logic behind these expressions (here $\overline{X}$ and $\overline{P}$ are given by (10.5)).

Exercise 10.12 Write $\frac{dS}{dt} = -\beta M$ in the form $\frac{dS}{dt} = -\beta C(N)\frac{SI}{N}$ and show that

$$C(N) = \frac{1 + 2\nu N - \sqrt{1 + 4\nu N}}{2\nu N} = \frac{2\nu N}{1 + 2\nu N + \sqrt{1 + 4\nu N}}. \qquad (10.8)$$

Show that (10.8) indeed has the desired properties (cf. the elaboration of Exercise 10.3).

Our bookkeeping was relatively simple since we had only two types of individuals, S and I. When we also consider immune individuals and introduce a class R, we get six categories of pairs: {SS}, {SI}, {SR}, {II}, {IR}, {RR}. But we would still have the system (10.4) to describe the dynamics of pairs and singles without paying attention to the S, I or R label that the individuals carry. Instead of the system (10.6), we would have

$$\frac{dS}{dt} = -\beta M, \quad \frac{dI}{dt} = \beta M - \alpha I, \quad \frac{dR}{dt} = \alpha I,$$

where M refers to {SI} pairs. Exactly as in Exercise 10.11, we would obtain the analogue of (10.7) and, in particular, the relation

$$\overline{M} = 2\overline{P}\frac{S}{N}\frac{I}{N}.$$

In other words, hardly anything changes and the resulting system on the long time scale would be

$$\frac{dS}{dt} = -\beta C(N)\frac{SI}{N}, \quad \frac{dI}{dt} = \beta C(N)\frac{SI}{N} - \alpha I, \quad \frac{dR}{dt} = \alpha I, \tag{10.9}$$

with $C(N)$ given by (10.8) and now $N = S + I + R$.

Likewise, one can incorporate host demography on the long time scale and still work with C as given by (10.8). Our derivation focused on the notationally simplest case, but the result is much more generally valid.

Exercise 10.13 Repeat Exercise 10.9 under the additional assumption that single individuals can revert to a resting state in which they are not active in searching for partners. Let ε_1 and ε_0 be the probabilities per unit of time of entering the resting state and returning to the active state respectively. Pairings dissociate into two active singles (a debatable assumption!). You can check your result by showing that the appropriate limit leads back to (10.5).

Exercise 10.14 With hindsight, we can now also give a Holling-type argument as in Exercise 10.4. This time we take into account that both individuals taking part in a contact are time-limited ('Holling squared', a procedure and name invented by J.A.J. Metz). We think again of singles and pairs as in the previous exercises and use a mix of notation from 10.9 and 10.4. Let $Y := Z/T$. Give the analogy with the Holling argument that leads to

$$Y = \frac{\rho X}{1 + \nu X},$$

where $\nu = \rho T_h$. In contrast to the prey in the Holling argument, the singles here are time-limited, in the sense that they are not all available for contacts. Convince yourself that

$$X = N\left(\frac{T - ZT_h}{T}\right).$$

Rewrite this in terms of Y. Express the steady-state density $\overline{P}$ of complexes in N, Y and T_h and substitute the expression for Y to find the same expression for $\overline{P}$ as in Exercise 10.9 and therefore the same expression for $C(N)$.

Remark: In this heuristic derivation we have not used the assumption that duration of pairings is exponentially distributed with some parameter σ; we have only used the average duration T_h. This suggests that the assumption of exponentiality in the formal derivation—although mathematically convenient— is not necessary.

To conclude, we show that if we do assume instantaneous contacts, the expression (10.8) for $C(N)$ indeed collapses to a constant times N (i.e. to transmission according to mass action), as is to be expected. One has to take care that the right procedure for taking limits is adopted. This exercise makes clear that it is worthwhile to distinguish between pairings and transmission opportunities during pairings. The obvious first thing to do would be to let the average duration of a pairing, $1/\sigma$, tend to zero (i.e. to take the limit $\sigma \to \infty$). Check that if we do this without adapting other quantities then the transmission rate becomes zero. The reason is that if the pairing duration becomes shorter and shorter, there comes a point beyond which the transmission opportunity during a pairing is negligible. The solution is to let the transmission opportunity tend to infinity 'equally fast' as the pairing duration tends to zero.

Exercise 10.15 Argue that the probability of transmission per pairing is given by

$$\frac{\beta}{\beta + \sigma}.$$

Hint: $1 - e^{-\beta t}$ is the probability that transmission occurs within the first t time units of the existence of the pairing; $\sigma e^{-\sigma t}$ is the probability density function for pairing duration.

Exercise 10.16 Show that the limit $\sigma \to \infty, \beta \to \infty$ with β/σ and ρ constant collapses C into a linear function of N.

As a third possibility for variation in contact duration, one has that the time scale of formation and dissociation of pairings is not fast compared with the demographic time scale or the time scale of infection. For this we refer to the literature.[4] For example, with regard to sexually transmitted diseases, it can be relevant to take the formation of longer-lasting (sexual) partnerships into account. We have seen some of the flavour of the resulting pair formation models when we calculated R_0 for examples from this class of models in Section 5.6.

10.3   Consistency conditions

In earlier chapters, we have repeatedly encountered situations in which the symmetry of contact implied that modelling ingredients should satisfy certain consistency relations. For instance if, according to our model, in a certain period of time 3251 females have sexual contact with a male then the number of sexual contacts of males

---

[4] K. Dietz & K.P. Hadeler: Epidemiological models for sexually transmitted diseases. *J. Math. Biol.*, **26** (1988), 1-25; For the calculation of R_0 for pair formation models see K. Dietz, O. Diekmann & J.A.P. Heesterbeek: The basic reproduction ratio for sexually transmitted diseases, Part 1: theoretical considerations. *Math. Biosci.*, **107** (1991), 325-339.

with females should also number 3251 in that period. In this subsection we address this consistency issue more systematically.

We consider a population in a demographic steady state. We assume that the h-state is static (this is not essential for the considerations below, but it becomes essential if one combines bookkeeping considerations with an infectivity submodel to compute the kernel $A(\tau, x, \xi)$). We distinguish finitely many h-states, which we number $1, ..., n$ (this is not at all essential; a continuum of h-states only requires a notational adaptation). We concentrate on *symmetric* contacts (so we exclude 'contacts' such as through blood transfusions, through a common water supply or through infected potato salad at a party). We adopt the counting convention that when two j-individuals have contact, there are *two* contacts (more generally, when a j-individual and a k-individual have contact, we count this as a kj-contact and a jk-contact).

Let

$$\phi_{ij} \quad := \quad \text{total number of contacts per unit time} \qquad (10.10)$$
$$\text{between } j\text{-individuals and } i\text{-individuals.}$$

Then, because of the imposed symmetry, consistency requires that

$$\phi_{ij} = \phi_{ji}. \qquad (10.11)$$

Therefore, out of the n^2 elements of the matrix Φ, only $\frac{1}{2}n(n+1)$ can be freely chosen.

In particular, when considering sexual contacts, the h-state often refers to a (sexual) activity level. More specifically, assume that j-individuals have, on average, c_j contacts per unit of time. Let N_j denote the number of j-individuals. Then Φ, c and N cannot be completely arbitrary. They should satisfy the relations

$$\sum_{i=1}^{n} \phi_{ij} = c_j N_j, \qquad j = 1, ..., n. \qquad (10.12)$$

Therefore, if one considers c and N as 'given', the number of degrees of freedom in the choice of Φ reduces to $\frac{1}{2}n(n-1)$. A particular choice for the elements of Φ is the *proportionate mixing* expression

$$\phi_{ij} = \frac{c_i c_j N_i N_j}{\sum_{k=1}^{n} c_k N_k} \qquad (10.13)$$

(in the terminology of C. Castillo-Chavez and co-workers,[5] this is called a Ross solution). However, other choices satisfying (10.11) and (10.12) are clearly possible. Indeed, the following result derives from work of Castillo-Chavez and various co-authors.

Theorem 10.17 *Let c and N be given. Let Ψ be a symmetric $n \times n$ matrix. Define*

$$\theta = \left(\sum_{k=1}^{n} c_k N_k \right)^2 - \sum_{i,j=1}^{n} c_i c_j N_i N_j \psi_{ij} \qquad (10.14)$$

---

[5] C. Castillo-Chavez, J.X. Velasco-Hernandez & S. Fridman: Modeling contact structures in biology. In: *Frontiers in Mathematical Biology*. S.A. Levin (ed.), Springer-Verlag, Berlin, 1994, pp. 454-491.

and

$$\alpha_j = \sum_{k=1}^{n} c_k N_k - \sum_{i=1}^{n} c_i N_i \psi_{ij}. \tag{10.15}$$

Then the matrix Φ defined by

$$\phi_{ij} = \frac{c_i c_j N_i N_j}{\sum_{k=1}^{n} c_k N_k} \left(\frac{\alpha_i \alpha_j}{\theta} + \psi_{ij} \right) \tag{10.16}$$

satisfies relations (10.11) and (10.12). (In order for this to be meaningful in our context, both Ψ and the right-hand sides of (10.14) and of (10.15) should be non-negative. A sufficient condition for this to be the case is $0 \le \psi_{ij} \le 1$ for all i, j.).

Conversely, if Φ is such that (10.11) and (10.12) hold for its elements then

$$\phi_{ij} = \frac{c_i c_j N_i N_j}{\sum_{k=1}^{n} c_k N_k} \psi_{ij}, \tag{10.17}$$

where the matrix Ψ is symmetric and the elements are such that

$$\sum_{i=1}^{n} c_i N_i \psi_{ij} = \sum_{k=1}^{n} c_k N_k \tag{10.18}$$

for all j. This shows that a representation of the form (10.16) is possible.

Proof Clearly, the matrix Φ defined by (10.16) is symmetric. Moreover,

$$\sum_{i=1}^{n} \phi_{ij} = \frac{c_j N_j}{\sum_{k=1}^{n} c_k N_k} \left(\frac{\sum_{i=1}^{n} c_i N_i \alpha_i}{\theta} \alpha_j + \sum_{i=1}^{n} c_i N_i \psi_{ij} \right).$$

By (10.15), the second factor on the right-hand side equals

$$\frac{\sum_{i=1}^{n} c_i N_i \alpha_i}{\theta} \alpha_j + \sum_{k=1}^{n} c_k N_k - \alpha_j.$$

Also by (10.15), we have that

$$\sum_{i=1}^{n} c_i N_i \alpha_i = \left(\sum_{i=1}^{n} c_i N_i \right)^2 - \sum_{i,j=1}^{n} c_i c_j N_i N_j \psi_{ij},$$

which, by (10.14), equals θ. Hence the second factor reduces to $\sum_{i=1}^{n} c_i N_i$ and the product with the first factor to $c_j N_j$, which then yields (10.12).

It is straightforward that θ and α_j are non-negative when $0 \le \psi_{ij} \le 1$.

If, finally, the matrix Φ is given and its elements satisfy (10.11) and (10.12) then a matrix Ψ with elements defined by

$$\psi_{ij} = \frac{\sum_{k=1}^{n} c_k N_k}{c_i c_j N_i N_j} \phi_{ij}$$

is certainly symmetric and, in addition,

$$\sum_{i=1}^{n} c_i N_i \psi_{ij} = \frac{\sum_{k=1}^{n} c_k N_k}{c_j N_j} \sum_{i=1}^{n} \phi_{ij} = \sum_{k=1}^{n} c_k N_k.$$

This ends the proof of Theorem 10.17.

On the one hand, Theorem 10.17 provides us with an algorithm to generate, for given c and N, a matrix Φ satisfying the constraints (10.11) and (10.12). It turns out that a symmetric matrix Ψ can be freely chosen, except for sign constraints on the right-hand sides of (10.14) and of (10.15), which themselves depend on c and N. On the other hand, the second part of the theorem tells us that there is no loss of generality in restricting Φ to have elements of the form (10.16).

Exercise 10.18 Let π_{ij} be the probability of transmission, given a contact between a susceptible i-individual and an infectious j-individual. Express the incidence among i-individuals in terms of π_{ij}, ϕ_{ij}, S_i and I_j for $j = 1, ..., n$.

The constraint (10.12) derives from the interpretation of c_j as an, on average, *realised* number of contacts per unit of time. If we relax the interpretation and characterise individuals by a *tendency* to make contacts, the rigid constraint disappears. Thus one can use a multi-type version of the time-scale arguments in Section 10.2 to derive ϕ_{ij} from a mass-action submodel.

The literature on consistency constraints is quite substantial. Yet we do not think that the present short summary misses any essential points!

Exercise 10.19 Consider a metapopulation of seal colonies. Number the colonies 1,2,..., n. Let ρ_{ij} denote the fraction of individuals of colony j that at any particular low tide sunbathe at the haul-out spot of colony i (we assume that such individuals are 'chosen' randomly from among all j-individuals at every low tide). Derive an expression for ϕ_{ij}.

10.4   Effects of subdivision

10.4.1   Aggregation

Suppose we have a population where the individuals are divided into n groups $G_1, ..., G_n$, of sizes $N_1, ..., N_n$ respectively, with different contact patterns. In this subsection we study the effect on the spread of infection of ignoring more and more of the heterogeneity in the population. Suppose we combine the n groups into k larger groups $H_1, ..., H_k$, with $k < n$. We are interested in the effect that such aggregation has on both R_0 and the final size of an epidemic. For the former we were inspired by results of F.R. Adler; for the latter we refer to work by H. Andersson and T. Britton[6] and to the end of Section 2.2. From Section 2.2, we already know that effects on R_0 and the final size need not be the same since in the heterogeneous case R_0 and the final size are not necessarily related in a monotone way. A small core group with high

---

[6] H. Andersson & T. Britton: Heterogeneity in epidemic models and its effect on the spread of infection. *J. Appl. Prob.*, **35** (1998), 651-661.

activity level can contribute substantially to R_0 (see Section 10.3.3), but may have little impact on the final size.

Let c_{ij} be the per couple contact rate of individuals from group i with individuals from group j. We consider direct contacts only; we think of, for example, vector-transmitted infections or sexually transmitted infections. Consistency then requires that

$$c_{ij} N_i N_j = c_{ji} N_j N_i$$

(see Section 10.3 (where $\phi_{ij} = c_{ij} N_i N_j$)), and the contact matrix $C = (c_{ij})_{1 \leq i,j \leq n}$ should therefore be symmetric.

Now partition the indices $\{1, ..., n\}$ into k non-overlapping subsets $s_1, ..., s_k$, and define new larger groups and group sizes

$$H_l := \cup_{i \in s_l} G_i, \qquad \widehat{N}_l := \sum_{i \in s_l} N_i.$$

Exercise 10.20 Let mixing between groups be proportional to their size and activity, i.e. assume that $c_{ij} = v_i v_j$. Check that this mixing pattern is preserved between the aggregated groups, i.e. show that $\widehat{c}_{lr} = \widehat{v}_l \widehat{v}_r$, where $\widehat{c}_{lr}$ is the contact rate between group H_l and group H_r. Hint: The total number of contacts between the aggregated groups l and r is $\sum_{i \in s_l} \sum_{j \in s_r} c_{ij} N_i N_j$.

If we include the probability p_{ij} that, upon contact between a susceptible of group i and an infective from group j, transmission is successful, we will, as a rule, have $p_{ij} \neq p_{ji}$. Furthermore, let T_j be the average length of the infectious period for infected individuals in group j. In analogy with the homogeneous case, equation (2.2), we can calculate elements m_{ij} of the next-generation matrix, defined as the expected number of new cases in group i caused by an infective from group j, if group i consists of susceptibles only. Instead of a single infection function $A(\tau)$, we then have functions $A_{ij}(\tau)$. We find, if we assume that all variability between infected individuals is in the length of the infectious period or, somewhat more generally, if infectivity during the infectious period is independent of the length of the period,

$$m_{ij} = N_i \int_0^\infty A_{ij}(\tau) \, d\tau = N_i c_{ij} p_{ij} T_j.$$

From Chapter 5, we know that R_0 is the dominant eigenvalue of the matrix M.

Now consider the special case that C is symmetric and primitive (i.e. that there is a power k of C such that all elements of C^k are strictly positive) and that $p_{ij} T_j = d$, for all i and j. This effectively means that we take out any influence of variability in infectious output and look at contact structure only. An argument analogous to the one given above gives a matrix $\widehat{M}$ and a dominant eigenvalue $\widehat{R}_0$ for the aggregated groups. We then have the following result by F. Adler:[7]

$$\widehat{R}_0 \leq R_0, \tag{10.19}$$

---

[7] F.R. Adler: The effects of averaging on the basic reproduction ratio, *Math. Biosci.* **111** (1992), 89-98.

or, in words, less heterogeneity (more aggregation) leads to a lower value for R_0 if infectivity is the same in all groups and if the contact pattern is symmetric and primitive. Note that in Section 2.2 we illustrated, with a numerical example, this relation for the case of two groups clustered into a single group, in the case that one has proportionate mixing.

Exercise 10.21 i) In Section 2.2 we used c_i to denote the average number of contacts per unit time of a type-i individual. Presently we are working with per-couple contact rates c_{ij}. The aim of this first part of the present exercise is to clarify the relationship between the various cs. In particular, show that

$$c_i = \sum_{j=1}^{n} c_{ij} N_j.$$

Moreover, show that a fraction

$$q_i = \frac{c_{ii} N_i}{c_i}$$

of the contacts of a type-i individual is within its own group.

ii) Using the notation from Section 2.2, prove (10.19) for the case of two groups clustered into a single group, in the case of proportionate mixing. Assume, for convenience, that $p_{ij} T_j = 1$ for all i and j.

ii) Prove the relation (10.19) in the case of two groups that are clustered into one group without assuming proportionate mixing. Use the notation of Section 2.2, i.e. show that

$$\bar{c} := \frac{c_1 N_1 + c_2 N_2}{N_1 + N_2} \leq R_0,$$

where R_0 is the dominant eigenvalue of the next-generation matrix

$$\begin{pmatrix} q_1 c_1 & (1 - q_2)c_2 \\ (1 - q_1)c_1 & q_2 c_2 \end{pmatrix}$$

for the case of two groups, and by using the consistency condition

$$(1 - q_1)c_1 N_1 = (1 - q_2)c_2 N_2,$$

where q_1 is the fraction of the contacts of group 1 individuals with others in group 1, et cetera, as above.

Exercise 10.22 For a counterexample in the case that infectivity varies between groups, take the same situation as in Exercise 10.21 (i.e. $n = 2$, $k = 1$), and let $p_{ij} T_j = d_j$, where we choose $d_1 = 1$ and $d_2 = 3$ respectively. To make life easier, assume that $N_1 = N_2 = N$ and $c_1 = c_2 = c$ (note that this implies $p_1 = p_2 = p$, equal for both groups). We have to make an assumption about how to define the infectivity for the aggregated group. Here simply take the arithmetic average (so the value is 2). Show that $\widehat{R}_0 > R_0$ for all $p < \frac{1}{2}$.

10.4.2   What is a core group?

From the book *Gonorrhea Transmission Dynamics and Control* by H.W. Hethcote &
J.A. Yorke (1984), we quote (p.35): 'The population in the groups with high prevalence
(with prevalences at least 20%) are lumped together and called the *core*'. This is
an operational definition (although it presupposes a splitting of the population into
subgroups) and the suggestion is that the core causes gonorrhea to remain endemic
and, more generally, determines to a large extent the endemic prevalence level. The
notion of a core is appealing as well as helpful in organising our thoughts about the role
of heterogeneity in determining overall levels of transmission. Yet, we claim, a good
definition is lacking. The present subsection offers a, not totally successful, attempt
at such a definition. We hope that it serves to stimulate our readers to give the issue
some thought and to come up with better ideas.

Consider finitely many h-states numbered 1,2,..., n. The next-generation matrix M
has as its entries m_{ij}, the expected number of secondary cases of type i due to one
primary case of type j, when the agent is introduced in a virgin host population. The
basic reproduction ratio R_0 is the dominant eigenvalue of the next-generation matrix
M (recall Section 5.1). The diagonal elements m_{ii} describe transmission within the
subpopulation of the same type (the own group). If we were to consider this type
in isolation, eliminating all other types, while assuming that this would not increase
the 'within-group' contacts, then the corresponding R_0 would be m_{ii}. In general, the
contacts outside the own group help to increase R_0 and therefore, as can be proved
mathematically as well, $R_0 \geq m_{ii}$ for all i, so $R_0 \geq \max_i m_{ii}$.

It is now tempting to call the subpopulation with h-state i a *core group* if m_{ii} and R_0
have the same order of magnitude while m_{jj} is substantially smaller for all $j \neq i$. More
generally, we would call the subpopulation with h-state $i \in C$, where C is a (small)
subset of $\{1, 2, ..., n\}$, a core group if for all $i \in C$ the quantity m_{ii} has the same order
of magnitude as R_0 while for all $j \notin C$ the quantity m_{jj} is substantially smaller. This
'definition' has (at least) two weak points. The first is that 'order of magnitude' and
'substantially' are subjective and it is a matter of taste how one should compare, for
example, 6 with 0.5. The second weak point is that loops, occurring for example when
considering the influence of prostitutes on STD prevalence, are ignored.

The second weakness can be remedied by looking at the diagonal elements of M^2
(and, more generally, higher powers of the matrix M). When $n = 2$, we have

$$M^2 = \begin{pmatrix} m_{11}^2 + m_{12}m_{21} & m_{11}m_{12} + m_{12}m_{22} \\ m_{21}m_{11} + m_{22}m_{21} & m_{21}m_{12} + m_{22}^2 \end{pmatrix},$$

and we see the product $m_{12}m_{21}$ appearing on the diagonal. In general (i.e. for all n),
the inequality

$$R_0 \geq \sqrt{\sum_{i=1}^{n} m_{ij}m_{ji}}$$

holds for all j. We therefore might use the term jk-*core loop* if $m_{kj}m_{jk}$ is of the same
order of magnitude as R_0, while all diagonal elements m_{ii} as well as all products
$m_{li}m_{il}$ with $(i, l) \neq (j, k)$ are substantially smaller. Of course we can also consider
loops of length three or greater, et cetera.

The title of this subsection is formulated as a question. We have serious doubts about the appropriateness and the completeness of our answer, and we welcome all suggestions that readers have to offer.

10.5 Network models (an idiosyncratic survey)

10.5.1 Reflections: what do we want and why?

We recall an observation made at the beginning of Section 5.6: in deterministic models in which, by assumption, all subpopulation sizes are infinite, two individuals never have contact twice. A quite different type of model conceives a population as a *network*, with connections between individuals that are either fixed or that can be formed and broken in the course of time. Each connection symbolises a relationship between individuals that involves repeated contacts, and therefore transmission of an infective agent proceeds along connections. In pure network models, transmission is restricted to connections, but of course one can formulate hybrid models that incorporate a second transmission route through uniform/random contacts.

In network models, populations are represented as *graphs* (static or dynamic) and one can use a colour coding to denote the status of individuals with respect to some infectious disease (e.g., an infective individual could be denoted by a red pixel in a population that is represented by a two-dimensional square lattice). Thus the results of a computer implementation and subsequent simulation can be visualised, and one can literally watch the agent spread. Such experiments catalyse the sharpening of our intuition.

We shall briefly indicate some characteristic features, while providing pointers to the literature (i.e. we try to give you a taste of the flavour, without serving the dish, but do give directions to the restaurant where the dish can be ordered). Much attention has been paid in ecology[8] and evolution, and increasingly in epidemiology, to the *interacting particle systems*. As described above, the basic idea is to represent space as a lattice of cells, to allow each cell to be in a number of different states, and to specify rules according to which the state of a cell at time $t + 1$ is determined (we consider only the discrete-time version, but continuous-time formulations are possible). If we think of an infectious agent of plants, the cells could, for example, be individual plants in a field, or fields in a region. Suppose each cell can be in either of three different states: susceptible (0), infected (1) or immune/dead (2). Part of the specification of the reconfiguration rules involves making a list of the cells that can, with their state at time t, influence the possible change in state of a given cell x between t and $t + 1$. Those cells are often called the 'neighbourhood' of x. One possible choice, in a square lattice, is to take only the four cells that share a boundary of positive measure with x, i.e. its nearest neighbours. This is called the von Neumann neighbourhood, after the mathematician who originally introduced the idea of a cellular automaton. (Different choices of neighbourhood can of course have a large effect on the dynamics.) One could specify that the infection probability for cell x is β times the number of infected cells in the neighbourhood of x and that the removal probability is α for each infected

---

[8] R. Durrett & S. Levin: Stochastic spatial models: a user's guide to ecological applications. *Phil. Trans. Roy. Soc. Lond. B*, **343** (1993), 329-350.

cell. In this situation one can show[9] that, starting from one infected cell, under certain conditions on β and α, we obtain a wave of infection travelling outward over the lattice with constant velocity and asymptotically a fixed wave front (roughly circular). This is similar to results we discussed in Chapter 8.

The conditions on β and α will be shaped as a condition such as $R_0 > 1$ in the situation that contacts can be made with all individuals in the population (that we focussed on for most of the book). Here, however, one should realise that the arrangement of the cells in the lattice determines what the threshold will be above which an infection will take off (loops!—see below). To determine analytically (as opposed to experimentally/numerically) how this threshold depends on the spatial arrangements in the lattice, neighbourhood choices and reconfiguration rules is a very difficult open mathematical problem.[10] (Incidentally, in this chapter we shall always have networks or graphs in mind that describe existing contact structures, on which transmission is then superimposed. A somewhat different approach is to consider the graph as 'constructed' by the transmission events and to use results from random graph theory to deduce information about the spread of the agent.[11])

Regular spatial lattices are easy to describe, characterise and represent in a computer. Yet we may want to introduce other types of graphs when modelling, for example, networks of sexual interaction between individuals. The ideal is to be able to describe a graph in terms of recognisable/observable *spatial patterns* such as

- *degree distribution* (for each vertex (i.e. individual), count how many edges (i.e. connections) attach to it);
- *loops* (starting in a vertex and following edges, what is the minimal length of a round trip? And how many alternative routes exist?);
- *clusters* (does the graph decompose into loosely coupled components?).

One furthermore wants to add some simple rule for *reconfiguration*, which determines the *temporal pattern*, and finally from such a description infer certain consequences for the spread of an infectious disease. In other words, the ideal is to deduce relevant characteristics of epidemic spread from a *statistical description* of the spatio-temporal patterns in a network. Obviously there is a long way to go before this ideal can be achieved, but once we can make the right connection, we can—in order to derive conclusions about the spread of the agent—rather directly use *observations* about quantities like the following:

- the frequency of forming new connections (e.g. the fequency of acquiring a new sexual partner);
- the length of time that a connection exists;
- the degree distribution (e.g. simultaneous partnerships);

---

[9] See e.g. R. Durrett: Spatial epidemic models. In: Mollison (1995), pp. 187-201; D. Mollison: Dependence of epidemic and population velocities on basic parameters. *Math. Biosci.*, **107** (1991), 255-287.

[10] See e.g. R. Durrett: Stochastic growth models: bounds on critical exponents. *J. Appl. Prob.*, **29** (1992), 11-20; S.A. Levin & R. Durrett: From individuals to populations. *Phil. Trans. Roy. Soc. Lond. B*, **351** (1996), 1615-1621.

[11] B. Bollobás: *Random Graphs*. Academic Press, London, 1985.

- types of connections (e.g. to distinguish casual from serious partnerships, wifes from mistresses, ...).

A second class that is widely studied is that of graph models. Usually, individuals are thought of as vertices, and if a connection between two individuals exists (i.e. if they can have contacts), one says that the two corresponding vertices are connected by an edge. An important class of graphs for epidemic application are the random graphs, where one can explicitly model birth and death of new vertices and forming and breaking up of edges.[12]

But, is there a good reason to expect any effect of incorporating spatio-temporal pattern? Yes, there is! The key point is that, from the point of view of the infective agent, contacts between two infected individuals are *wasted*. When contacts are repeated and/or when a graph has many short cycles, the probability of contacts between two infectives is higher than in a randomly mixing population.

To measure the effect, we should investigate how such indicators as

- the basic reproduction ratio R_0,
- the intrinsic growth rate r,
- the size of a major outbreak,
- the probability of a minor outbreak,

or, when incorporating host demography,

- the probability of extinction after a first major outbreak,
- the critical community size,
- the endemic level,
- the effort needed for eradication

depend on the spatio-temporal pattern. A characteristic (but rather annoying) difficulty is that we want to gauge the models that we compare in such a way that the contact 'intensity' is the same, while the contact 'pattern' differs, but that there often is not a unique well-defined way to specify 'intensity'.

As an alternative to the direct statistical description of pattern, we can postulate *mechanisms* (or, at least, phenomenological rules) that govern the formation and reconfiguration of the network. Here we can think, for example, of the tendency for individuals to form a new partnership when single or when having already some given number of partners (in principle, one can include other characteristics, such as age). The advantage of *simulation* models is that one can specify rules directly in terms of a computer program (although an interaction between two symmetric individuals remains a subtle modelling challenge; recall Section 10.3 on consistency conditions). The disadvantage is, of course, that it is difficult to extract insight and understanding from the myriad of simulation runs. With *analytical* models it is usually the other way around: they are, as a rule, intractable, but IF we manage to analyse them, we obtain qualitative insight into the relationship between pattern and the spread of the agent.

---

[12] A.D. Barbour & D. Mollison: Epidemics and random graphs. In: Gabriel et al. (1990), pp. 86-89; Ph. Blanchard, G.F. Bolz & T. Krüger: Modelling AIDS-epidemics or any venereal disease on random graphs. In: Gabriel et al. (1990), pp. 104-117.

We conclude that the quest for tractable analytical network models is a valuable endeavour!

If one aims to carry out all good plans for a textbook, it will never get finished. In this light it is a pity that we could not expand on important recent developments in, for example, explicit spatial epidemic and population dynamic models (e.g. techniques such as pair approximation) and contact processes (e.g. incorporating household structure). For the first we refer to the review paper of D. Rand.[13] For the second we refer to Ball et al. (1997), Becker & Dietz (1995) and Ball (1999), and the references given there.[14]

10.5.2 A network with hardly any structure

We saw that, apart from demographic stochasticity, there are two key features that distinguish network models from standard deterministic models, viz.

- repeated contacts between the same individuals,
- local correlations due to short cycles in the graph.

In this subsection, we focus on a caricature of a network that displays the first of these features but hardly anything of the second. One could say that we sacrifice most of the second in order to achieve tractability.

We assume that a connection is symmetric, and we shall call individuals that are connected *acquaintances* of each other. The number of acquaintances of an individual is a stochastic variable k with probability distribution $\{\mu_k\}$. That is, with probability μ_k, an individual has k acquaintances. We shall then say that the individual is of type k (or a k-individual). We assume that the mean

$$\overline{k} := \sum_{k=1}^{\infty} k\mu_k \qquad (10.20)$$

is finite.

The crucial assumption of the model is that acquaintances are a random sample of the population. By this we mean two things:

(1) An acquaintance of some individual x is, with probability

$$\nu_k := \frac{k\mu_k}{\overline{k}}, \qquad (10.21)$$

of type k, irrespective of the type of x itself (so the probability that an arbitrary acquaintance is of type k is proportional to the occurrence of individuals of this

---

[13] D.A. Rand: Correlation equations and pair approximations for spatial ecologies. In: McGlade (1999), pp. 100-142.

[14] F.G. Ball, D. Mollison & G. Scalia-Tomba: Epidemics with two levels of mixing. *Ann. Appl. Prob.*, **7** (1997), 46; N.G. Becker & K. Dietz: The effect of the household distribution on transmission and control of highly infectious diseases. *Math. Biosci.*, **127** (1995), 207; F.G. Ball: Stochastic and deterministic models for SIS epidemics among a population partitioned into households. *Math. Biosci.*, **156** (1999), 41-67.

type, with constant of proportionality k; the $\bar{k}$ in the denominator serves to make ν into a probability distribution, i.e. it serves to guarantee that the ν_k sum to one).

(2) The probability that two particular individuals in a population of size N are acquaintances of each other is $O(N^{-1})$. In particular, the probability that acquaintances have other acquaintances in common equals zero in the limit of an infinite population.

As we shall indeed restrict our considerations to the infinite-population limit, it is this second assumption that eliminates short cycles.

Having thus specified the network structure (or, rather, the lack of it), we now turn to transmission. We assume that the contact intensity between two acquaintances is independent of the types k_1 and k_2 of these two individuals (allowing dependence would create either asymmetry or inconsistency or both). So, individuals with high k have a high overall contact intensity.

Infected individuals may differ in the amount of infectious material that they disseminate to their acquaintances. Depending on what we want to compute, we have to describe this variability in more or less detail. We will use this as a guiding principle for the order in which we present the results, the key idea being that we add detail only when needed to proceed.

Let $\bar{q}$ denote the probability that an individual picked at random will transmit the infective agent to any one of its acquaintances, when infected itself. Then

$$R_0 = \bar{q} \sum_{k=1}^{\infty} (k-1)\nu_k. \tag{10.22}$$

Exercise 10.23 Explain the rationale behind the expression (10.17).

Exercise 10.24 Rewrite (10.17) to obtain

$$R_0 = \bar{q}\left(\bar{k} - 1 + \frac{\text{variance}(k)}{\bar{k}}\right). \tag{10.23}$$

This should remind you of an earlier exercise. Which one?

In line with earlier conclusions (Section 2.2. Exercise 5.10, Section 10.3.2), we see that heterogeneity increases R_0. But what is the effect of the network structure?

To focus on that, let us eliminate the heterogeneity by assuming that every individual has the same number of acquaintances $\bar{k}$ (so $\mu_k = 0$ for $k \neq \bar{k}$, and $\mu_{\bar{k}} = 1$, for some integer $\bar{k}$). Then (10.18) reduces to

$$R_0 = \bar{q}(\bar{k} - 1). \tag{10.24}$$

The effect of repeated contacts between the same individuals is hidden in $\bar{q}$ (it will be exposed if we calculate $\bar{q}$ from a submodel for infectivity; see below). The effect of the loop between two acquaintances (recall that we assumed the acquaintance relation to be symmetric!) is in the '−1'. Indeed, a derivation ignoring the network structure would argue that R_0 equals the product of the number of acquaintances $\bar{k}$ (during the infectious period, say) and the probability $\bar{q}$ of transmission, and would therefore lead to

$$R_0 = \bar{q}\bar{k}. \qquad (10.25)$$

The difference is that in this second way of reasoning it is ignored that the individual that caused the infection of the individual we consider is still among the $\bar{k}$ acquaintances.

To illustrate the effect of the '−1', we note that in order to have $R_0 = \bar{q}(\bar{k} - 1) > 1$, we require

$$\bar{k} > 1 + \frac{1}{\bar{q}} > 2. \qquad (10.26)$$

And indeed, when every individual has exactly two acquaintances (so for $\bar{k} = 2$), the 'network' is a chain. As we allow for nearest-neighbour transmission only, spread in any of the two directions will stop as soon as one link fails (by chance) to transmit the infection. Thus major outbreaks are precluded.

Our next aim is to justify the expression

$$s_\infty = \sum_{k=1}^{\infty} \mu_k (1 - \bar{q} + \pi_b \bar{q})^k \qquad (10.27)$$

for the final size s_∞, where π_b is the unique root (which exists when we assume $R_0 > 1$) in the interval $(0, 1)$ of the equation

$$\pi = g_b(\pi), \qquad (10.28)$$

where

$$g_b(z) := \sum_{k=1}^{\infty} \nu_k (1 - \bar{q} + z\bar{q})^{k-1} \qquad (10.29)$$

This smells like a branching process, and that is indeed what is lurking behind it: the 'b' refers to 'backward' and we shall motivate the expression (10.19) by interpreting s_∞ as the probability of extinction of the so-called backward (or dual) branching process when starting (or, rather, ending) with a randomly chosen individual.

The idea is as follows. For each individual x we ask for each of its acquaintances y: will x be infected by y, should y itself become infected? Here we ignore the fact x may already have been infected earlier by one of its other acquaintances. If the answer to the question is 'yes' for a given y, we call y a child of x. Note that this explains the name 'backward', since we turn the parent-offspring relation, defined by 'infection', upside down.

For this process of producing children, the distribution of type at birth is given by ν as well. Each individual begets offspring according to a binomial distribution with probability parameter $\bar{q}$ and a type-dependent parameter for the maximum number of offspring. In fact, this maximum number equals $k - 1$ for a k-type individual, once the process is started, since one of its k acquaintances is the 'mother' of x, and therefore cannot be a 'daughter' of x as well.

Exercise 10.25 Now check that $g_b(z)$ defined in (10.21) is the generating function. Verify that $g_b'(1) = R_0$.

The root π_b of equation (10.20) therefore equals the probability of extinction for this backward process if we start with one 'typical' individual. However, our starting point is somewhat different. To begin our thought experiment, we pick one individual at random, which gives it type k with probability μ_k (and not ν_k!). This first individual, if it is of type k, can beget k offspring (rather than $k - 1$!).

Exercise 10.26 Our aim is to interpret the right-hand side of (10.19) as the probability of extinction of the backward process when starting from (read: ending with) a randomly chosen individual. Finish the argument.

It remains to relate this probability of extinction to the final size s_∞. To do so, we imagine that a large outbreak has occurred. The collection of victims then forms a so-called *giant component* of the graph.[15] Pick an arbitrary individual and ask what the probability is that it belongs to the giant component. By definition, this probability is $1 - s_\infty$. The key observation now is that the individual belongs to the giant component if and only if the backward branching process starting from this individual does NOT go extinct. Or, equivalently, the individual does not belong to the giant component if and only if the backward branching process starting from this individual does go extinct. The point is that only members of the giant component have an infinite 'family' tree. Therefore s_∞ equals the probability of extinction and we have arrived at (10.19). We have deliberately kept this derivation somewhat heuristic, but it can be made precise.[16]

Exercise 10.27 Show that for R_0 only slightly bigger than 1, we have

$$1 - s_\infty = \frac{2\bar{k}}{\bar{q}\sum_{k=3}^{\infty}(k-1)(k-2)\nu_k}(R_0 - 1) + \text{h.o.t.} \qquad (10.30)$$

Hint: Assume $\pi_b = 1 - \eta$ with η small. First expand the identity $\pi_b = g_b(\pi_b)$ in powers of η; next do the same with the expression (10.22) for s_∞ and, finally, combine these results to eliminate η from the lowest-order terms.

Now recall, from Exercise 1.19-ii, that for the standard final-size equation we have

$$1 - s_\infty = 2(R_0 - 1) + \cdots.$$

So the difference resides in the factor

$$\frac{\bar{k}}{\bar{q}\sum_{k=3}^{\infty}(k-1)(k-2)\nu_k}.$$

To investigate this factor, we employ a (rough) approximation: we replace $k - 2$ in the denominator by $\bar{k} - 2$ (note that this is exact when μ is concentrated in one point, i.e.

---

[15] B. Bollobás: *Random Graphs*. Academic Press, London, 1985.

[16] O. Diekmann, M.C.M. de Jong & J.A.J. Metz: A deterministic epidemic model taking account of repeated contacts between the same individuals. *J. Appl. Prob.* **35** (1998), 448-462; H. Andersson: Epidemic in a population with social structures. *Math. Biosci.*, **140** (1997), 79-84; H. Andersson: Limit theorems for a random graph epidemic model. *Ann. Appl. Prob.*, **8** (1988), 1331-1349.

in $\bar{k}$), while extending the sum to include $k = 2$. Using the expression (10.17) for R_0 and our assumption that R_0 is only slightly bigger than one, we find that the factor is approximately equal to

$$\frac{\bar{k}}{\bar{k} - 2}.$$

For relatively small values of $\bar{k}$, this is substantially bigger than one. Our conclusion is therefore that the 'steepness' of the relation between $1 - s_\infty$ and $R_0 - 1$ is much enhanced by the network structure. In other words, the all $(R_0 > 1)$ or nothing $(R_0 < 1)$ dichotomy, which in general is an exaggeration, is closer to the truth in a network setting than in a situation with random mixing.

It remains to characterise the real-time growth rate r and the probability of a minor outbreak. To do so, however, we need to be more specific about the submodel for infectivity. Assume that the dissemination of infectious material is determined by a stochastic variable ξ, the distribution of which is given by a probability measure m on a set Ω. If we restrict attention to a generation perspective, as we do when calculating the probability of a minor outbreak, we only need to specify the overall probability of transmission $q(\xi)$. The quantity $\bar{q}$, introduced earlier, is related to $q(\xi)$ through the formula

$$\bar{q} = \int_\Omega q(\xi) m(d\xi). \tag{10.31}$$

Again we shall consider a branching process, but now the *forward* process (hence the index f for π below) in which the parent-offspring relation corresponds to true transmission of infection. Hence, if we pick an individual whose infectivity is characterised by ξ, all its susceptible acquaintances experience that infectivity. (Note the difference with the backward branching process, where it was the infectivity 'at the other end of the edge' that mattered; in that case we have independence, whereas now, when looking forward, we have dependence.)

Exercise 10.28 Show that the probability of a minor outbreak equals

$$\sum_{k=1}^{\infty} \mu_k \int_\Omega \left(1 - q(\xi) + \pi_f q(\xi)\right)^k m(d\xi),$$

where, for $R_0 > 1$, π_f is the unique root in the interval $(0, 1)$ of the equation

$$\pi = g_f(\pi),$$

with

$$g_f(z) := \sum_{k=1}^{\infty} \nu_k \int_\Omega \left(1 - q(\xi) + zq(\xi)\right)^{k-1} m(d\xi).$$

Exercise 10.29 Check that $g_f'(1) = R_0$.

Before turning to the equation for r, we focus on the submodel for infectivity. As an example, consider the case of an infectious period that is exponentially distributed with parameter α. Assume that within this period each susceptible acquaintance has probability β per unit of time of becoming infected. This implies that the probability $\mathcal{F}(\tau, \xi)$ that a susceptible acquaintance of a type-ξ infectious individual is still uninfected at time τ after the start of the infectious period satisfies

$$\frac{d\mathcal{F}}{d\tau} = -\beta\mathcal{F}, \quad 0 \le \tau \le \xi,$$
$$\mathcal{F}(0, \xi) = 1,$$

and hence we have

$$\mathcal{F}(\tau, \xi) = e^{-\beta \min(\tau, \xi)} \tag{10.32}$$

and

$$q(\xi) = 1 - \mathcal{F}(\infty, \xi) = 1 - \mathcal{F}(\xi, \xi) = 1 - e^{-\beta \xi}. \tag{10.33}$$

To calculate $\bar{q}$, we average $q(\xi)$ with respect to the measure m, which has density $\alpha e^{-\alpha \xi}$ (compare Exercise 10.11):

$$\bar{q} = \int_0^\infty (1 - e^{-\beta \xi}) \alpha e^{-\alpha \xi} \, d\xi = 1 - \frac{\alpha}{\alpha + \beta} = \frac{\beta}{\alpha + \beta}. \tag{10.34}$$

It is now possible to determine the effect of the repetition of contacts. If we disregard loops, our calculation yields (see (10.20))

$$R_0 = \bar{q}\bar{k} = \frac{\beta \bar{k}}{\alpha + \beta}, \tag{10.35}$$

whereas a calculation that, in addition, disregards repetitions of contacts would give R_0 as the product of the rate $\beta\bar{k}$ of making effective contacts and the expected length $1/\alpha$ of the infectious period; i.e. it would yield

$$R_0 = \frac{\beta \bar{k}}{\alpha}. \tag{10.36}$$

We conclude that allowing for repetition reduces R_0 by a factor $\frac{\alpha}{\alpha+\beta}$.

Exercise 10.30 As an alternative submodel for infectivity, assume that the infectious period lasts from $\tau = 1$ to $\tau = \xi$, where ξ is uniformly distributed over the interval $[2, 3]$. Calculate $q(\xi)$ and $\bar{q}$. Determine the reduction factor of R_0 due to the repetition of contacts.

Exercise 10.31 Returning to the exponentially distributed infectious period, calculate the mean $\overline{\mathcal{F}}(\tau)$ of $\mathcal{F}(\tau, \xi)$ with respect to ξ.

Exercise 10.32 Explain that the initial real-time growth rate r is determined by the equation

$$1 = \sum_{k=1}^{\infty} \nu_k (k - 1) \int_0^\infty e^{-r\tau} \frac{d}{d\tau} (1 - \overline{\mathcal{F}}(\tau)). \tag{10.37}$$

We conclude that, for this particular quasi-network, one can calculate (or at least characterise) all indicators of infectivity for a demographically closed host population. In particular, one can disentangle and quantify the effects of repeated contacts on the one hand, and of short loops on the other hand. Reluctantly, we admit that this is an exceptional situation and that, as a rule, network models are not at all amenable to mathematical analysis. We hope this statement provokes our readers and that we are able to extend this chapter considerably in a second edition of the book.

Part III

The hard part: elaborations to (almost) all exercises

11

Elaborations for Part I

11.1 Elaborations for Chapter 1

Exercise 1.1 Let c denote the number of blood meals a mosquito takes per unit of time. Suppose a human receives k bites per unit of time. Consistency requires that $kD_{\text{human}} = cD_{\text{mosquito}}$. Our assumption is that c is a given constant. Hence necessarily

$$k = c\frac{D_{\text{mosquito}}}{D_{\text{human}}}.$$

Exercise 1.2 cT_1p_1.

Exercise 1.3 kT_2p_2.

Exercise 1.4 Consider one infected mosquito. It is expected to infect cT_1p_1 humans, each of which is expected to infect kT_2p_2 mosquitoes. So, going full circle, we have a multiplication factor

$$cT_1p_1kT_2p_2 = c^2T_1T_2p_1p_2\frac{D_{\text{mosquito}}}{D_{\text{human}}}.$$

When this multiplication factor is below one, an initial infection will die out in a small number of 'generations'. If, however, it is above one then most likely (see Section 1.2.2 for a qualitative and quantitative elaboration of how likely this actually is) an avalanche will result.

Exercise 1.5 i) Direct substitution. ii) $g(1) = \sum_{k=0}^{\infty} q_k$ and, since $\{q_k\}$ should be a probability distribution, we require that $\sum_{k=0}^{\infty} q_k = 1$. iii) By term-by-term differentiation, we find $g'(z) = \sum_{k=0}^{\infty} kq_k z^{k-1} = \sum_{k=1}^{\infty} kq_k z^{k-1}$ and therefore $g'(1) = \sum_{k=1}^{\infty} kq_k = R_0$. iv) All $q_k \geq 0$ (by their interpretation) and $\sum_{k=0}^{\infty} q_k = 1$ guarantee that at least one of the q_k is strictly positive; the expression for $g'(z)$ derived in 1.5-iii then implies $g'(z) > 0$. v) $g''(z) = \sum_{k=0}^{\infty} k(k-1)q_k z^{k-2}$, and the argument is now identical to that used in proving 1.5-iv.

Exercise 1.6 Draw in one picture the graph $y = g(z)$ and the $45°$line $y = z$; see Figure 11.1.

Recall the assumption that $q_0 > 0$ and note that consequently $g(0) = q_0 > 0$. Graphically, we can construct the sequence z_n as in the picture, that is, by going to the graph of g, then going to the $45°$ line, then going to the graph of g, etc. We then

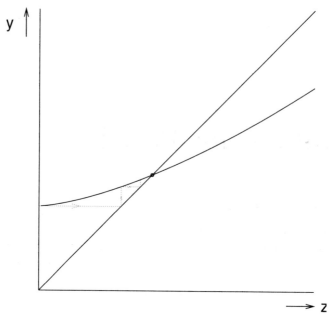

Figure 11.1

observe that z_n converges to the intersection of the two graphs that is nearest to the origin, or, in other words, to the smallest root of $z = g(z)$.

This graphical argument can be elaborated analytically. Since g is increasing, $g(q_0) > g(0) = q_0$ or $z_1 > z_0$. This in turn implies $z_2 = g(z_1) > g(z_0) = z_1$, and inductively we find that the sequence z_n is increasing. Since $g(z) \leq g(1) = 1$, the sequence is also bounded. A bounded increasing sequence has a limit, which we shall call z_∞. By taking the limit $n \to \infty$ in the recurrence relation $z_n = g(z_{n-1})$, we obtain $z_\infty = g(z_\infty)$ (here we use that g is continuous). Let now w denote the smallest root of the equation $z = g(z)$; then $w > 0$ and hence $w = g(w) > g(0) = q_0 = z_0$. Applying g repeatedly, we find $w > z_n$ for all n. By taking the limit $n \to \infty$ we deduce that $w \geq z_\infty$. But since w is defined to be the smallest root, we must have equality.

The interpretation as a consistency condition is as follows. The probability to go extinct equals the probability to go extinct immediately plus, summed over k, the probability to go extinct conditional on infecting k other individuals. In mathematical terms this sentence reads

$$z_\infty = q_0 + \sum_{k=1}^{\infty} q_k z_\infty^k = g(z_\infty),$$

where we have used the independence to write the probability that all k 'branches', or 'lines', go extinct as z_∞^k.

Exercise 1.7 The function $z - g(z)$ can have at most two zeros, since its second derivative equals $-g''(z) < 0$ (and if a function has three, or more, zeros, its derivative has two (or more) zeros and consequently its second derivative has at least one zero). If $R_0 < 1$, this function is negative for z slightly larger than zero and for z slightly less than one (recall Exercise 1.5-iii: $g'(1) = R_0$), so either $z - g(z)$ has no zero at all on the interval $[0, 1)$ or at least two. Since $z = 1$ is a root, the last case would imply that altogether three roots exist, which is impossible. Hence $z = 1$ is the smallest root. If, on the other hand, $R_0 > 1$ then, since $g'(1) = R_0$, the function $z - g(z)$ is positive for z slightly less than one, so a zero less than one must exist.

Exercise 1.8 For $R_0 = 1$ we have $z_\infty = 1$. Indeed, since $g''(1) > 0$, the graph of g lies above the $45°$ line for z slightly less than one, and the same argument as used in Exercise 1.7 for the case $R_0 > 1$ applies.

Exercise 1.9 If k individuals are contacted, the probability that m of them are actually infected equals

$$\binom{k}{m} p^m (1 - p)^{k-m}$$

(the first factor $\binom{k}{m} = \frac{k!}{m!(k-m)!}$ equals the number of ways one can choose m individuals out of k individuals). Hence the probability that m individuals are infected equals

$$\sum_{k=m}^{\infty} \binom{k}{m} p^m (1 - p)^{k-m} \frac{(c\Delta T)^k}{k!} e^{-c\Delta T}$$

$$= \frac{p^m}{m!} (c\Delta T)^m e^{-c\Delta T} \sum_{k=m}^{\infty} \frac{(1 - p)^{k-m}}{(k - m)!} (c\Delta T)^{k-m}$$

$$= \frac{p^m}{m!} (c\Delta T)^m e^{-c\Delta T} \sum_{l=0}^{\infty} \frac{(1 - p)^l}{l!} (c\Delta T)^l$$

$$= \frac{p^m}{m!} (c\Delta T)^m e^{-c\Delta T} e^{(1-p)c\Delta T}$$

$$= \frac{(pc\Delta T)^m}{m!} e^{-pc\Delta T}.$$

Exercise 1.10 For given ΔT, the probability that m individuals are infected equals (Exercise 1.9)

$$\frac{(pc\Delta T)^m}{m!} e^{-pc\Delta T}.$$

The probability density function for ΔT is, by assumption, $\alpha e^{-\alpha \Delta T}$. Hence

$$q_m = \alpha \int_0^\infty e^{-\alpha \Delta T} \frac{(pc\Delta T)^m}{m!} e^{-pc\Delta T} \, d(\Delta T)$$

and

$$g(z) = \sum_{m=0}^{\infty} q_m z^m = \alpha \int_0^\infty e^{-\alpha \Delta T} e^{pc\Delta T(z-1)} \, d(\Delta T)$$

(here we have interchanged the summation and the integration and subsequently evaluated the sum, i.e. we have applied (1.6) with $\lambda = pc\Delta T$). Evaluation of the integral yields the expression

$$g(z) = \frac{\alpha}{\alpha - pc(z - 1)}.$$

Exercise 1.11 i) If $g(z) = \frac{\alpha}{\alpha - pc(z-1)}$ then $g'(z) = \frac{\alpha pc}{(\alpha - pc(z-1))^2}$ and hence $R_0 = g'(1) = \frac{\alpha pc}{\alpha^2} = \frac{pc}{\alpha}$.

ii) If ΔT is exponentially distributed with parameter α then its expected value equals $1/\alpha$ (if you don't know this by heart, you have to compute $\alpha \int_0^\infty \Delta T e^{-\alpha \Delta T} d(\Delta T)$ and find, by using partial integration, that this integral equals $1/\alpha$). During this period contacts are made with rate (= probability per unit of time) c; hence the expected number of contacts equals $\frac{c}{\alpha}$. Finally, a contact results in transmission with probability p, so the expected number of secondary cases R_0 equals $\frac{pc}{\alpha}$.

Exercise 1.12 The equation is $z = \frac{\alpha}{\alpha - pc(z-1)}$, which, as can be verified by direct substitution, has roots $z = 1$ and $z = \frac{\alpha}{pc}$. As our earlier geometric argument in Exercise 1.7 showed, there are at most two roots, so we know all of them. Alternatively, one can multiply the equation by the denominator on the right-hand side and solve the resulting quadratic equation in z by using the explicit formula for the roots.

Exercise 1.13 A new case at time t results from a contact with an infective that was itself infected some time earlier, say at time $t - \tau$ for some positive τ. In order to be infective, τ should actually be restricted to $T_1 \leq \tau \leq T_2$. Infective individuals make contacts at rate c and, given a contact, transmit the agent with probability p. Hence, if we break down the incidence (i.e. the number of new cases per unit of time) according to the possibilities for the value of τ of the infector (i.e. the individual that is responsible for the transmission), we find (1.9).

Exercise 1.14 For $r = 0$ the right-hand side equals $pc \int_{T_1}^{T_2} d\tau = pc(T_2 - T_1) = R_0$. Furthermore,

$$\frac{d}{dr}\left(pc \int_{T_1}^{T_2} e^{-r\tau} \, d\tau\right) = -pc \int_{T_1}^{T_2} \tau e^{-r\tau} \, d\tau < 0.$$

For $r \to \infty$, $e^{-r\tau} \to 0$ uniformly for $T_1 \leq \tau \leq T_2$ and hence the limit of the integral equals zero. Likewise $e^{-r\tau} \to +\infty$ for $r \to -\infty$, uniformly for $T_1 \leq \tau \leq T_2$ (if $T_1 > 0$; if $T_1 = 0$, restrict to $\varepsilon \leq \tau \leq T_2$ for some fixed small ε and the same argument applies) and the limit of the integral equals $+\infty$.

Exercise 1.15 The key idea is that if you produce many 'children', but only at a very advanced age, population growth may be relatively slow! If $T_2^* - T_1^* = T_2^{**} - T_1^{**}$ then $R_0^* = R_0^{**}$. But if, for instance, $T_i^* = T_i^{**} + S$, for some $S > 0$, then $r^* < r^{**}$, since, for given r, $e^{-r(\tau+S)} = e^{-rS}e^{-r\tau} < e^{-r\tau}$ and hence the right-hand side of (1.10) is, for the same r, smaller in the * case than in the ** case (now draw a picture and you see at once that $r^* < r^{**}$). If S is large then actually r^* is much smaller than r^{**}. If we subsequently make T_2^* a bit bigger, we make sure that $R_0^* > R_0^{**}$ while still $r^* < r^{**}$ (since both R_0 and r are continuous functions of T_1 and T_2).

Exercise 1.16 The question is for which time T_d we have $\exp(rT_d) = 2$. By taking logarithms, we find $T_d = \frac{\ln 2}{r}$.

Exercise 1.18 The function $h(s) = \ln s - R_0(s-1)$ satisfies $h(1) = 0$, $h'(1) = 1 - R_0$, $h''(s) = -\frac{1}{s^2} < 0$ (for $0 < s \leq 1$). Moreover, $h(s) \downarrow -\infty$ for $s \downarrow 0$ and $h(s) \downarrow -\infty$ for $s \to \infty$. The property $h''(s) < 0$ guarantees that there are at most two roots. Since

$h(1) = 0$ and $h'(1) = 1 - R_0$, we know that $h(s) > 0$ for s slightly larger than one when $R_0 < 1$, and for s slightly less than one when $R_0 > 1$. The continuity of h and the fact that $h(s) \downarrow -\infty$ for both $s \downarrow 0$ and $s \to \infty$ then imply that for $R_0 \neq 1$ there is always a second zero, which is larger than one (and hence irrelevant) when $R_0 < 1$, but smaller than one (and then called $s(\infty)$) for $R_0 > 1$. For the exceptional case $R_0 = 1$ we have that $s = 1$ is a double zero.

Exercise 1.19 For notational convenience we omit the argument ∞ of s in the following. We are studying roots of the equation (1.11) as a function of R_0, in particular the root that is different from one (and the preceding exercise already showed that there exists exactly one such root).

For i) it is easiest to take the exponential on both sides of (1.11) to obtain

$$s = e^{R_0 s} e^{-R_0}.$$

Assuming that $R_0 s \to 0$ for $R_0 \to \infty$, we come to the consistent conclusion that

$$\lim_{R_0 \to \infty} s e^{R_0} = 1,$$

which is what we mean when we write $s \sim e^{-R_0}$ for $R_0 \to \infty$. To justify the assumption, we multiply the equation by $R_0 e^{-R_0 s}$, give $R_0 s$ a new name, say ξ, and write the equation in the form

$$h(\xi) = h(R_0),$$

where we now define $h(\xi) := \xi e^{-\xi}$. The graph of h is depicted in Figure 11.2, the relevant properties being that h is strictly increasing on $[0, 1)$, attaining its maximum at one, and is strictly decreasing towards zero on $(1, \infty)$.

Graphically, we find ξ from R_0 by going from R_0 on the horizontal axis vertically up to the graph of h, then horizontally to find the second point of the graph with the same value of h, and finally down to find the corresponding value ξ on the horizontal axis (see Figure 11.2). The construction immediately shows that ξ is a decreasing function of R_0 with limit zero for $R_0 \to \infty$. Recalling that ξ is just a name for $R_0 s$, we conclude that our assumption is valid.

ii) The assertion is obtained by formal Taylor expansion near $R_0 = 1$ and is justified by the implicit function theorem. Introducing $\sigma = s - 1$, we can rewrite equation (1.11) as

$$\frac{\ln(\sigma + 1)}{\sigma} = R_0.$$

For small σ we have $\ln(\sigma + 1) = \sigma - \frac{1}{2}\sigma^2 + O(\sigma^3)$. (For the sake of completeness, we recall that $O(\sigma^3)$ denotes a quantity that after division by σ^3 remains bounded when we let σ approach zero.) So the left-hand side equals $1 - \frac{1}{2}\sigma + O(\sigma^2)$. Bringing the 1 to the other side and multiplying by -2 we obtain

$$\sigma + O(\sigma^2) = -2(R_0 - 1)$$

which is the required result (recall that $\sigma = s - 1$ and that we want to consider σ as a function of $R_0 - 1$, rather than the other way around; but this relation shows that σ and $R_0 - 1$ are of the same order, so, when changing perspective, we may replace $O(\sigma^2)$ by $O((R_0 - 1)^2)$. For the important implicit function theorem, which justifies

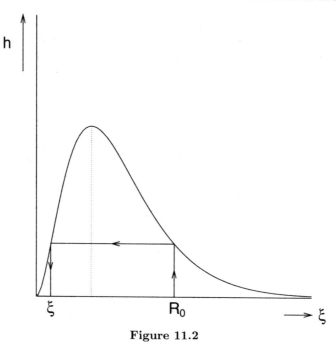

Figure 11.2

those formal calculations, we refer to any introductory analysis book (e.g. Dieudonné (1969)) or to advanced calculus books.

Exercise 1.21 i) The expected duration of the infectious period is $1/\alpha$. During the infectious period, a force of infection β is exerted on all members of the susceptible population, which, in the initial phase, has size N. Hence $R_0 = \beta N/\alpha$. To determine r, the initial rate of increase of the I subpopulation, we replace in the equation for dI/dt the variable S by the constant N (this amounts to linearisation in the steady state $(S, I) = (N, 0)$). From $\frac{dI}{dt} = (\beta N - \alpha)I$, we conclude that I initially grows with rate $\beta N - \alpha$. Hence $r = \beta N - \alpha$. Clearly R_0 passes the value one exactly when r passes the value zero, whatever parameter α, β or N we vary. Concentrating on the dependence on N, we see that below the threshold value $N_{\text{th}} := \alpha/\beta$ we have $R_0 < 1$ while above it we have $R_0 > 1$.

ii) We have

$$\frac{dI}{dS} = \frac{\beta SI - \alpha I}{-\beta SI} = -1 + \frac{\alpha}{\beta}\frac{1}{S}$$

$$\Rightarrow \quad dI = \left(-1 + \frac{\alpha}{\beta}\frac{1}{S}\right) dS$$

$$\Rightarrow \quad \int^{I(t)} d\xi = \int^{S(t)} \left(-1 + \frac{\alpha}{\beta}\frac{1}{\sigma}\right) d\sigma$$

$$\Rightarrow \quad I(t) = -S(t) + \frac{\alpha}{\beta} \ln S(t) + C$$

$$\Rightarrow \quad \frac{\alpha}{\beta} \ln S(t) - S(t) - I(t) \text{ is independent of } t.$$

iii) In the context of models (and perhaps in reality sometimes as well) the first case of an infection is always a bit mysterious. In deterministic models we have lost the concept of 'one individual'. What we have in mind here is that when we go backwards in time we approach (asymptotically for $t \to -\infty$) the point $(S, I) = (N, 0)$, which describes the situation before the infective agent appeared. What we leave completely open and unspecified is how and when the agent appeared, simply because such a discrete event affecting one individual cannot be incorporated in our deterministic description.

Since $R_0 > 1$, I at first increases. In general, S can only decrease. Once $\beta S - \alpha < 0$, I decreases too. So $I(t)$ has a limit for $t \to +\infty$ and so has $S(t)$. Limits must be steady states. Looking at (1.12), ignoring the R variable, which is redundant, we see that all points on the I axis are steady states and these are the only ones. So, $I(+\infty) = 0$. Since $\frac{\alpha}{\beta} \ln S(t) - S(t) - I(t)$ is independent of t, its values at $t = \pm\infty$ must be equal, that is

$$\frac{\alpha}{\beta} \ln S(+\infty) - S(+\infty) = \frac{\alpha}{\beta} \ln N - N.$$

Rewriting this identity as

$$\ln \frac{S(+\infty)}{N} = \frac{\beta N}{\alpha} \left(\frac{S(+\infty)}{N} - 1 \right),$$

we only have to realise that the fraction $s(\infty)$ introduced earlier is precisely $\frac{S(+\infty)}{N}$, to conclude that we have derived (1.11).

iv) A brute force solution to the difficulty of modelling the start of the epidemic is to specify at some particular time instant (which without loss of generality we take to be zero) values for S, I and R that add up to N. And if we really intend the time instant to be the 'start', it makes sense to take $R(0) = 0$ and $I(0)$ small. But anyhow we know from ii) that

$$\frac{\alpha}{\beta} \ln S(t) - S(t) - I(t) = \frac{\alpha}{\beta} \ln S(0) - S(0) - I(0).$$

Subtracting $\ln N$ from both sides and rearranging, we find, using $s(t) := S(t)/N$,

$$\ln s(t) = \ln s(0) + \frac{\beta N}{\alpha} \left(s(t) - \frac{S(0) + I(0)}{N} + \frac{I(t)}{N} \right).$$

Arguing exactly as in iii), we deduce that $I(t) \to 0$ for $t \to +\infty$. So if we choose $R(0) = 0$ or, equivalently, $S(0) + I(0) = N$ then, by taking the limit $t \to \infty$, we obtain

$$\ln s(\infty) = \ln s(0) + R_0(s(\infty) - 1).$$

Note that $s(0) < 1$, so $\ln s(0) < 0$. The arguments of Exercise 1.18 carry over and show that the equation $\ln s = \ln s(0) + R_0(s - 1)$ has exactly two roots, one bigger than one and (the relevant) one in the interval (0,1). See Figure 11.3, where the functions $y = \ln s$ and $y = \ln s(0) + R_0(s - 1)$ are plotted for $R_0 < 1$ and $R_0 > 1$.

For $R_0 < 1$ the relevant root is only slightly less than $s(0)$ when $s(0)$ itself is near one (and hence $\ln s(0)$ is almost zero). In fact it converges to 1 for $s(0) \uparrow 1$. This consolidates our conclusion that for $R_0 < 1$ the size of the epidemic is so small that the name 'epidemic' doesn't even apply.

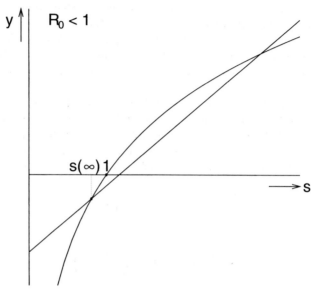

Figure 11.3a

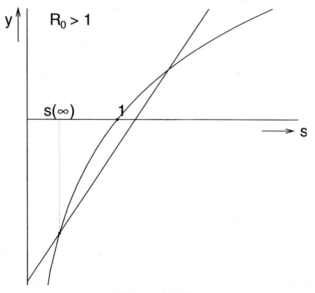

Figure 11.3b

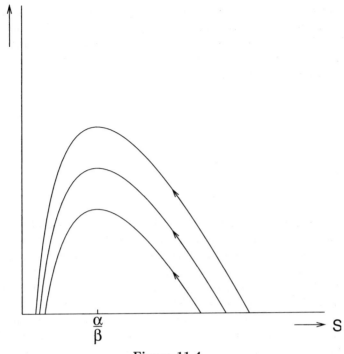

Figure 11.4

In contrast, for $R_0 > 1$ and small $\ln s(0)$, the relevant root is slightly smaller than the (relevant) root of (1.11), and for $s(0) \uparrow 1$ we have convergence towards that root. So, indeed, for $R_0 > 1$ we observe a sizeable epidemic, no matter how small we make the initial introduction of infectives.

v) See Figure 11.4.

vi) A necessary condition for I to be maximal is that $\frac{dI}{dt} = 0$. Since $\frac{dI}{dt} = (\beta S - \alpha)I$ and $I \neq 0$, this requires that $S = \frac{\alpha}{\beta}$. And, indeed, in the phase portrait of Figure 11.4 we see all curves reaching their maximum at this value of S.

If we 'freeze' S at a particular value, the argument presented in i) predicts that a case, on average, will produce $\beta S/\alpha$ new cases. As long as this number exceeds one, we will observe an increase in the infected subpopulation, as soon as it falls below one, in contrast, we will observe a decrease. This is exactly what Figure 11.4 shows.

vii) In the elaboration of Exercise 1.18 we have already observed that $s(\infty)$ is a decreasing function of R_0. Since $R_0 = \beta N/\alpha$ is an increasing function of N, this observation suffices.

In vi) we noted that I starts to decrease as soon as S falls below the level α/β. Furthermore, we saw that, for S below this level, a case produces less than one new case, on average. The point, however, is that there may be very many cases around at the moment that S passes this level. So, even though one case is expected to produce less than one new case, all of these together may still produce a lot of new cases.

How many cases there are when S reaches the critical level α/β depends on the population size. This is clearly demonstrated in the phase portrait in Figure 11.4. The further to the right on the S axis we start, the higher the peak of the I level will be, and the lower we 'end' on the S axis. So by 'overshoot phenomenon' we mean that there will be many new cases after the size of the susceptible subpopulation has dropped below the critical level, simply because there are many infectives around.

viii) Let a small letter correspond to the quantity denoted by the corresponding capital letter divided by N. Dividing both sides of (1.12) by N, we get, in this notation, the system

$$\frac{ds}{dt} = -\beta N si,$$

$$\frac{di}{dt} = \beta N si - \alpha i,$$

$$\frac{dr}{dt} = \alpha i.$$

So the structure is preserved, but the effect is that β becomes multiplied by N.

The force of infection has dimension (time)$^{-1}$ and equals, in this model, βI. The dimension of β thus depends on the dimension of I, which we have deliberately left unspecified. To make the connection with the quantities we introduced earlier, it helps to write $\beta I = \beta N \frac{I}{N}$ and to realise that $\frac{I}{N}$ is the fraction of the population that is infective, i.e., in the deterministic setting and assuming random 'mixing', the probability that a contact is with an infective. We may then compare βN with the product pc, i.e. with the number of contacts per unit of time multiplied by the probability that transmission occurs given a contact between a susceptible and an infectious individual. If we take β to be constant, this means that c is proportional to N.

Warning In the text of Section 1.3.2 following Exercise 1.21 the symbol i has a different meaning.

Exercise 1.22 If we integrate both sides of

$$\frac{s'(t)}{s(t)} = c \int_0^\infty A(\tau) s'(t - \tau) \, d\tau$$

from $-\infty$ to $+\infty$, we find

$$\begin{aligned}
\ln \frac{s(\infty)}{s(-\infty)} &= \int_{-\infty}^{+\infty} \frac{s'(t)}{s(t)} \, dt = c \int_{-\infty}^{+\infty} \int_0^\infty A(\tau) s'(t - \tau) \, d\tau \, dt \\
&= c \int_0^\infty A(\tau) \int_{-\infty}^{+\infty} s'(t - \tau) \, dt \, d\tau \\
&= c \int_0^\infty A(\tau) \{ s(+\infty) - s(-\infty) \} \, d\tau.
\end{aligned}$$

(Here we have interchanged the order of the two integrations. The monotonicity of s, and hence the existence of $s(\pm\infty)$, is compatible with (1.16), is required by the interpretation and can be proved rigorously.[1] But since the proof would force us to

---

[1] O. Diekmann: Limiting behaviour in an epidemic model. *Nonl. Anal. Theory, Meth. Appl.*, **1** (1977), 459-470.

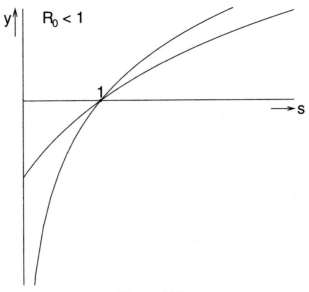

Figure 11.5a

discuss the way we specify the initial condition as well as the existence and uniqueness of a solution, we take the monotonicity of s for granted here.) Since $s(-\infty) = 1$ we can rewrite the identity as

$$\ln s(\infty) = c \int_0^\infty A(\tau)d\tau \, (s(+\infty) - 1)$$

which is exactly (1.11) since $c \int_0^\infty A(\tau)d\tau = cp(T_2 - T_1) = R_0$.

Exercise 1.23 Whenever we have $c = \theta N(t)$ and mixing is at random, i.e. the probability that a contact is with a susceptible equals the fraction $S(t)/N(t)$ of susceptibles in the population, we have equation (1.18) and the dependence on $N(t)$ cancels, allowing us to derive (1.11). So the correct answer is: no, it does not!

Exercise 1.24 The fraction that got infected but survived is $n(\infty) - s(\infty)$. The fraction that got infected is $1 - s(\infty)$. So if f denotes the probability of surviving infection then (1.20) should hold.

Exercise 1.25 i) Taking logarithms of (1.21), we arrive at

$$\ln n(\infty) = \frac{1 - f}{R_0} \ln s(\infty).$$

By (1.20), we have

$$\ln n(\infty) = \ln(s(\infty) + f(1 - s(\infty))).$$

The combination of these two identities implies (1.22).

ii) Consider the graphs in Figure 11.5a and Figure 11.5b, where the functions $y = \ln s$ and $y = \frac{R_0}{1-f} \ln(f + (1 - f)s)$ respectively.

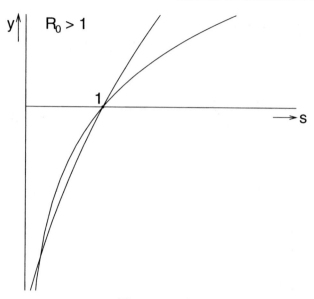

Figure 11.5b

A first key element in the underpinning of these graphs is the computation of the derivatives in $s = 1$ of both curves:

$$\frac{d}{ds}\ln s = \frac{1}{s}, \quad \text{so this derivative equals 1 for } s = 1;$$

$$\frac{d}{ds}\frac{R_0}{1-f}\ln(f + (1-f)s) = \frac{R_0}{1-f}\frac{1-f}{f + (1-f)s} = \frac{R_0}{f + (1-f)s},$$

so this derivative equals R_0 for $s = 1$. The intersection of the two curves for the cases $R_0 > 1$ and $R_0 < 1$ is therefore correctly represented in the two graphs.

For $s = 0$ we have $\frac{R_0}{1-f}\ln(f + (1-f)s) = \frac{R_0}{1-f}\ln f < 0$ but finite (whereas $\ln s \downarrow -\infty$ for $s \downarrow 0$). The ordering of the values at $s = 0$ is therefore also correctly represented in the two graphs.

Next we look at the derivative of the difference:

$$\frac{d}{ds}\left\{\ln s - \frac{R_0}{1-f}\ln(f + (1-f)s)\right\} = \frac{1}{s} - \frac{R_0}{f + (1-f)s},$$

which tends to $+\infty$ for $s \downarrow 0$ and equals $1 - R_0$ for $s = 1$. Moreover, in order for the right-hand side to equal zero we should have $f + (1-f)s = R_0 s \Rightarrow s = \frac{f}{R_0 - 1 + f}$. The point is that this derivative has at most one zero in $(0,1)$. So for $R_0 < 1$, when the derivative is positive at $s = 1$ and near $s = 0$, there can be no zero at all for this derivative: it is positive on the entire interval $(0,1)$. Consequently, the function $\ln s - \frac{R_0}{1-f}\ln(f + (1-f)s)$ cannot have two or more zeros on $(0,1)$ and the picture for $R_0 < 1$ is confirmed. For $R_0 > 1$, on the other hand, the derivative is negative at $s = 1$ and positive near $s = 0$, so the derivative has exactly one zero. But then the function $\ln s - \frac{R_0}{1-f}\ln(f + (1-f)s)$ cannot have three or more zeros on $(0,1)$, which confirms our picture for the case $R_0 > 1$.

To summarise: 1 is always a root of (1.22) and for $R_0 < 1$ this is the only root in $[0,1]$, but for $R_0 > 1$ there is exactly one additional root $s(\infty) \in (0,1)$. For $R_0 = 1$ the root 1 is a double root.

iii) We use the fact that $\ln(1 + x) = x + O(x^2)$ for $x \to 0$. Writing $f = 1 - \varepsilon$ and assuming $\varepsilon \downarrow 0$, we deduce $\ln(f + (1 - f)s) = \ln(1 - \varepsilon(1 - s)) = -\varepsilon(1 - s) + O(\varepsilon^2)$. Inserting this into equation (1.22), we find

$$\ln s(\infty) = \frac{R_0}{\varepsilon}\{-\varepsilon(1 - s(\infty)) + O(\varepsilon^2)\} = R_0(s(\infty) - 1) + O(\varepsilon),$$

and in the limit $\varepsilon \downarrow 0$ we indeed arrive at (1.11).

iv) For $f \uparrow$ we have $(1 - f) \downarrow$ and hence $\frac{1}{1-f} \uparrow$. Moreover, for $f \uparrow$ we have, for $0 < s < 1$, that $(s + f(1 - s)) \uparrow$ and hence that $\ln(s + f(1 - s)) = \ln(f + (1 - f)s) \uparrow$. We conclude that the right-hand side of (1.22) is an increasing function of f. A graphical argument (see Figure 11.5b for $R_0 > 1$) now shows that $s(\infty)$ is a monotonically increasing function of f. An analytical elaboration can be based on the implicit function theorem and involves sign considerations about derivatives with respect to s, as presented in ii) above.

v) When $f \downarrow 0$ the right-hand side of (1.22) tends to $R_0 \ln s$ for any $s > 0$. So, since $\ln s < 0$ for $0 < s < 1$, the right-hand side is less than the left-hand side when $R_0 > 1$, for any $s \in (0,1)$, and f sufficiently small (depending on R_0 and s). Hence (see Figure 11.5b), $s(\infty) \downarrow 0$ for $f \downarrow 0$. Next look at (1.20) and conclude that also $n(\infty) \downarrow 0$ for $f \downarrow 0$.

Exercise 1.26 The relation (1.20) implies that

$$f = \frac{n(\infty) - s(\infty)}{1 - s(\infty)}.$$

Taking logarithms of (1.21), we find, after a little rearrangement,

$$R_0 = (1 - f)\frac{\ln s(\infty)}{\ln n(\infty)} = \frac{1 - s(\infty) - n(\infty) + s(\infty)}{1 - s(\infty)}\frac{\ln s(\infty)}{\ln n(\infty)},$$

giving the desired result.

Exercise 1.27 i) The parameter γ is the product of the expected number of contacts per unit of time and the probability of transmission given a contact between an infective and a susceptible. The product $\gamma\frac{SI}{N}$ may now be read as $\gamma\frac{S}{N}I$, in which case we consider the contacts made by the infectives and say that the probability that such a contact is with a susceptible equals the fraction $\frac{S}{N}$ of susceptibles in the population. We can also read the product as $\gamma\frac{I}{N}S$, in which case we consider the contacts made by the susceptibles and say that the probability that such a contact is with an infective equals the fraction $\frac{I}{N}$ of infectives in the population. Note that consistency is guaranteed. When we take γ to be a given constant (parameter), the number of contacts per unit of time is indeed independent of population size N.

A special feature of the model is the way in which we have incorporated mortality. The more usual way is to add a term $-\nu I$ to the equation for dI/dt, which is then interpreted as infectives being exposed to an additional force of mortality ν. A consequence is that the length of the infectious period depends on the parameter ν.

The present system expresses a different point of view: the length of the infectious period is exponentially distributed with parameter α, but whether or not the infectious period ends with fatality is determined by another chance process, which gives a probability f to survive.

ii) When $S = N$, the expected number of secondary cases per primary case equals γ/α and the real-time growth of the infected subpopulation is governed by $\frac{dI}{dt} = (\gamma - \alpha)I$.

iii) We start from

$$\frac{dS}{dN} = \frac{\gamma \frac{SI}{N}}{(1-f)\alpha I} = \frac{\gamma S}{(1-f)\alpha N} = \frac{R_0 S}{(1-f)N}.$$

Now separate variables, $\frac{dS}{S} = \frac{R_0}{1-f}\frac{dN}{N}$, then integrate from σ to t,

$$\ln \frac{S(t)}{S(\sigma)} = \frac{R_0}{1-f}\ln \frac{N(t)}{N(\sigma)},$$

and finally take exponentials,

$$\frac{S(t)}{S(\sigma)} = \left(\frac{N(t)}{N(\sigma)}\right)^{\frac{R_0}{1-f}}.$$

For $\sigma \downarrow -\infty$ both $S(\sigma)$ and $N(\sigma)$ tend to the initial population size. So, by taking the limits $\sigma \downarrow -\infty$ and $t \uparrow \infty$, we arrive at (1.21).

iv) Following the hint, we have

$$N(\infty) - N(-\infty) = -(1-f)\alpha \int_{-\infty}^{+\infty} I(\tau)\, d\tau.$$

Since $I(t) \downarrow 0$ for both $t \downarrow -\infty$ and $t \uparrow \infty$, we find by integrating $\frac{dI}{dt} = -\frac{dS}{dt} - \alpha I$ that $0 = -S(\infty) + S(-\infty) - \alpha \int_{-\infty}^{+\infty} I(\tau)\, d\tau$. Next we eliminate $\int_{-\infty}^{+\infty} I(\tau)\, d\tau$ from these two identities and arrive at $N(\infty) - N(-\infty) = (1-f)(S(\infty) - S(-\infty))$. Now divide by $N(-\infty)$, realise that $S(-\infty) = N(-\infty)$ and obtain $n(\infty) - 1 = (1-f)(s(\infty) - 1)$, which is just another way of writing (1.20).

v) Start from

$$\left.\begin{array}{l} \frac{1}{S}\frac{dS}{dt} = -\gamma\frac{I}{N} \\[2mm] \frac{1}{N}\frac{dN}{dt} = -(1-f)\alpha\frac{I}{N} \end{array}\right\} \quad \frac{1}{N}\frac{dN}{dt} = \frac{(1-f)\alpha}{\gamma}\frac{1}{S}\frac{dS}{dt} = \frac{(1-f)}{R_0}\frac{1}{S}\frac{dS}{dt},$$

then integrate with respect to time from σ to t and take exponentials.

Exercise 1.28 i) With c equal to the number of contacts per unit time and $A(\tau)$ equal to the product of the probability to be alive τ units of time after infection and the probability that transmission takes place, given a contact with a susceptible, τ units of time after infection we have, by definition, $R_0 = c\int_0^\infty A(\tau)\, d\tau$.

ii) We derive a second identity involving $\int_0^\infty A(\tau)\, d\tau$ as follows: $f - 1 = \mathcal{F}(\infty) - 1 = \int_0^\infty \mathcal{F}'(\tau)\, d\tau = -\int_0^\infty \mu(\tau)\mathcal{F}(\tau)\, d\tau = -q\int_0^\infty a(\tau)\mathcal{F}(\tau)\, d\tau = -q\int_0^\infty A(\tau)\, d\tau$. If we eliminate $\int_0^\infty A(\tau)d\tau$ from this identity and the relation for R_0, we find

$$q = c\frac{1-f}{R_0}.$$

Exercise 1.29 Rewriting (1.26) as

$$N(t) = S(t) - \int_{-\infty}^t S'(\tau)\mathcal{F}(t-\tau)\,d\tau,$$

we obtain by differentiation

$$N'(t) = S'(t) - S'(t)\mathcal{F}(0) - \int_{-\infty}^t S'(\tau)\mathcal{F}'(t-\tau)\,d\tau.$$

The first two terms on the right-hand side cancel each other, and $\mathcal{F}'(t-\tau) = -\mu(t-\tau)\mathcal{F}(t-\tau) = -qa(t-\tau)\mathcal{F}(t-\tau) = -qA(t-\tau)$, so we have in fact derived that

$$N'(t) = q\int_{-\infty}^t S'(\tau)A(t-\tau)\,d\tau = q\int_0^\infty S'(t-\tau)A(\tau)\,d\tau,$$

and consequently

$$\frac{d}{dt}\ln N(t) = \frac{N'(t)}{N(t)} = \frac{q\int_0^\infty S'(t-\tau)A(\tau)\,d\tau}{N(t)}.$$

We can then rewrite (1.25) as

$$\frac{d}{dt}\ln S(t) = \frac{S'(t)}{S(t)} = \frac{c\int_0^\infty S'(t-\tau)A(\tau)\,d\tau}{N(t)} = \frac{c}{q}\frac{d}{dt}\ln N(t).$$

If we now integrate and use the expression for q derived in Exercise 1.28, we obtain (1.32).

Exercise 1.30 We rewrite, as above, (1.26) as

$$N(t) = S(t) - \int_{-\infty}^t S'(\tau)\mathcal{F}(t-\tau)\,d\tau.$$

For $t \to \infty$ we have $\mathcal{F}(t-\tau) \to f$ (uniformly on every bounded τ interval), so, by taking the limit $t \to \infty$, we find

$$N(\infty) = S(\infty) - f\int_{-\infty}^\infty S'(\tau)\,d\tau = S(\infty) - f(S(\infty) - N(-\infty)),$$

which, after division by $N(-\infty)$, is precisely (1.20).

Exercise 1.31 The differences are irrelevant if population size is fixed. They pertain to the dependence of contact intensity on population size/density. In the foregoing section we assumed that the combination of individual behaviour and the way we

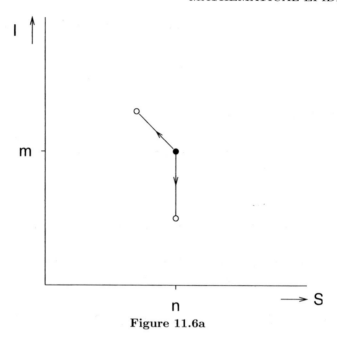

Figure 11.6a

measure population size is such that it is reasonable to assume that the number of contacts per unit time per individual is proportional to population size.

Even if infection may lead to fatalities, we can ignore the effect on population size in the very early phase of an epidemic, when there are just a few cases. This means that the determination of R_0 and r for an infection in a particular population is not influenced at all by the difference between the two models.

However, when comparing the effect of the introduction of the infection in two (or more) populations that differ in size, the difference between the two models will manifest itself.

In the model of the preceding section we have $R_0 = \frac{\beta N}{\alpha}$, $r = \beta N - \alpha$ and $s(\infty)$ is determined by (1.11). We see that both R_0 and r increase with N, while $s(\infty)$ decreases with N.

In the model of the present section we have $R_0 = \frac{\gamma}{\alpha}$, $r = \gamma - \alpha$ and $s(\infty)$ is determined by (1.22). All of these quantities are independent of population size.

In both cases the probability that the introduction of one initial case does not lead to an epidemic is given, according to (1.8), by $1/R_0$. So, again, this is independent of population size for the model in this section, but decreases with population size for the model of the preceding section.

Exercise 1.32 It is reasonable to take this as a constant, independent of population size. This clearly leaves out certain subtleties: people who want to be very sexually active may be attracted to big cities, and a small number of whales in a large ocean may have difficulty finding each other for mating.

Exercise 1.33 i) The rate of leaving state (n, m) equals $\beta nm + \alpha m$. State (n, m) is entered from state $(n, m + 1)$ at rate $\alpha(m + 1)$ and from state $(n + 1, m - 1)$ at rate $\beta(n + 1)(m - 1)$; see Figure 11.6a and Figure 11.6b.

ii) Initially, there are $N - 1$ susceptibles and 1 infective.

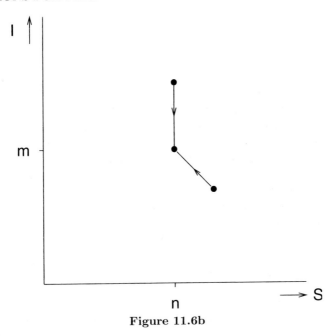

Figure 11.6b

iii) An absorbing state is a state where, once the system attains it, the system will stay fixed forever. In a diagram such as Figures 11.6a and 11.6b, we can spot absorbing states by looking for points that have no outgoing arrows. If $m = 0$, there are no infectives and so, according to the interpretation, m should stay zero. We can verify this mathematically:

$$\frac{d}{dt}P_{(n,0)}(t) = \alpha P_{(n,1)}(t)$$

(recall the convention introduced at the beginning of the exercise), which tells us that there is only a positive contribution to the rate of change of $P_{(n,0)}$, i.e. one can only enter the state $(n,0)$ but not leave it. There are no other absorbing states because every other state has $m > 0$ and therefore the corresponding differential equation for $P_{(n,m)}(t)$ always contains a negative contribution $-\alpha m P_{(0,m)}(t)$ caused by an infective being removed.

iv) For $m > 0$, there is a positive probability per unit of time of moving 'down' due to removal, which cannot be compensated, in the long run, by the oblique movement due to infection (just look at the diagram, restricted to the feasible triangular region $n \geq 0, m \geq 0, n + m \leq N$). So $\lim_{t \to \infty} P_{(n,m)}(t) = 0$ for any $m > 0$. This 'intuitive' argument is easily elaborated formally by analysing the system of differential equations.

v) The probability that there are k victims in total is given by $P_{(N-k,0)}(\infty)$.

vi) For $m > 0$, the rate of leaving state (n,m) equals $\beta nm + \alpha m$, so the sojourn time is exponentially distributed with mean $(\beta nm + \alpha m)^{-1}$. The next state will be either $(n-1, m+1)$ or $(n, m-1)$. The ratio of the probabilities equals the ratio of the rates, that is how one arrives at the expressions. More formally, if

$$\frac{d\mathcal{F}}{dt} = -(\lambda_A + \lambda_B)\mathcal{F}, \quad \mathcal{F}(0) = 1$$

$$\frac{dP_A}{dt} = \lambda_A \mathcal{F}$$

$$\frac{dP_B}{dt} = \lambda_B \mathcal{F}$$

then $P_A(\infty) = \frac{\lambda_A}{\lambda_A + \lambda_B}$ and $P_B(\infty) = \frac{\lambda_B}{\lambda_A + \lambda_B}$.

vii) Again Figures 11.6a and 11.6b are helpful. But note carefully that for $m = 1$ there is no incoming arrow from the south-east, as there can be no new cases after all infectives have been removed. This observation, together with the results of vi), allow us to reformulate the diagrams in Figures 11.6a and 11.6b in the form of recurrence relations (note that a factor cancels in the expressions for the probabilities). The expression for the initial condition is self-evident.

viii) $Q_{(n,0)}(l)$ does not occur on the right-hand side of any of the recurrence relations.

ix) If everybody gets infected, we have the maximal number of events. These are $N - 1$ transmissions and N removals, so $2N - 1$ events in total.

x)

$$\sum_{l=1}^{2N-1} Q_{(N-k,0)}(l) = \Pr\{k \text{ victims in total}\}.$$

Note that this is easily evaluated numerically by programming the recurrence relations and the initial condition. In fact, however, all Q contributions in the sum are zero except for one. In order to go from state $(N - 1, 1)$ to state $(N - k, 0)$, there have to occur $k - 1$ transmissions and k removals, so $2k - 1$ events in total (stated differently: there are many routes from $(N - 1, 1)$ to $(N - k, 0)$, but they all involve $2k - 1$ steps). So $Q_{(N-k,0)}(l)$ is non-zero iff (meaning 'if and only if') $l = 2k - 1$.

xi) Multiply both the numerator and the denominator by N/α and use $R_0 = \beta N/\alpha$.

Exercise 1.34 We do not attempt to draw a three-dimensional analogue of Figures 11.6a and 11.6b, but instead write in self-explanatory notation

$$(n, m, k) \rightarrow \begin{cases} (n - 1, m + 1, k) & \text{with rate } \frac{\gamma mn}{n+m+k}, \\ (n, m - 1, k) & \text{with rate } (1 - f)\alpha m, \\ (n, m - 1, k + 1) & \text{with rate } f\alpha m. \end{cases}$$

These rates add up to $\frac{\gamma mn}{n+m+k} + \alpha m$. We can now compute the probabilities with which each of the three possible outcomes occurs, as the relative contributions of the rates to the total rate (exactly as we did in Exercise 1.33-vi).

For $m > 1$ there are three ways to arrive at (n, m, k): 1) from $(n + 1, m - 1, k)$ by a transmission; 2) from $(n, m + 1, k)$ by a removal that is fatal; 3) from $(n, m + 1, k - 1)$ by a removal that is non-fatal. These observations, and the fact that $R_0 = \gamma/\alpha$, show that the stated recurrence relation does the right bookkeeping. The absorbing states are now $(n, 0, k)$ with $n + k \leq N$. The maximum number of events is still $2N - 1$. And

$$\sum_{l=1}^{2N-1} Q_{(N-r,0,k)}(l) = \Pr\{r \text{ individuals in total are infected,}$$

$$k \text{ of which survived the disease caused by the infection}\}.$$

But, as in Exercise 1.33-x, it follows that $Q_{(N-r,0,k)}(l)$ is non-zero iff $l = 2r - 1$.

Exercise 1.35 This is indeed the case when (1.11) applies (see Exercise 1.19-i). In the case of (1.22), a second parameter f is involved and we have to take possible differences in f into account when determining $s(\infty)$.

Exercise 1.36 Under the vaccination regime, a fraction q of all contacts is 'wasted' on protected individuals. So the expected number of secondary cases per primary case equals $(1 - q)R_0$ and in order that this quantity lies below one we should have $q > 1 - \frac{1}{R_0}$.

Exercise 1.37 If $z_\infty > \frac{1}{2}$, it is quite likely that, even though $R_0 > 1$, the introduction of one infective into a virgin population leads to a minor outbreak only. Therefore, if one observes a few minor outbreaks, the conclusion that evidently $R_0 < 1$ is NOT warranted.

Exercise 1.38 By assumption, the chains of infection develop independently as long as there are only a few cases. So the probability that all chains are of the 'minor outbreak' type equals $(z_\infty)^k$.

Exercise 1.39 The first infected individual is expected to generate R_0 new cases; therefore, after one generation we expect $1 + R_0$ infected individuals. The R_0 first-generation cases are also expected to make R_0 new cases each in the second generation, etc. We therefore expect a total of

$$1 + R_0 + R_0^2 + R_0^3 + \cdots = \frac{1}{1 - R_0}$$

cases for $R_0 < 1$.

Exercise 1.40 i)

$$I_{k+1} + S_{k+1} - \frac{N}{R_0} \ln S_{k+1}$$

$$= S_k \left(1 - \exp\left(-R_0 \frac{I_k}{N} \right) \right) + S_k \exp\left(-R_0 \frac{I_k}{N} \right) - \frac{N}{R_0} \left(\ln S_k - R_0 \frac{I_k}{N} \right)$$

$$= I_k + S_k - \frac{N}{R_0} \ln S_k.$$

ii) In the relation

$$I_k + S_k - \frac{N}{R_0} \ln S_k = I_l + S_l - \frac{N}{R_0} \ln S_l$$

we take the limit $k \to +\infty$ and $l \to -\infty$ and use that both I_k and I_l tend to zero, while S_l tends to N:

$$S_\infty - \frac{N}{R_0} \ln S_\infty = N - \frac{N}{R_0} \ln N.$$

Multiplying this relation by $\frac{R_0}{N}$ we find, after a little rearrangement, the familiar final size equation (1.11).

iii) We have nothing to add to what we said in the main text.

iv) In discrete-time formulations one should multiply (i.e. work with fractions) rather than subtract. It is quite possible that $\beta I_k S_k$ exceeds (for large β and suitable I_k, S_k)

the quantity S_k, in which case we get negative numbers for quantities that ought to be positive.

Note that *for small I_k* we have

$$(1 - e^{-R_0 I_k/N}) = R_0 \frac{I_k}{N} + o(I_k),$$

so the quadratic term shows up by Taylor expansion for small numbers of infectives.

Exercise 1.41 Here μ is the probability per unit of time of dying and $a(t)$ is the expected infectivity at disease age t, conditional on survival.

If we integrate the differential equation for I, we find $I(t) = I_0 \exp(-\int_0^t \mu(\sigma)\, d\sigma)$, and so in the limit for $t \to \infty$: $I(\infty) = f I_0$. Integrating the differential equation for N we obtain $N(\infty) - N_0 = -\int_0^\infty \mu(\sigma) I(\sigma)\, d\sigma = \int_0^\infty \frac{dI}{dt}(\sigma)\, d\sigma = I(\infty) - I_0 = (f-1)I_0$. Considering $\frac{dS}{dN}$ and using separation of variables, we arrive, exactly as in Exercise 1.27-iii, at $S(t)/S_0 = (N(t)/N_0)^{R_0/(1-f)}$.

Now we have to translate this information into the recurrence relation that corresponds to the 'next-generation' procedure. The relation

$$I_{k+1} = S_k - S_{k+1}$$

holds by definition. The relation $N(\infty) - N_0 = (f-1)I_0$ derived above translates into

$$N_{k+1} = N_k - (1-f)I_k.$$

Combining this with $S_{k+1} = S_k(N_{k+1}/N_k)^{R_0/(1-f)}$ (which is the straightforward limit $t \to +\infty$ of an equation derived above), we deduce that

$$S_{k+1} = S_k \left(1 - (1-f)\frac{I_k}{N_k}\right)^{R_0/(1-f)}.$$

(So we can compute I_{k+1} even though S_{k+1} occurs on the right-hand side of the equation for I_{k+1}.)

It is actually helpful to also rewrite the relation between $S(t)/S_0$ and $N(t)/N_0$, derived above, in the form

$$\frac{S_{k+1}}{S_k} = \left(\frac{N_{k+1}}{N_k}\right)^{R_0/(1-f)},$$

because, by induction, this implies that

$$\frac{S_\infty}{S_0} = \left(\frac{N_\infty}{N_0}\right)^{R_0/(1-f)},$$

which, with $I_0 + S_0 = N_0$ and in the limit $I_0 \downarrow 0$, is (1.21) in minor disguise. Now define $R_0 = 0$ and $R_{k+1} = R_k + f I_k$; then we shall prove, by induction, that $N_k = S_k + I_k + R_k$. For $k = 0$ this is true by assumption. Suppose it holds for $k = l$; then $S_{l+1} + I_{l+1} + R_{l+1} = S_l + R_l + f I_l$, while $N_{l+1} = N_l - I_l + f I_l = S_l + I_l + R_l - I_l + f I_l = S_l + R_l + f I_l$, so the relation holds for $k = l + 1$. Likewise, we prove by induction that $R_{k+1} = f(N_0 - S_k)$. Again this is true for

$k = 0$ by the assumption $R_0 = 0, N_0 = S_0 + I_0$. Suppose it is true for $k = l$. Then $R_{l+1} = R_l + f I_l = f(N_0 - S_{l-1}) + f I_l = f(N_0 - S_{l-1} + I_l) = f(N_0 - S_l)$, which shows that the relation is true for $k = l + 1$. In the limit for $k \to \infty$ we find $N_\infty = S_\infty + R_\infty$ and $R_\infty = f(N_0 - S_\infty)$, which, in combination, and after division by N_0, yields (1.20).

Exercise 1.42 Recall from calculus (or take our word for it) that

$$\lim_{y \to \infty} \left(1 + \frac{x}{y} \right)^y = e^x.$$

In the present context take $y = -\frac{1}{1-f}$ in the recurrence relation for S and use the outcome in the recurrence relation for I.

11.2   Elaborations for Chapter 2

Exercise 2.1 The $-\alpha I$ term in the equation for $\frac{dI}{dt}$ represents that the length of the infectious period is exponentially distributed with parameter α. In other words, if $P_I(\tau)$ denotes the probability to be still infectious at time τ after infection then

$$\frac{dP_I}{d\tau} = -\alpha P_I, \quad P_I(0) = 1,$$

and consequently $P_I(\tau) = e^{-\alpha \tau}$.

Assume that death does not occur and that removed individuals are immune. We should read the term βSI as $\beta N \frac{S}{N} I$. If we consider just one infected individual, we delete the factor I. The factor $\frac{S}{N}$ is the fraction of susceptibles in the total population, and is therefore declared to be the probability that a contact is with a susceptible (rather than with an infected or immune individual). The factor βN is in fact the product of two quantities: the number of contacts per unit of time and the probability of transmission, given a contact between an infectious and a susceptible individual. The number of contacts per unit of time we called c. So the probability of transmission, given a contact between an infectious individual and a susceptible individual, equals $\beta N / c$.

Let $A(\tau)$ be the probability that a susceptible is infected when it comes into contact with another individual that was itself infected τ units of time ago. Then

$$A(\tau) = \frac{\beta N}{c} P_I(\tau) = \frac{\beta N}{c} e^{-\alpha \tau}.$$

When death does occur, but we are still in the initial phase where the influence of the agent on population size can be neglected, the factor $\beta N / c$ is motivated exactly as above, but $P_I(\tau)$ should be interpreted as the probability to be alive and infectious at time τ after infection.

Exercise 2.2 i) Let $P_E(\tau)$ denote the probability to be still in the latency period at time τ after infection and $P_I(\tau)$ the probability to be in the infectious period at that time. Then

$$\frac{dP_E}{d\tau} = -\theta P_E, \qquad P_E(0) = 1,$$

$$\frac{dP_I}{d\tau} = \theta P_E - \alpha P_I, \qquad P_I(0) = 0,$$

from which it follows that $P_E(\tau) = e^{-\theta\tau}$ and

$$P_I(\tau) = \int_0^\tau e^{-\alpha(\tau-\sigma)}\theta P_E(\sigma)\,d\sigma$$

(this last formula is an instance of the variation-of-constants formula from the theory of ordinary differential equations (see e.g. Hale (1969)), and has exactly the interpretation outlined in the hint). After some algebra, this expression becomes

$$P_I(\tau) = \frac{\theta}{\alpha - \theta}(e^{-\theta\tau} - e^{-\alpha\tau}).$$

In the preceding exercise it was already shown that

$$A(\tau) = \frac{\beta N}{c} P_I(\tau).$$

Combination of these two identities yields the desired result.

ii) We find that

$$R_0 = \beta N \frac{\theta}{\alpha - \theta}\left(\frac{1}{\theta} - \frac{1}{\alpha}\right) = \frac{\beta N}{\alpha},$$

and see that time delay does not influence R_0, which is a 'generation' quantity. Time delays do, however, influence r, which is a 'real-time' quantity.

iii) The formulation of this exercise is a bit vague. The idea is the following. Let some $A = A(\tau)$ be given. Extend the domain of definition of the function A by the convention that $A(\tau) = 0$ for $\tau < 0$.

Now consider a one-parameter family of 'shifted' versions of A that at τ assume the value $A(\tau - T_L)$, where the parameter $T_L \geq 0$ corresponds to the latency period (see Figure 11.7).

Then, on the one hand,

$$\int_0^\infty A(\tau - T_L)\,d\tau = \int_{T_L}^\infty A(\tau - T_L)\,d\tau = \int_0^\infty A(\sigma)\,d\sigma$$

and so the value of R_0 is independent of the parameter T_L. On the other hand, we have

$$\int_0^\infty e^{-r\tau}A(\tau - T_L)\,d\tau = e^{-rT_L}\int_0^\infty e^{-r\sigma}A(\sigma)\,d\sigma$$

(so for fixed r this is a decreasing function of T_L, which implies that the root of the equation

$$1 = ce^{-rT_L}\int_0^\infty e^{-r\sigma}A(\sigma)\,d\sigma$$

shifts to the left, and so becomes smaller, if we increase T_L).

Exercise 2.3 The change is incorporated as a factor c_{reduced}/c, so, for example,

$$A(\tau) = \begin{cases} pc_{\text{reduced}}/c & \text{if } T_1 \leq \tau \leq T_2, \\ 0 & \text{otherwise.} \end{cases}$$

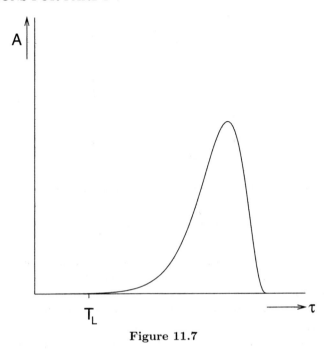

Figure 11.7

Then A is no longer a conditional probability, but also incorporates aspects of the contact process.

Exercise 2.4 i) When we use the word 'expected' in relation to $A(\tau)$, we refer to all that might happen after some individual is infected, and, more precisely, we should take into account all that can occur in the interval of length τ after the moment of infection.

Only host individuals that are alive take part in the contact process and can transmit the agent to another host. So we have to incorporate the survival probability $\exp(-\mu\tau)$ as a factor in $A(\tau)$.

ii) The idea is to redo Exercise 2.2-iii, but now with

$$A(\tau) = e^{-\mu\tau}\widetilde{A}(\tau).$$

When 'shifting' to incorporate the effect of a latency period, we shift $\widetilde{A}$ while keeping the survival probability unchanged. That is, we consider the one-parameter family of functions that at time τ take the value $e^{-\mu\tau}\widetilde{A}(\tau - T_L)$, with $\widetilde{A}$ being defined to be zero for negative values of the argument. Then

$$R_0 = R_0(T_L) = c\int_0^\infty e^{-\mu\tau}\widetilde{A}(\tau - T_L)\,d\tau = e^{-\mu T_L}c\int_0^\infty e^{-\mu\sigma}\widetilde{A}(\sigma)\,d\sigma.$$

iii) Starting from

$$\frac{dP_I}{d\tau} = -(\alpha + \mu)P_I \quad \Rightarrow \quad P_I(\tau) = e^{-(\alpha+\mu)\tau},$$

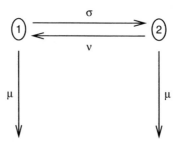

Figure 11.8

the arguments presented in the elaboration of Exercise 2.1 yield

$$R_0 = \beta N \int_0^\infty e^{-(\alpha+\mu)\tau} d\tau = \frac{\beta N}{\alpha + \mu}.$$

iv) The probability to survive the latency period and to enter the infectious period equals $\frac{\theta}{\theta+\mu}$ (see Exercise 1.33-vi). If that happens, we are in the situation just considered in 2.4-iii.

v) The time scale of the infectious period is $1/\alpha$ and the time scale of life is $1/\mu$. When $\frac{1}{\mu} \gg \frac{1}{\alpha}$, we can ignore death (here '$\gg$' means 'is much larger than'). We can also express this condition as $\mu \ll \alpha$. (When there is a latency period we also require that $\frac{1}{\mu} \gg \frac{1}{\theta}$, i.e. $\mu \ll \theta$, where $1/\theta$ is the expected length of the latency period.)

Exercise 2.5 The picture in Figure 11.8 gives a useful symbolic representation of the assumptions.

i) When $\mu = 0$ but $\nu > 0$, infected individuals will either always be infectious (if $\sigma = 0$) or will 'visit' the state 1 infinitely often. Each visit has an exponentially distributed duration with parameter σ. So the expected total time spent in 1 equals infinity.

ii) If you don't follow the hint, you need to do some calculations, which are presented here in self-explanatory notation.

$$\frac{dP_1}{d\tau} = -\mu P_1 - \sigma P_1 + \nu P_2, \qquad P_1(0) = 1,$$

$$\frac{dP_2}{d\tau} = -\mu P_2 + \sigma P_1 - \nu P_2, \qquad P_2(0) = 0,$$

$$\Rightarrow \frac{d}{d\tau}(P_1 + P_2) = -\mu(P_1 + P_2) \Rightarrow P_1(\tau) + P_2(\tau) = e^{-\mu\tau}.$$

Hence

$$\frac{dP_1}{d\tau} = -(\mu + \sigma + \nu)P_1 + \nu e^{-\mu\tau}, \qquad P_1(0) = 1.$$

Instead of solving this equation, we integrate it from 0 to $+\infty$ (note that one can deduce, with a bit of effort, from the equation that $P_1(\tau) \to 0$ for $\tau \to \infty$, as should be the case according to the interpretation) to obtain

$$-1 = -(\mu + \sigma + \nu) \int_0^\infty P_1(\tau) \, d\tau + \frac{\nu}{\mu}.$$

It follows readily that

$$\int_0^\infty P_1(\tau)\, d\tau = \frac{\mu + \nu}{\mu(\mu + \sigma + \nu)},$$

and the definition of h_1 is meant to imply the stated identity from this.

Exercise 2.6 Consider a continuous-time Markov chain with transition rate matrix G ($= \Sigma - D$ in the text). Then $e^{\tau G}$ is the fundamental matrix solution; see e.g. Hale (1969). When applied to the unit vector e_j, this gives the vector of probabilities to be in the various states at time τ, given that we started in state j. Now integrate with respect to τ from 0 to ∞, and use that $\int_0^\infty e^{\tau G}\, d\tau = -G^{-1}$ whenever G is defective, in the sense that $\exp(\tau G)$ decreases exponentially at ∞. So if we take the ith component of $-G^{-1}e_j$, we obtain the expected total time spent in state i, given that we started in state j. But this is exactly $(-G^{-1})_{ij}$.

Exercise 2.7 i) The scheme of Figure 11.8 is represented by the matrix

$$\Sigma - D = \begin{pmatrix} -(\mu + \sigma) & \nu \\ \sigma & -(\mu + \nu) \end{pmatrix}.$$

Hence

$$-(\Sigma - D)^{-1} = \frac{1}{(\mu + \sigma)(\mu + \nu) - \nu\sigma} \begin{pmatrix} \mu + \nu & \nu \\ \sigma & \mu + \sigma \end{pmatrix}.$$

The starting vector is $\begin{pmatrix} 1 \\ 0 \end{pmatrix}$ and the infectivity vector is $(h_1\ 0)^\top$, so

$$-h \cdot (\Sigma - D)^{-1}\Theta = \frac{h_1(\mu + \nu)}{(\mu + \sigma)(\mu + \nu) - \nu\sigma} = \frac{h_1(\mu + \nu)}{\mu(\mu + \nu + \sigma)}.$$

Another way of arriving at the same conclusion is to observe that the 11-element of $-(\Sigma - D)^{-1}$ is the expected time spent in state 1, given that one starts in state 1.

ii) As a function of ν, the factor $\int_0^\infty A(\tau)\, d\tau$ of R_0 increases from the value $\frac{h_1}{\mu + \sigma}$ at $\nu = 0$ to the limit $\frac{h_1}{\mu}$ for $\nu \to \infty$.

Exercise 2.8 Look at Figure 2.2. The matrix $\Sigma - D$ is now given by

$$\begin{pmatrix} -\sigma_1 & 0 & 0 \\ \sigma_1 & -\sigma_2 & 0 \\ 0 & \sigma_2 & -\alpha \end{pmatrix},$$

and one can use linear algebra to compute the inverse. The interpretation allows us, however, to find a more efficient procedure to calculate the relevant quantities. We start with certainty in state 1, but are also sure to arrive in state 2 (when exactly does not matter for an R_0 calculation!). The expected duration of the period spent in state 2 equals $1/\sigma_2$ and we are certain to go to state 3 from there. The expected duration of the period spent in state 3 equals $1/\alpha$. So, from the values given for h_i, we find that

$$\int_0^\infty A(\tau)\, d\tau = \frac{1}{\sigma_2} + \frac{2}{\alpha}.$$

Exercise 2.9 The force of mortality that an individual experiences is proportional to the force of infection that it exerts. Now that we are working with a state concept at the i-level, this should hold for every state separately. Taken together, this translates into $\mu = qh$.

Exercise 2.10 In the deterministic case

$$\int_\Omega f(\xi)m(d\xi) = f(\bar\xi) = f\left(\int_\Omega m(d\xi)\right)$$

for every f, since m is the unit measure concentrated at the point $\bar\xi$. So, in particular, this holds for

$$f(\xi) = \exp\left(c\int_0^\infty a(\sigma;\xi)\,d\sigma\,(z-1)\right).$$

Thus we find that

$$g(z) = e^{c\int_0^\infty a(\sigma;\bar\xi)\,d\sigma\,(z-1)},$$

which we rewrite as $g(z) = e^{R_0(z-1)}$, since, according to (2.5), $A(\tau) = a(\tau;\bar\xi)$ and, according to (2.2), $R_0 = \int_0^\infty A(\tau)d\tau$.

Exercise 2.11 i) Jensen's inequality (see e.g. Rudin (1974)) tells us that for a convex function φ we have

$$\varphi\left(\int_\Omega f(\xi)m(d\xi)\right) \le \int_\Omega (\varphi \circ f)(\xi)m(d\xi),$$

when m is a positive measure and f is integrable with respect to m. We apply this here by taking $\varphi(x) = \exp(x(z-1))$ and $f(\xi) = c\int_0^\infty a(\sigma;\xi)\,d\sigma$. By Fubini's theorem (about conditions for interchanging the order of integration, see e.g. Kolmogorov & Fomin (1975)) and (2.5), we have

$$\int_\Omega f(\xi)m(d\xi) = c\int_0^\infty \int_\Omega a(\sigma;\xi)m(d\xi)\,d\sigma = c\int_0^\infty A(\sigma)\,d\sigma = R_0,$$

so the left-hand side equals $e^{R_0(z-1)}$ while the right-hand side is, according to (2.6), equal to $g(z)$.

ii) The easiest deduction is the graphical argument in Figure 11.9, where we have drawn typical graphs of $y = g(z)$ and the graph of $y = e^{R_0(z-1)}$ and their intersection with the line $y = z$.

iii) Yes. When we repeat the 'experiment' of introducing the agent many times, the probability that we observe a major outbreak is less for the situation where there is variability in infectivity. More or less intuitively obvious: the very first infective may have low infectivity. (Once we have many cases the mean is decisive.) In the initial phase, the low-infectivity part of the distribution weighs more heavily than the high-infectivity part in determining whether or not a major outbreak occurs.

Exercise 2.12 We have

$$\frac{\partial R_0}{\partial c_1} = \frac{2c_1 N_1(c_1 N_1 + c_2 N_2) - N_1(c_1^2 N_1 + c_2^2 N_2)}{(c_1 N_1 + c_2 N_2)^2}$$

$$= \frac{(c_1^2 N_1 + c_2(2c_1 - c_2)N_2)N_1}{(c_1 N_1 + c_2 N_2)^2}.$$

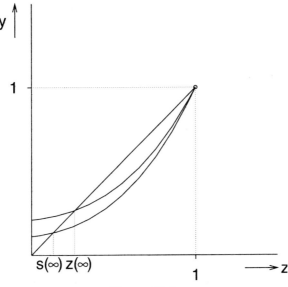

Figure 11.9

Hence the condition for $\partial R_0 / \partial c_1 < 0$ is

$$\frac{N_1}{N_2} < \left(\frac{c_2}{c_1} - 2 \right) \frac{c_2}{c_1},$$

or, equivalently,

$$\frac{c_2}{c_1} > 1 + \sqrt{1 + \frac{N_1}{N_2}}.$$

Exercise 2.13 New cases occur in the ratio

$$\frac{c_1 N_1}{c_1 N_1 + c_2 N_2} : \frac{c_2 N_2}{c_1 N_1 + c_2 N_2} = 1 : 10.$$

The ratio among those responsible for transmission can be read from the two terms in the formula for R_0, and so equals

$$\frac{c_1^2 N_1}{c_1 N_1 + c_2 N_2} : \frac{c_2^2 N_1}{c_1 N_1 + c_2 N_2} = 1 : 1.$$

Exercise 2.14 Suppose the contact pattern (i.e. the distribution of contacts over the various types) is τ-independent, but that infectivity is a τ-dependent scalar multiplication factor. Then the distribution of new cases with respect to type will depend on the contact pattern only (while the number may depend on τ), and so will be the same for the generation and real-time perspective.

11.3   Elaborations for Chapter 3

Exercise 3.1 Of course we can linearise the full system (3.3) around the 'virgin' state and compute the eigenvalues. But this kind of biological invasibility problem always

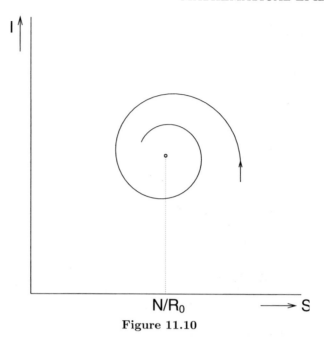

Figure 11.10

allows a short cut. The point is that the right-hand side of the differential equation for I has a factor I and a factor $\beta S - \mu - \alpha$. Linearisation amounts to replacing S by N in the second factor, and leads to the decoupled linear equation

$$\frac{dI}{dt} = (\beta N - \mu - \alpha)I.$$

Hence we have stability if $\beta N - \mu - \alpha < 0$ (equivalently: $R_0 < 1$) and instability if $\beta N - \mu - \alpha > 0$ (equivalently: $R_0 > 1$).

Exercise 3.2 For $I \neq 0$, $\frac{dI}{dt} = 0$ requires $\beta \overline{S} - \mu - \alpha = 0 \Rightarrow \overline{S} = \frac{\mu+\alpha}{\beta}$. We can rewrite this as $\frac{\overline{S}}{N} = \frac{\mu+\alpha}{\beta N} = \frac{1}{R_0}$. The same observation also shows that $S = \frac{\mu+\alpha}{\beta} = \frac{N}{R_0}$ is an isocline, so along orbits the variable I takes its maxima and minima on this line. In particular, a steady state has to lie on this line. This is not at all surprising, since in steady state a case has to produce, on average, one secondary case and the expected number of secondary cases is R_0 multiplied by the reduction factor $\overline{S}/N$, which takes into account that only a fraction $\overline{S}/N$ of contacts is with susceptibles. See also Figure 11.10, where we have drawn part of a trajectory of the system (3.3).

Note that if we can estimate $\overline{S}/N$ (from blood samples taken at random), we can estimate $R_0 = N/\overline{S}$.

Exercise 3.3 Putting $\frac{dS}{dt} = 0$ in (3.3) we find

$$\overline{I} = \frac{B - \mu\overline{S}}{\beta\overline{S}} = \frac{\mu}{\beta}\left(\frac{B}{\mu\overline{S}} - 1\right) = \frac{\mu}{\beta}\left(\frac{N}{\overline{S}} - 1\right) = \frac{\mu}{\beta}(R_0 - 1).$$

Exercise 3.4 We calculate the Jacobi matrix and evaluate the elements for $S = \overline{S}$

and $I = \bar{I}$:

$$\frac{\partial}{\partial S}(B - \beta SI - \mu S) = -\beta I - \mu = -\beta \bar{I} - \mu,$$

$$\frac{\partial}{\partial I}(B - \beta SI - \mu S) = -\beta S = -\beta \bar{S},$$

$$\frac{\partial}{\partial S}((\beta S - \mu - \alpha)I) = \beta I = \beta \bar{I},$$

$$\frac{\partial}{\partial I}((\beta S - \mu - \alpha)I) = \beta S - \mu - \alpha = 0.$$

The corresponding matrix

$$\begin{pmatrix} -\beta \bar{I} - \mu & -\beta \bar{S} \\ \beta \bar{I} & 0 \end{pmatrix}$$

has trace $T = -\beta \bar{I} - \mu < 0$ and determinant $D = \beta^2 \overline{SI} > 0$, and so both roots of the characteristic equation $\lambda^2 - T\lambda + D = 0$ have negative real part (see e.g. Hale (1969)). So both eigenvalues of the matrix have negative real part and, according to the principle of linearised stability, the endemic steady state is (locally asymptotically) stable.

Exercise 3.5 As we saw in the preceding exercise, the characteristic equation is $\lambda^2 + (\beta \bar{I} + \mu)\lambda + \beta^2 \overline{SI} = 0$. Dividing by μ^2 and using (3.4)-(3.6), we rewrite this as

$$\left(\frac{\lambda}{\mu}\right)^2 + (R_0 - 1 + 1)\frac{\lambda}{\mu} + (R_0 - 1)\frac{\beta}{\mu}\frac{N}{R_0},$$

$$= \left(\frac{\lambda}{\mu}\right)^2 + R_0\frac{\lambda}{\mu} + \frac{\alpha + \mu}{\mu}(R_0 - 1) = 0.$$

When $\frac{1}{\mu} \gg \frac{1}{\alpha}$, we have $\frac{\alpha}{\mu} \gg 1$, and consequently we approximate the last term by $\frac{\alpha}{\mu}(R_0 - 1)$. The equation

$$y^2 + R_0 y + \frac{\alpha}{\mu}(R_0 - 1) = 0$$

has roots

$$y = \frac{-R_0 \pm \sqrt{R_0^2 - 4\frac{\alpha}{\mu}(R_0 - 1)}}{2}.$$

Using once more that $\frac{\alpha}{\mu} \gg 1$, we see that the expression under the square root is negative (we are considering the endemic steady state, so implicitly we have assumed that $R_0 > 1$) and that the roots are, in the first approximation,

$$\frac{\lambda}{\mu} = y = -\frac{R_0}{2} \pm i\sqrt{\frac{\alpha}{\mu}(R_0 - 1)}.$$

So the relaxation time $\frac{1}{|\text{Re}\lambda|}$ equals $\frac{2}{\mu R_0}$, while the frequency equals $\sqrt{\alpha\mu(R_0 - 1)}$, both in first approximation with respect to the small parameter $\frac{\mu}{\alpha}$. For $\mu \ll \alpha$, the relaxation time is of the order of $\frac{1}{\mu}$ but the period is of the order, of $\frac{1}{\sqrt{\mu}}$, so the ratio between the two goes to infinity for $\mu \downarrow 0$. This means that we shall see many oscillations while the deviation from steady state is damping out.

Exercise 3.6 In by now familiar notation, we have

$$\frac{dP_S}{da} = -(\Lambda + \mu)P_S, \qquad P_S(0) = 1,$$

$$\frac{dP_I}{da} = \Lambda P_S - \cdots.$$

From the first equation, we have $P_S(a) = e^{-(\mu+\Lambda)a}$, and consequently the probability per unit time that the individual is infected at age a equals $\Lambda P_S(a) = \Lambda e^{-(\mu+\Lambda)a}$. Hence

$$\bar{a} = \frac{\int_0^\infty a\Lambda P_S(a)\, da}{\int_0^\infty \Lambda P_S(a)\, da} = \frac{1}{\mu + \Lambda}.$$

Since $\Lambda = \beta\bar{I}$, we have, by (3.6), that $\Lambda = \mu(R_0 - 1)$ and so $\bar{a} = \frac{1}{\mu R_0}$.

Exercise 3.8 For $\bar{a}$ ranging from 4 to 5 the period ranges, when computed from the formula, from 2.28 to 2.55 years. This is the right order of magnitude and, in fact, surprisingly close to the observed period of two years. (You should realise that seasonal effects, such as weather and the school system, force the actual period to be an integer, while in our model no such restriction or effect has been built in).

Exercise 3.9 This is really an exercise for people who like to do calculations (and the authors have to admit that, consequently, it is not one of their favourites).

i) This is indeed already done: Exercise 2.4-iv.

ii) Putting $\frac{dI}{dt} = 0$, we find $\bar{E} = \frac{\alpha+\mu}{\theta}\bar{I}$, while $\frac{dE}{dt} = 0$ leads to $\beta\overline{SI} = (\mu + \theta)\bar{E}$. Combining these two relations, we obtain, after dividing out a factor $\bar{I}$ (remember we are determining the endemic steady state, so we are not interested in $\bar{I} = 0$) that $\beta\bar{S} = \frac{1}{\theta}(\mu + \theta)(\alpha + \mu)$ or, equivalently, $\bar{S} = N/R_0$. From $\frac{dS}{dt} = 0$, we find that $\bar{I} = \frac{B-\mu\bar{S}}{\beta\bar{S}} = \frac{\mu}{\beta}(R_0 - 1)$.

iii) The Jacobi matrix is derived analogously to the derivation in Exercise 3.4. To deduce the characteristic equation from the matrix is straightforward but tedious.

iv) This is an invitation, not an exercise!

Exercise 3.10 ii) In steady state, $\bar{\Lambda} = \frac{c}{N}\int_0^\infty A(\tau)\, d\tau\, \overline{\Lambda S} = \frac{R_0}{N}\overline{\Lambda S}$, so if $\bar{\Lambda} \neq 0$ then necessarily $\bar{S} = \frac{N}{R_0}$.

iii) The variation-of-constants formula says that the differential equation $\frac{dx}{dt}(t) = m(t)x(t) + f(t)$ has as a solution that satisfies the initial condition $x(t_0) = x_0$ the expression

$$x(t) = e^{\int_{t_0}^t m(\sigma)\, d\sigma}x_0 + \int_{t_0}^t e^{\int_\tau^t m(\sigma)\, d\sigma}f(\tau)\, d\tau.$$

By letting $t_0 \to -\infty$, while assuming that x tends to zero at $-\infty$, we obtain the limiting form

$$x(t) = \int_{-\infty}^t e^{\int_\tau^t m(\sigma)\, d\sigma}f(\tau)\, d\tau,$$

which is what we applied here, with f the constant function assuming the value B, with $m(t) = -\mu - \Lambda(t)$ and $x(t) = S(t)$.

iv) In an endemic steady state we must have $B - \mu\overline{S} - \overline{\Lambda}\overline{S} = 0$, which gives $\overline{\Lambda} = \frac{B - \mu\overline{S}}{\overline{S}} = \mu(R_0 - 1)$. If we identify $\overline{\Lambda}$ with $\beta\overline{I}$ in (3.6), this is exactly the same identity.

v) To derive the equation for $y(t)$, we only have to expand

$$\Lambda(t - \tau)S(t - \tau) = \overline{\Lambda}\overline{S} + \overline{\Lambda}x(t - \tau) + \overline{S}y(t - \tau) + \text{h.o.t.}$$

('h.o.t.' means 'higher-order terms'). To derive the equation for $x(t)$, write

$$e^{-\int_{t-\sigma}^{t} \Lambda(s)\, ds} = e^{-\overline{\Lambda}\sigma} e^{-\int_{t-\sigma}^{t} y(s)\, ds} = e^{-\overline{\Lambda}\sigma}\left(1 - \int_{t-\sigma}^{t} y(s)\, ds + \text{h.o.t.}\right)$$

and apply partial integration to arrive at the form as stated.

If we substitute as an Ansatz the given expressions for $x(t)$ and $y(t)$, we find, after dividing out a factor $\exp(\lambda t)$, the system of equations

$$x_0 = -\frac{N}{R_0}\int_0^\infty e^{-(\mu R_0 + \lambda)\sigma}\, d\sigma\, y_0,$$

$$y_0 = \frac{c}{R_0}\int_0^\infty A(\tau)e^{-\lambda\tau}\, d\tau\, y_0 + \frac{c}{N}\mu(R_0 - 1)\int_0^\infty A(\tau)e^{-\lambda\tau}\, d\tau\, x_0,$$

which is, for given λ, a homogeneous linear system of two equations in two unknowns. So, in order for there to be a non-trivial solution, the corresponding matrix should be singular. The characteristic equation is precisely the condition for the matrix to be singular.

vi) On the one hand, for $R_0 = 1$ the characteristic equation reads

$$1 = c\overline{A}(\lambda).$$

On the other hand, we can rewrite $R_0 = 1$ as $c\overline{A}(0) = 1$. So $\lambda = 0$ is a root. When λ is restricted to be real, $\overline{A}$ is a monotonically decreasing function of λ, so in particular $\overline{A}(\lambda) < 1$ for $\lambda > 0$. Now suppose $c\overline{A}(\lambda) = 1$. Then necessarily $\text{Im}\overline{A}(\lambda) = 0$ and

$$c\overline{A}(\lambda) = c\int_0^\infty e^{-\text{Re}\lambda\tau}\cos(\text{Im}\lambda\tau)\, A(\tau)\, d\tau.$$

But then in fact, under very mild conditions on A, we have that $c|\overline{A}(\lambda)| < c\overline{A}(\text{ Re }\lambda)$ if $\text{Im }\lambda \neq 0$, which implies that no root can lie in the right half-plane or on the imaginary axis (other than $\lambda = 0$).

vii) The monotonicity of $\lambda \mapsto \overline{A}(\lambda)$ for real λ shows that the root $\lambda = 0$ must shift to the left when we increase R_0 a bit. So, for R_0 slightly bigger than R_0, all roots lie in the left half-plane.

viii) In fact $\overline{A}(\lambda) \to 0$ if $|\lambda| \to \infty$ while λ is restricted to the right half-plane. This is immediately clear if $\text{Re }\lambda \to +\infty$ and follows from the Riemann-Lebesgue lemma (see e.g. Rudin (1974); the key idea behind this lemma is that oscillations become so rapid that positive and negative contributions cancel each other) if $\text{Re }\lambda$ stays bounded.

ix) We argue by contradiction. That is, we assume that $\lambda = i\omega$ is a root. Then necessarily

$$1 = c\frac{|\overline{A}(i\omega)|}{R_0}\left|\frac{i\omega + \mu}{i\omega + \mu R_0}\right|.$$

But in fact the two inequalities

$$c\left|\overline{A}(i\omega)\right| = c\left|\int_0^\infty A(\tau)e^{-i\omega\tau}\,d\tau\right| \leq c\int_0^\infty A(\tau)\left|e^{-i\omega\tau}\right|\,d\tau \leq R_0$$

and

$$\left|\frac{i\omega + \mu}{i\omega + \mu R_0}\right| = \frac{\omega^2 + \mu^2}{\omega^2 + \mu^2 R_0^2} < 1$$

show that the right-hand side is less than one—a contradiction.

x) This is a conclusion that follows by combining vii), viii) and ix) with the principle of linearised stability and the exponential decay of solutions of linear equations when all roots of the characteristic equation lie strictly to the left of the imaginary axis.

Exercise 3.11 If we calculate the derivative of V with respect to t we obtain

$$\begin{aligned}
\frac{dV}{dt} &= \frac{\partial V}{\partial S}\frac{dS}{dt} + \frac{\partial V}{\partial I}\frac{dI}{dt} \\
&= \left(1 - \frac{\overline{S}}{S}\right)(B - \beta SI - \mu S) + \left(1 - \frac{\overline{I}}{I}\right)I(\beta S - \alpha - \mu) \\
&= (S - \overline{S})\left(\frac{B}{S} - \beta I - \mu + \beta I - \beta\overline{I}\right) \\
&= (S - \overline{S})\left(\frac{B}{S} - \frac{B}{\overline{S}}\right) \\
&= -\frac{B}{S\overline{S}}(S - \overline{S})^2 = -\frac{\mu R_0}{S}\left(S - \frac{N}{R_0}\right)^2,
\end{aligned}$$

where we have used $\overline{S} = \frac{\alpha + \mu}{\beta}$ and $B - \beta\overline{S}\overline{I} - \mu\overline{S} = 0$. We conclude that $\frac{dV}{dt} < 0$ except on the line $S = \frac{N}{R_0}$, where $\frac{dV}{dt} = 0$. At this line (which is an I-nullcline) we have $\frac{dI}{dt} = 0$ and $\frac{dS}{dt} = B - (\mu + \beta I)\frac{N}{R_0}$. So clearly $\frac{dS}{dt} > 0$ for $I < \overline{I}$ and $\frac{dS}{dt} < 0$ for $I < \overline{I}$ (with $\overline{I}$ the steady state value for I given explicitly by (3.6)). We conclude that orbits cannot 'stay' on the line, unless we consider the steady state. The LaSalle invariance principle (see Hale (1969)), which is a very useful generalisation of Lyapunov's stability theorem, then allows us to conclude that all orbits that stay bounded do converge to the steady state. The boundedness of orbits, on the other hand, is a direct consequence of our modelling assumptions (a constant population birth rate, while the per capita death rate is constant). Mathematically this is reflected in the invariance of the region $\{(S, I) : S \geq 0, I \geq 0 \text{ and } S + I \leq N\}$ (note that $\frac{d}{dt}(S + I) = B - \mu(S + I) - \alpha I \leq B - \mu(S + I)$).

Exercise 3.12 We first explain 'Hence $\pi_j = \frac{\tau_j}{L}$'. The point is that the fractions π_j sum to one and that the constant of proportionality does not depend on j (this is somewhat implicit in the formulation of the microcosm principle), so the mean sojourn times should sum to L. The strength of the principle is that it yields relationships of a general nature that are independent of the details of the model specification. That we apply it here to the simple models is meant as a training in the use of the principle. Once you understand it, you may use it in far more complex situations where explicit calculations are often cumbersome (to put it mildly). Before starting with the elaboration, we make one more remark: sometimes it is helpful to number the i-states, and so j refers to a number, at other times it is helpful to indicate i-states by letters that carry information, and then j may take 'values' like S, I, E. This should not lead to confusion.

For the system (3.3) we have $\pi_S = \frac{\bar{S}}{N} = \frac{1}{R_0}$ while $\tau_S = \frac{1}{\mu + \beta \bar{I}} = \frac{1}{\mu + \mu(R_0 - 1)} = \frac{1}{\mu R_0}$, so indeed $\pi_S = \frac{\tau_S}{L}$. Next consider $j = I$ and observe that $\pi_I = \frac{\bar{I}}{N} = \frac{\mu(R_0 - 1)}{\beta N} = \frac{\mu(R_0 - 1)}{(\alpha + \mu) R_0}$. Now comes a subtlety. We have to distinguish between the mean time spent as an infective and the mean time spent as an infective, *given* that the individual does indeed become infected. The latter equals $\frac{1}{\mu + \alpha}$, while the former equals $\frac{\beta \bar{I}}{\mu + \beta \bar{I}} \frac{1}{\mu + \alpha}$, where the first factor is the probability that a newborn susceptible individual is infected during its life (see Exercise 1.33-vi for explanation, if needed). In the microcosm principle as formulated in the exercise the quantity τ_I refers to a 'mean' computed over newborn individuals. So we should check that

$$\frac{1}{L} \frac{\beta \bar{I}}{\mu + \beta \bar{I}} \frac{1}{\mu + \alpha} = \pi_I,$$

which is straightforward (use (3.6)).

Next consider the system (3.11). Again $\pi_S = \frac{\bar{S}}{N} = \frac{1}{R_0}$ while $\tau_S = \frac{1}{\mu + \beta \bar{I}} = \frac{1}{\mu R_0}$, so $\pi_S = \frac{\tau_S}{L}$. Concerning the i-state E we have

$$\pi_E = \frac{\bar{E}}{N} = \frac{\alpha + \mu}{\theta} \frac{\bar{I}}{N} = \frac{\alpha + \mu}{\theta} \frac{\mu(R_0 - 1)}{\beta N},$$

which is indeed equal to $\frac{\beta \bar{I}}{\mu + \beta \bar{I}} \frac{1}{\theta + \mu} \mu$, since $\beta \bar{I} = \mu(R_0 - 1)$. Finally, $\pi_I = \frac{\bar{I}}{N} = \frac{\mu(R_0 - 1)}{\beta N}$. To arrive in state I, two conditions should be met, so we have two factors

$$\tau_I = \frac{\beta \bar{I}}{\mu + \beta \bar{I}} \frac{\theta}{\theta + \mu} \frac{1}{\mu + \alpha}.$$

Again the identity $\beta \bar{I} = \mu(R_0 - 1)$ is now all that is needed to verify that $\pi_I = \frac{\tau_I}{L}$.

Exercise 3.13 The underlying idea is that, quite generally under steady-state conditions *rate of inflow × expected sojourn time = amount*. Three remarks are in order: i) it does not matter whether we measure 'amount' in numbers or in density, as long as we compute the rate in the corresponding units; ii) when individuals can return to the state under consideration, like to the S-state in an SIS model, we should take the sojourn time of a *single* visit; iii) steady state is too strong a condition, in periodic conditions one can formulate an analogue in terms of averages over the period (but in growing populations the rate of population growth has to be taken into account).

In this exercise we have to check that

$$\frac{\beta \overline{SI}}{\bar{I}} = \alpha + \mu,$$

which is indeed the case (combine (3.5) and (3.4)).

Exercise 3.14 We consider the system

$$\frac{dS}{dt} = (1-q)B - \mu S - \beta SI,$$

$$\frac{dI}{dt} = -\mu I + \beta SI - \alpha I,$$

but we keep defining R_0 as the expected number of secondary cases per primary case in the unvaccinated population, that is, we still take

$$R_0 = \frac{\beta N}{\alpha + \mu} = \frac{\beta B}{\mu(\alpha + \mu)}.$$

Putting $\frac{dI}{dt} = 0$, we find that in endemic steady state

$$\overline{S} = \frac{\mu + \alpha}{\beta} = \frac{N}{R_0},$$

so $\overline{S}/N$ does not change at all (which is not surprising, since the argument given in Exercise 3.2 still applies). Putting $\frac{dS}{dt} = 0$, we find, with a bit of manipulation,

$$\overline{I} = \frac{\mu}{\beta}((1-q)R_0 - 1).$$

So the force of infection $\Lambda = \beta\overline{I} = \mu((1-q)R_0 - 1)$ and the mean age at infection $\overline{a}$ (see Excercise 3.6) is given by

$$\overline{a} = \frac{1}{\mu + \Lambda} = \frac{1}{(1-q)\mu R_0}.$$

We see that the force of infection decreases and the mean age at infection increases (by a factor $1/(1-q)$) with q. For rubella this is a serious complicating factor that has to be taken into consideration when comparing various vaccination strategies.

The linearisation around the steady state has the characteristic equation of Exercise 3.5 with everywhere R_0 replaced by $(1-q)R_0$. Our analysis of Exercise 3.5 remains valid as long as $(1-q)R_0 > 1$. Both the relaxation time and the period of the oscillations increase with increasing q.

Exercise 3.15 In general, if the incidence of a lethal disease is substantial, the demography will be influenced (e.g. AIDS in part of Africa), and one should not model the inflow of newborn susceptibles as a constant.

Exercise 3.16 We number the years with the integers and denote by $n-$ the moment just before the reproduction event and by $n+$ the moment just after the reproduction event. We first describe the demography in as simple a manner as we can (constant number of newborns, fixed probability to survive one year):

$$S(n+) = S(n-) + B,$$
$$S((n+1)-) = S(n+)e^{-\mu},$$

which combine into

$$S((n+1)+) = S(n+)e^{-\mu} + B,$$

and show convergence towards the globally stable steady state

$$\overline{S}_+ = \frac{B}{1 - e^{-\mu}};$$

make a cobweb picture (see e.g. Edelstein-Keshet (1988)) for yourself, plotting in the (x, y) plane the lines $y = x$ and $y = e^{-\mu}x + B$. Next we put the infectious agent onto the stage. The reproduction event is now described by

$$
\begin{aligned}
S(n+) &= S(n-) + B, \\
I(n+) &= I(n-),
\end{aligned}
$$

which means that all newborns are susceptible (no vertical transmission) and that the disease has no influence on fertility. To model the infection process in between reproduction events, we introduce the continuous time variable t, which runs from 0 to 1 (and so which is reset to zero at every reproduction event; in other words, t measures time within a year just as a calendar does), and suppose that

$$
\begin{aligned}
\frac{dS}{dt} &= -\mu S - \beta SI, \\
\frac{dI}{dt} &= -\mu I + \beta SI - \alpha I.
\end{aligned}
$$

The notion of R_0 does not make sense in this setting, since even when we have a stroboscopic steady state $\overline{S}_+$, the quantity S depends on time t and is not constant. Yet we can easily derive an invasibility criterion (by which we mean a criterion that decides whether or not the agent will spread when initially rare). To this end, we consider

$$\frac{dI}{dt} = (\beta \overline{S}(t) - \mu - \alpha)I,$$

with

$$\overline{S}(t) = \frac{B}{1 - e^{-\mu}} e^{-\mu t},$$

and note that I will increase from year to year (when we take stock at the reproduction time) if and only if

$$\beta \int_0^1 \overline{S}(t)\, dt - \mu - \alpha > 0$$

or, equivalently,

$$\frac{\beta B}{\mu(\mu + \alpha)} > 1.$$

It is possible to do some more work on this problem (but in the limited amount of time devoted to it, the authors did not arrive at clear and meaningful conclusions). We refer to Roberts & Kao (1998)[2] for related work. Moreover, we mention that a version of the numerical continuation and bifurcation package CONTENT (Kuznetsov

---

[2] M.G. Roberts & R.R. Kao: The dynamics of an infectious disease in a population with birth pulses. *Math. Biosci.*, **149** (1988), 23-36.

& Levitin $(1997)^3$) exists that can deal with hybrid models. We fear that analytical methods are not powerful enough and that one needs such computational tools to proceed, and we therefore advocate their use. Likewise, we advocate hybrid models: we think they are unjustly neglected in the literature.

Exercise 3.17 Consider

$$\frac{dS}{dt} = -\beta SI + \gamma I,$$
$$\frac{dI}{dt} = \beta SI - \gamma I.$$

Then $\frac{d}{dt}(S+I) = 0$; so $S + I = \text{constant} = N$, and we can eliminate one variable, for instance S, and deduce

$$\frac{dI}{dt} = (\beta N - \gamma - \beta I)I,$$

which is the famous logistic equation. An endemic steady state exists provided that

$$R_0 = \frac{\beta N}{\gamma} > 1,$$

and is globally stable.

When there is temporary immunity we may put

$$\frac{dS}{dt} = -\beta SI \qquad + \gamma R,$$
$$\frac{dI}{dt} = \beta SI - \alpha I,$$
$$\frac{dR}{dt} = \qquad +\alpha I \quad - \gamma R,$$

for which $S + I + R = \text{constant} = N$. The system can therefore be reduced to the two-dimensional system

$$\frac{dS}{dt} = -\beta SI \qquad + \gamma(N - S - I),$$
$$\frac{dI}{dt} = \beta SI - \alpha I.$$

We find a positive endemic state

$$(\overline{S}, \overline{I}) = \left(\frac{\alpha}{\beta}, \frac{\gamma(N - \frac{\alpha}{\beta})}{\gamma + \alpha} \right),$$

provided that $R_0 = \frac{\beta N}{\alpha} > 1$. The corresponding Jacobi matrix

$$\begin{pmatrix} -\beta\overline{I} - \gamma & -\beta\overline{S} - \gamma \\ \beta\overline{I} & 0 \end{pmatrix}$$

---

[3] Y.A. Kuznetsov & V.V. Levitin (1997): CONTENT. A multiplatform environment for analysing dynamical systems. Dynamical Systems Laboratory, Centre for Mathematics and Computer Science, Amsterdam, ftp.cwi.nl/pub/CONTENT

has negative trace and positive determinant, so the endemic state is (locally asymptotically) stable.

Exercise 3.18 We consider the following system:

$$\frac{dS}{dt} = B\frac{S + (1-\theta)I + R}{N} - \beta SI - \mu S,$$

$$\frac{dI}{dt} = \theta B\frac{I}{N} + \beta SI - \mu I - \alpha I,$$

$$\frac{dR}{dt} = -\mu R + \alpha R,$$

the idea being that a fraction θ of the offspring of an infectious individual is infected at birth. Here we use both B and N in the formulation. They are not independent parameters, however, but rather related to each other by $B = \mu N$. There are now two different transmission routes, and accordingly the interpretation yields the expression

$$R_0 = \frac{\beta N}{\mu + \alpha} + \frac{\theta B}{(\mu + \alpha)N} = \frac{\beta N}{\mu + \alpha} + \frac{\theta \mu}{\mu + \alpha},$$

which is the sum of two contributions. Likewise, we find, by looking at the equation for dI/dt, that

$$r = \frac{\theta B}{N} + \beta N - \mu - \alpha.$$

It follows from $\frac{d}{dt}(S+I+R) = B - \mu(S+I+R)$ that the set $\{(S,I,R) : S+I+R = N\}$ is attracting, and so we restrict our further analysis to that set. This means that we study the system

$$\frac{dS}{dt} = B(1 - \theta\frac{I}{N}) - \beta SI - \mu S,$$

$$\frac{dI}{dt} = \theta B\frac{I}{N} + \beta SI - \mu I - \alpha I;$$

that is

$$\frac{dS}{dt} = B - \theta\mu I - \beta SI - \mu S,$$

$$\frac{dI}{dt} = \theta\mu I + \beta SI - \mu I - \alpha I.$$

The endemic state has $\overline{S} = \frac{\mu + \alpha - \theta\mu}{\beta}$ (put $dI/dt = 0$), and $\overline{I} = \frac{B - \mu\overline{S}}{\theta\mu + \beta\overline{S}}$ (put $dS/dt = 0$) with corresponding Jacobi matrix

$$\begin{pmatrix} -\beta\overline{I} - \mu & -\theta\mu - \beta\overline{S} \\ \beta\overline{I} & 0 \end{pmatrix},$$

which has negative trace and positive determinant (provided $\overline{S} > 0$ and $\overline{I} > 0$, that is, provided $R_0 > 1$). So the endemic state is (locally asymptotically) stable. We refer to Busenberg & Cooke (1993) for a wealth of information on vertical transmission and its modelling.

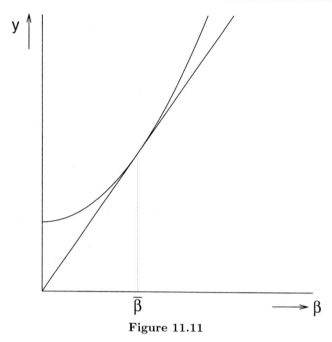

Figure 11.11

Exercise 3.19 This exercise touches the surface of an important and not yet fully explored problem area. See Section 3.5 and the references given there.

We first note that (from (3.5)) $\frac{\overline{S}}{N} = \frac{1}{R_0^1}$ and next that, consequently, the expected number of secondary cases per primary case of infection by agent 2 will be $\frac{\overline{S}}{N} R_0^2 = \frac{R_0^2}{R_0^1}$, which will exceed one iff $R_0^2 > R_0^1$.

The next question is: if $R_0 = \frac{\beta N}{\alpha + \mu}$ (see (3.4)), with β a free parameter, but α related to β as depicted in Figure 3.1, for what value of β will R_0 be maximal? A straightforward computation shows that $dR_0/d\beta = 0$ is equivalent to $\alpha + \mu = \frac{d\alpha}{d\beta}\beta$. Now note that in general the tangent line (see Figure 11.11, where we have drawn the graph of $y = \alpha(\beta) + \mu$) to the graph of $y = \alpha(\beta) + \mu$ at the point $(\overline{\beta}, \alpha(\overline{\beta}) + \mu)$ is given by $y = \alpha(\overline{\beta}) + \mu + \frac{d\alpha}{d\beta}(\overline{\beta})(\beta - \overline{\beta})$ and that this tangent line runs through the origin iff $\alpha(\overline{\beta}) + \mu = \frac{d\alpha}{d\beta}(\overline{\beta})\overline{\beta}$.

So the graphical criterion for extremality of R_0 is that the tangent line runs through the origin. Graphically, we also see that the extremum is a maximum: on the straight tangent line we have that $\frac{\beta}{y}$ is constant, say c, and the fact that the graph lies above the line for $\beta \neq \overline{\beta}$ implies the inequality $\frac{\beta}{\alpha(\beta) + \mu} < c$ for $\beta \neq \overline{\beta}$, with equality for $\beta = \overline{\beta}$.

Within this setting the statement is correct. But in fact we have forced the statement to be true by assuming the trade-off between α and β as depicted in Figure 11.11. So the real question is: does such a trade-off exist? What is its mechanistic underpinning?

Exercise 3.20 i) If we assume that, for a certain fixed density, the number of 'contacts' per individual is constant, we should use $\gamma\frac{SI}{N}$ when the variables refer to numbers. When, in contrast, we assume that the curiosity of mice is such that they explore every corner of the cage, no matter how large it is, we should take βSI. We think that the last assumption is not warranted and that the first gives a more reasonable description until we go down to a really small cage (i.e. a really small number).

ii) From $\frac{dI}{dt} = 0$ we find that $\overline{S} = \frac{\mu+\alpha}{\beta}$, and from $\frac{dR}{dt} = 0$ that $\overline{R} = \frac{f\alpha}{\mu}\overline{I}$. Now put $\frac{dS}{dt} = 0$, insert the expression for $\overline{S}$ and solve for $\overline{I}$. This yields $\overline{I} = \frac{B}{\mu+\alpha} - \frac{\mu}{\beta}$. All that remains to be done is to rearrange the resulting expression for $\overline{S} + \overline{I} + \overline{R}$ a little.

iii) This computation is slightly more complicated than the preceding one. Putting $\frac{dI}{dt} = 0$, we find $\overline{S} = \frac{\mu+\alpha}{\gamma}\overline{N}$, while $\frac{dR}{dt} = 0$ yields $\overline{R} = \frac{f\alpha}{\mu}\overline{I}$, and $\frac{dI}{dt} = 0$ yields $\overline{I} = \frac{B}{\mu+\alpha} - \frac{\mu}{\mu+\alpha}\overline{S}$. This last equation, together with $\overline{S} = \frac{\mu+\alpha}{\gamma}\overline{N} = \frac{\mu+\alpha}{\gamma}(\overline{S}+\overline{I}+\frac{f\alpha}{\mu}\overline{I})$, can be considered as a system of two linear equations with two unknowns, $\overline{S}$ and $\overline{I}$. Standard manipulation to eliminate $\overline{I}$ then gives one equation for $\overline{S}$, with solution

$$\overline{S} = \left(1 - (1-f)\frac{\alpha}{\gamma}\right)^{-1} \frac{B}{\gamma}\left(1 + \frac{f\alpha}{\mu}\right).$$

Rather than computing $\overline{I}$ and $\overline{R}$ from this, we now recall that $\overline{N} = \frac{\gamma}{\mu+\alpha}\overline{S}$ and find the expression for $\overline{N}$ that is stated in the text.

Exercise 3.21 We assume that in between vaccination events, the susceptible density develops according to

$$\frac{dS}{dt} = B - \mu S$$

when the population is free from infection. Hence

$$S(t) = N + (S_0 - N)e^{-\mu t},$$

with N the total population size, i.e. $N = B/\mu$. If we vaccinate a fraction p of the susceptible population at time T, we have that, after vaccination,

$$S(T) = (1-p)\left\{N + (S_0 - N)e^{-\mu T}\right\}.$$

So the infection-free dynamics is described by the recurrence relation

$$S(nT) = (1-p)\left\{N + (S((n-1)T) - N)e^{-\mu T}\right\},$$

which has the attracting fixed point

$$\overline{S} = \frac{(1-p)N(1-e^{-\mu T})}{1-(1-p)e^{-\mu T}}$$

(found by solving the fixed-point equation that is obtained by putting both $S(nT)$ and $S((n-1)T)$ equal to $\overline{S}$). In between two jumps, we accordingly then have

$$S(t) = N + (\overline{S} - N)e^{-\mu t}, \quad (n-1)T \le t < nT,$$

with average value

$$\frac{1}{T}\int_0^T S(t)\,dt = N + (\overline{S} - N)\frac{1 - e^{-\mu T}}{\mu T}.$$

We now first compute the effort E. Per event, there are

$$p\left\{N + (\overline{S} - N)e^{-\mu T}\right\}$$

vaccinations, and to obtain E we have to divide this amount by T:

$$E = \frac{p}{T}\left\{N + (\overline{S} - N)e^{-\mu T}\right\}.$$

Now rewrite the fixed-point equation as

$$(\overline{S} - N)(1 - e^{-\mu T}) = -p\left\{N + (\overline{S} - N)e^{-\mu T}\right\},$$

and find

$$E = (N - \overline{S})\frac{1 - e^{-\mu T}}{T}.$$

Hence

$$\frac{1}{T}\int_0^T S(t)\, dt = N - \frac{E}{\mu},$$

which shows that the magnitude of the critical quantity $\frac{1}{T}\int_0^T S(t)\, dt$ depends on the vaccination effort E, but *not* on the vaccination period T. So, in particular, the vaccination effort required for elimination is independent of T, and is given by

$$E_{crit} = B\left(1 - \frac{1}{R_0}\right).$$

When we vaccinate a fraction q of all newborns, the vaccination effort is given by qB, and the condition for elimination $q > 1 - 1/R_0$ of Exercise 1.36 can be reformulated as

$$E = pB > B\left(1 - \frac{1}{R_0}\right).$$

We conclude that, for a constant transmission-rate constant β, pulse vaccination yields no advantage whatsoever. Of course, this may be different when, for instance, β is periodic and when the susceptible density fluctuates by other causes (and not only due to the vaccination campaigns). Indeed, the point is to level off the peaks in susceptible density as much as possible. We refer to Shulgin et al. (1998).[4]

Exercise 3.22 By adding the equations, we arrive at

$$\begin{aligned}
\dot{N} &= bS + bR - \mu N - (1 - f)\alpha I \\
&\Rightarrow \dot{N} = bN - bI - \mu N - (1 - f)\alpha I \\
&\Rightarrow \dot{N} = \{b - \mu - (b + \alpha(1 - f))y\}N.
\end{aligned}$$

Next we find, using $\frac{S}{N} = 1 - y - z$, that

$$\begin{aligned}
\dot{y} &= \frac{\dot{I}}{N} - \frac{I}{N}\frac{\dot{N}}{N} = \gamma(1 - y - z)y - \mu y - \alpha y - y\{b(1 - y - z) + bz - \mu - (1 - f)\alpha y\} \\
&\Rightarrow \dot{y} = y\{\gamma(1 - y - z) - \alpha + \alpha(1 - f)y + b(y - 1)\}
\end{aligned}$$

---

[4] B. Shulgin, L. Stone & Z. Agur: Pulse vaccination strategy in the SIR epidemic model. *Bull. Math. Biol.*, **60** (1988), 1123-1148.

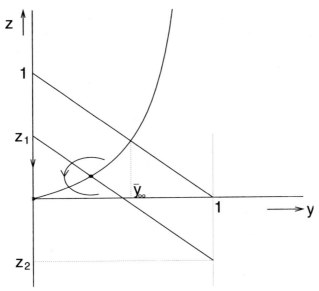

Figure 11.12

and

$$\dot{z} = \frac{\dot{R}}{N} - \frac{R}{N}\frac{\dot{N}}{N} = -\mu z + f\alpha y - z\{b(1 - y - z) + bz - \mu - (1 - f)\alpha y\}$$
$$\Rightarrow \quad \dot{z} = y\{f\alpha + \alpha(1 - f)z\} - bz(1 - y).$$

Exercise 3.23 Consider Figure 11.12, where we have drawn, for the system (3.14), the isoclines at which $\dot{y} = 0$ and isoclines at which $\dot{z} = 0$, as well as part of a trajectory.

The isoclines at which $\dot{y} = 0$ are $y = 0$ and $\gamma z = \gamma - \alpha - b + \{b - \gamma + \alpha(1 - f)\}y$, which is a straight line intersecting the line $y = 0$ at $z = z_1 = 1 - \frac{\alpha + b}{\gamma}$ and the line $y = 1$ at $z = z_2 = -\frac{\alpha f}{\gamma}$. Note that $z_1 < 1$ and $z_1 > 0$ iff $R_{0,\text{relative}} = \frac{\gamma}{\alpha + b} > 1$, while always $z_2 < 0$.

The triangle $\{(y, z) : 1 - y - z \geq 0, y \geq 0, z \geq 0\}$ is invariant (as it should be because of the interpretation!) since the z axis is invariant; at the y axis we have $\frac{dz}{dt} = f\alpha y > 0$, and on the line $z + y = 1$ we have (after writing out and simplifying)

$$\frac{d}{dt}(y + z) = b(y - 1) < 0 \qquad \text{for } y < 1.$$

The isocline at which $\dot{z} = 0$ is given by

$$z = \frac{f\alpha}{b\frac{1-y}{y} - (1 - f)\alpha},$$

which is an increasing function of y (since $(1 - y)/y$ is a decreasing function of y). So there is a unique intersection with the sloping $\dot{y} = 0$ isocline in the interior of the triangle whenever $R_{0,\text{relative}} > 1$. The $\dot{z} = 0$ isocline does not depend on γ. For fixed y the sloping $\dot{y}= 0$ isocline has a z value that increases with γ (differentiating the equation for the line for fixed y with respect to γ, we find $z + \gamma\frac{dz}{d\gamma} = 1 - y$, which implies $dz/d\gamma > 0$). Hence both the y and z coordinates of the steady state are increasing functions of γ. For $\gamma \to \infty$, the sloping $\dot{y}= 0$ isocline approaches the line $z = 1 - y$, and the limiting value of the y coordinate of the steady state is therefore found by solving the equation

$$1 - y = \frac{f\alpha}{b\frac{1-y}{y} - (1 - f)\alpha},$$

which is a quadratic equation in y that has $\bar{y}_\infty$ (see main text) as the root in the interval (0,1).

We first show that, for $R_{0,\text{relative}} = \frac{\gamma}{\alpha+b} < 1$, the origin $(y, z) = (0,0)$ is globally stable. Indeed, we then have

$$
\begin{aligned}
\dot{y} &= y\{\gamma(1 - y - z) - \alpha + \alpha(1 - f)y + b(y - 1)\} \\
&\leq y\{(\alpha + b)(1 - y - z) - \alpha + \alpha y - \alpha fy + b(y - 1)\} \\
&\leq y\{-(\alpha + b)z - \alpha fy\} \leq -\alpha fy^2,
\end{aligned}
$$

which shows that y decreases monotonically towards its limit zero. Since y tends to zero, the limiting equation for z is $dz/dt = -bz$, and consequently also z tends to zero for $t \to \infty$. In the rest of the elaboration we concentrate on the case $R_{0,\text{relative}} > 1$.

The Jacobi matrix of the non-trivial steady state is

$$
\begin{pmatrix}
y(-\gamma + \alpha(1 - f) + b) & -\gamma y \\
f\alpha + (1 - f)\alpha z + bz & (1 - f)\alpha y - b(1 - y)
\end{pmatrix},
$$

and we claim that its sign pattern is

$$
\begin{pmatrix}
- & - \\
+ & -
\end{pmatrix},
$$

which then implies that the trace $T < 0$ and the determinant $D > 0$, and this, in turn, implies the (local asymptotic) stability of the steady state. The signs of the anti-diagonal elements are immediately clear, so we concentrate on the diagonal elements, starting with position 11. The assumption $R_{0,\text{relative}} > 1$ implies $\gamma > b + \alpha$, or $-\gamma + b + \alpha < 0$, and hence certainly $-\gamma + b + \alpha - \alpha f < 0$. The element at position 22 is negative for small y, but changes sign at $y = b/((1 - f)\alpha + b)$. An easy calculation reveals that $\bar{y}_\infty$ is smaller than this value. Hence the 22-element is negative for the relevant range of y values.

We conjecture that the steady state is in fact globally stable (this should follow from limit-cycle theory for quadratic systems and Poincaré-Bendixson theory, but the authors have so far made no attempt to study the relevant literature, and they doubt that they will ever do so).

It remains to calculate the critical value γ_c at which the growth rate changes sign. To do this, we append to the two isocline equations the condition that $N' = 0$ for $N \neq 0$. In other words, we consider the three equations

$$
\begin{aligned}
\gamma(1 - y - z) - \alpha + \alpha(1 - f)y + b(y - 1) &= 0, \\
y(f\alpha + \alpha(1 - f)z) - bz(1 - y) &= 0, \\
b - \mu - (b + \alpha(1 - f))y &= 0,
\end{aligned}
$$

and try to solve for y, z and γ while considering α, f, b and μ as known. The third equation gives

$$y = \frac{b - \mu}{b + \alpha(1 - f)}.$$

This we substitute in the second equation and then solve for z to find

$$z = \frac{\alpha f(b - \mu)}{\mu(b + \alpha(1 - f))}.$$

The first equation gives

$$\gamma_c = \frac{\alpha - \alpha(1 - f)y + b(1 - y)}{1 - y - z},$$

and upon substitution of the expressions for y and z we arrive at the expression for γ_c as given.

Exercise 3.24 As long as N is exponentially increasing, we may neglect the term K in the denominator and the system is very similar to system (3.12).[5] Now we look for steady states. Putting $I' = 0$, we find $-\mu + \frac{\gamma S}{K+N} - \alpha = 0$, which we rewrite as $(\gamma - \alpha - \mu)S - (\alpha + \mu)I = (\alpha + \mu)K$. Putting $dI/dt = 0$ (and using the relation we already have) we obtain $(b_1 - \mu)S + (b_2 - \alpha - \mu)I = 0$. These are two linear equations in the two unknowns S and I. The solution is given by

$$\overline{S} = \frac{(\alpha + \mu)K}{\gamma - \alpha - \mu - \frac{(\alpha+\mu)(\mu-b_1)}{b_2-\alpha-\mu}},$$

$$\overline{I} = \frac{\mu - b_1}{b_2 - \alpha - \mu}\overline{S},$$

but in order to be biologically meaningful, both of these expressions should be positive. In order to have a growing population in the absence of the infective agent we assume $b_1 > \mu$. If an individual is infected right at birth, it is expected to produce $b_2/(\alpha + \mu)$ offspring. The condition that this is less than one is exactly the condition that the multiplication factor that relates $\overline{S}$ to $\overline{I}$ is positive. It remains to investigate the sign of the denominator in the expression for $\overline{S}$. The condition for this denominator to be positive turns out to be exactly the condition for the negativity of the joint growth rate of I and S in a model in which we omit K in the denominator (again we refer to Diekmann & Kretzschmar, loc. cit., for details; the joint growth rate is explicitly given by $\frac{\gamma b_2}{\gamma - b_1 - \alpha + b_2} - \mu - \alpha$, which becomes zero for $\gamma = \gamma_c = \frac{(\mu+\alpha)(b_1+\alpha-b_2)}{\mu+\alpha-b_2}$; to find the growth rate, one has to use a transformation that decouples the equations for relative quantities from the equation for population size, just as we did when introducing (3.13)).

Exercise 3.25 The function $C(N)$ should increase with N. In the foregoing exercise we had $C(N) = \frac{N}{K+N}$, which is bounded; in the next exercise we shall consider $C(N) = N$, which is unbounded. As we shall see, the difference matters. In the bounded case,

---

[5] For a detailed justification and elaboration see O. Diekmann & M. Kretzschmar: Patterns in the effects of infectious diseases on population growth. *J. Math. Biol.*, **29** (1991), 539-570.

everything proceeds essentially as in the foregoing exercise, except that it is more difficult (or, rather, impossible) to do explicit calculations when we do not have an explicit expression for the inverse C^{-1}.

Exercise 3.26 i) If we put $dI/dt = 0$, we obtain $\overline{S} = \frac{\mu+\alpha}{\beta}$, and substitution of this into the equation obtained by putting $dS/dt = 0$ leads to $(\mu + \alpha - b_2)\overline{I} = (b_1 - \mu)\overline{S}$. Because we assume that $b_1 > \mu$ (to have exponential growth when the agent is absent), positivity of $\overline{I}$ requires positivity of $\mu + \alpha - b_2$ or, equivalently $\frac{b_2}{\mu+\alpha} < 1$ (which has the interpretation that the expected number of offspring produced by an individual that is infected right at birth is less than one). The Jacobi matrix

$$\begin{pmatrix} b_1 - \mu - \beta\overline{I} & b_2 - \beta\overline{S} \\ \beta\overline{I} & 0 \end{pmatrix} \text{ has sign structure } \begin{pmatrix} - & - \\ + & 0 \end{pmatrix},$$

and hence has negative trace and positive determinant, and therefore the steady state is stable (to see the negativity of $b_1 - \mu - \beta\overline{I}$ rewrite it as $-b_2\overline{I}/\overline{S}$).

ii) We rewrite the equation for dS/dt as

$$\frac{dS}{dt} = \left(b_1 + b_2\frac{I}{S} - \mu - \beta I\right) S.$$

As long as $\frac{b_2}{S} - \beta > 0$, we can infer from this that $S(t)$ grows faster than $e^{(b_1-\mu)t}$. So, for large t, necessarily the opposite inequality has to hold and therefore $S(t) \geq \frac{b_2}{\beta}$. If that is the case then

$$\frac{dI}{dt} = (\beta S - \mu - \alpha)I \geq (b_2 - \mu - \alpha)I,$$

and when $b_2 > \mu + \alpha$ this implies exponential growth of I. Returning to the equation for dS/dt in the form above, we then see that $\frac{dS}{dt} \ll 0$ if $S > \frac{b_2}{\beta}$. Hence $S(t)$ converges to $\frac{b_2}{\beta}$ from above. Replacing S in the equation for dI/dt by this quantity, we infer that I grows exponentially at the rate $b_2 - \mu - \alpha$. In their pioneering work on host regulation, May & Anderson (1979)[6] investigated a variant of the system in this exercise.

Exercise 3.27 Didn't we advise you to wait until you had worked through Chapter 5? The elaboration is given as Exercise 5.44.

---

[6] R.M. Anderson & R.M. May: Population biology of infectious diseases, Part 1. *Nature*, **280** (1979), 361-367; Part 2, **280**, 455-461.

12

Elaborations for Part II

12.1 Elaborations for Chapter 5

Exercise 5.2 i) We have $Kx = c_1 K\psi^{(1)} + c_2 K\psi^{(2)} = c_1\lambda_1\psi^{(1)} + c_2\lambda_2\psi^{(2)}$ and, by induction, $K^n x = c_1\lambda_1^n\psi^{(1)} + c_2\lambda_2^n\psi^{(2)}$.

ii) We can rewrite $K^n x$ as

$$K^n x = c_1\lambda_1^n \left[\psi^{(1)} + \frac{c_2}{c_1}\left(\frac{\lambda_2}{\lambda_1}\right)^n \psi^{(2)}\right]$$

(provided $c_1 \neq 0$) and since $|\lambda_2/\lambda_1| < 1$, the second term within the square brackets approaches zero for $n \to \infty$. The influence of the initial condition (the zeroth generation) is restricted to the value of c_1. When $\lambda_1 > 1$ we observe exponential growth, when $0 < \lambda_1 < 1$ exponential decline.

Exercise 5.5 i) The kth element of the vector $K^n e_j$, where e_j denotes the jth unit vector (which has a one at position j and zeros everywhere else), is precisely $(K^n)_{kj}$. So if K is irreducible, there exists an n for which this quantity is positive, meaning that, via a chain of infections of length n, the infectious individual with h-state j 'at birth', produces cases with h-state k at birth, with positive probability. 'Eventually' thus means 'some generation ahead'. When K is primitive, not only can we choose the length n of the chain independent of j and k, but in addition $(K^n)_{kj}$ is positive for all m larger than this n. This means that for all generations after the nth, the expected number of cases with h-state k at birth is positive, no matter what the h-state at birth was of the ancestor that started the chain of infections. In that case we can take 'eventually' to mean 'after a certain number of generations'.

ii) Yes, you should agree. The key point is the precise specification of 'eventually'; otherwise the difference is just the context dependent interpretation of the positivity of $(K^n)_{kj}$.

iii) First let us concentrate on the sign structure and write $K = \begin{pmatrix} 0 & + \\ + & 0 \end{pmatrix}$.

It then follows that $K^2 = \begin{pmatrix} + & 0 \\ 0 & + \end{pmatrix}$ and $K^3 = \begin{pmatrix} 0 & + \\ + & 0 \end{pmatrix}$, and, by induction,

$K^{2n} = \begin{pmatrix} + & 0 \\ 0 & + \end{pmatrix}$ and $K^{2n+1} = \begin{pmatrix} 0 & + \\ + & 0 \end{pmatrix}$. The sign structure is periodic with period 2, and K is irreducible, but not primitive.

Now we will compute $\| K^n \|^{1/n}$. First we note that $K^2 = 10^3 I$, and consequently we have that $K^3 = 10^3 K$, $K^4 = 10^6 I$, $K^5 = 10^6 K$, ... So we alternate between K and I, modulo a multiplication factor. From this observation, the following follows readily: for $n = 1, 2, ..., 10$ we find respectively $\| K^n \|^{1/n} = 10^2, 10^{3/2}, 10^{5/3}, 10^{3/2}, 10^{8/5}, 10^{3/2}, 10^{11/7}, 10^{3/2}, 10^{14/9}, 10^{3/2}$.

iv) The subpopulation of homosexual males is isolated from the rest, and so is the subpopulation of homosexual females. By 'isolated' we mean that if the agent is introduced only in that subpopulation, it stays in that subpopulation and does not spread to other subpopulations. The reason is the contact structure: in the absence of bisexuality, homosexual males only have contacts within their own group. Likewise the subpopulation of heterosexual individuals, comprising both the males and females, is isolated.

So 'reducible' means that we can subdivide the population into isolated groups or, in more mathematical terms, that we can decompose the next-generation matrix into uncoupled lower-dimensional matrices (or matrices that are coupled in one direction only, as in $\begin{pmatrix} + & 0 \\ + & + \end{pmatrix}$, where spread from 1 to 2 is possible, but spread from 2 to 1 is not possible).

Exercise 5.7 The matrix

$$K = \begin{pmatrix} \rho_1 c_1 & (1 - \rho_2)c_2 \\ (1 - \rho_1)c_1 & \rho_2 c_2 \end{pmatrix}$$

has as its characteristic equation $\lambda^2 - (\rho_1 c_1 + \rho_2 c_2)\lambda + (\rho_1 + \rho_2 - 1)c_1 c_2 = 0$ and consequently

$$R_0 = \frac{1}{2}\left\{\rho_1 c_1 + \rho_2 c_2 + \sqrt{(\rho_1 c_1 + \rho_2 c_2)^2 - 4(\rho_1 + \rho_2 - 1)c_1 c_2}\right\}.$$

When $\rho_1 = \rho_2 = 1$, this, of course, yields $R_0 = \max\{c_1, c_2\}$. The other boundary extreme occurs when one of the ρs is equal to zero. Suppose that $c_1 N_1 < c_2 N_2$; then this happens for $\rho_1 = 0$, $\rho_2 = 1 - \frac{c_1 N_1}{c_2 N_2}$. For the numbers as given in Section 2.2 we have $\rho_2 = \frac{9}{10}$ and $R_0 \approx 15\frac{1}{2}$ (while $\max\{c_1, c_2\} = c_1 = 100$). We did not formally check that R_0 is maximal for $\rho_1 = \rho_2 = 1$ and that it achieves its minimum for $\rho_1 = 0$, $\rho_2 = 1 - \frac{c_1 N_1}{c_2 N_2}$ when $c_1 N_1 < c_2 N_2$, but we are quite confident that this is nevertheless true.

Exercise 5.8 The perturbed matrix

$$\begin{pmatrix} 10^3 & \varepsilon_{12} \\ \varepsilon_{21} & 1 \end{pmatrix}$$

has as its characteristic equation $\lambda^2 - 1001\lambda + (1000 - \varepsilon_{12}\varepsilon_{21}) = 0$, with roots

$$\lambda_\pm = \frac{1001}{2} \pm \frac{999}{2}\sqrt{1 + \frac{4\varepsilon_{12}\varepsilon_{21}}{998001}},$$

and therefore $R_0 = \lambda_d = \lambda_+ \approx 1000 + \varepsilon_{12}\varepsilon_{21}10^{-3}$. The eigenvector $x = (x_1, x_2)^\top$ has to satisfy

$$10^3 x_1 + \varepsilon_{12}x_2 = \lambda_+ x_1 \approx 10^3 x_1 + \varepsilon_{12}\varepsilon_{21}10^{-3},$$

from which it follows that

$$x_2 \approx \varepsilon_{21} 10^{-2} x_1.$$

Therefore if $x_1 + x_2 = 1$ then $x_1 = O(1)$ while $x_2 = O(\varepsilon_{21})$, and we conclude that new cases predominantly have h-state 1 at birth and only a very small fraction has h-state 2. The reason is that most effective contacts are within the 1-group and the coupling is a very loose one. The point of the remark is that we see this reflected in an order-of-magnitude difference between the components of the normalised eigenvector corresponding to R_0.

Exercise 5.9 When $k_{ij} = a_i b_j$ we have $(K\phi)_i = a_i \sum_{j=1}^{n} b_j \phi_j$, i.e. $K\phi = (\sum_{j=1}^{n} b_j \phi_j) a$ or, in words, for arbitrary ϕ the vector $K\phi$ is a multiple of the vector a (using a bit more jargon, we can also express this by saying that the range of K is spanned by a). Hence eigenvectors have to be a multiple of a. Since $Ka = (\sum_{j=1}^{n} b_j a_j) a$, the corresponding (one and only non-zero) eigenvalue equals $\sum_{j=1}^{n} b_j a_j$. By definition, this is also R_0.

If $k(\xi, \eta) = a(\xi) b(\eta)$, the range of K is spanned by the function a, and, following exactly the same line of argument, one arrives at the expression $R_0 = \int_\Omega b(\eta) a(\eta) \, d\eta$.

If $\Lambda(\eta)(\omega) = a(\omega) b(\eta)$, the range of K is spanned by the measure a, and hence $R_0 = \int_\Omega b(\eta) a(d\eta)$.

Exercise 5.10 We have

$$R_0 = \sum_{j=1}^{n} b_j a_j = \frac{\sum_{j=1}^{n} c_j^2 N_j}{\sum_{k=1}^{n} c_k N_k}.$$

Normalise the N_i such that $\sum_{k=1}^{n} N_k = 1$ (this keeps the expression for R_0 unchanged, since the N_i occur linearly in both numerator and denominator). Then $\bar{c} := \sum_{k=1}^{n} c_k N_k$ is the mean, while the variance is $\bar{\bar{c}} := \sum_{k=1}^{n} (c_k - \bar{c})^2 N_k = \sum_{k=1}^{n} c_k^2 N_k - (\bar{c})^2$. We find

$$R_0 = \frac{(\bar{c})^2 + \bar{\bar{c}}}{\bar{c}} = \bar{c} + \frac{\bar{\bar{c}}}{\bar{c}}.$$

Exercise 5.11 Under the assumption that $k_{ij} = a_i b_j + c_j \delta_{ij}$, the eigenvalue problem $K\psi = \lambda\psi$, written out in components, reads

$$a_i \sum_{j=1}^{n} b_j \psi_j = (\lambda - c_i)\psi_i.$$

Now multiply this identity by $b_i(\lambda - c_i)^{-1}$ and sum over i to obtain

$$\sum_{i=1}^{n} \frac{b_i a_i}{\lambda - c_i} \sum_{j=1}^{n} b_j \psi_j = \sum_{i=1}^{n} b_i \psi_i,$$

which shows that necessarily $\sum_{j=1}^{n} b_j \psi_j = 0$ or that

$$\sum_{i=1}^{n} \frac{b_i a_i}{\lambda - c_i} = 1.$$

For the dominant eigenvalue, the first possibility is ruled out, since all contributions to the sum are positive (see Theorem 5.6). As a function of a real variable λ, the sum $\sum_{i=1}^{n} \frac{b_i a_i}{\lambda - c_i}$ is strictly decreasing on the interval $(\max c_i, \infty)$. Let Q_0 denote the value of this sum for $\lambda = 1$. Then the sum assumes the value 1 to the right of $\lambda = 1$ if and only if $Q_0 > 1$. As the value of λ for which the sum equals 1 must be R_0, we have verified the threshold property.

A derivation of the third case goes as follows. In terms of $\Lambda(\eta)(\omega) = a(\omega)b(\eta) + c(\eta)\delta_\eta(\omega)$, the eigenvalue problem $Km = \lambda m$ reads

$$a(\omega) \int_\Omega b(\eta)m(d\eta) + \int_\omega c(\eta)m(d\eta) = \lambda m(\omega),$$

where $\omega \subset \Omega$. Now integrate the function $\xi \mapsto \frac{b(\xi)}{\lambda - c(\xi)}$ over Ω with respect to both the left- and right-hand sides to obtain

$$\int_\Omega \frac{b(\xi)}{\lambda - c(\xi)} a(d\xi) \int_\Omega b(\eta)m(d\eta) + \int_\Omega \frac{b(\xi)c(\xi)}{\lambda - c(\xi)} m(d\xi) = \lambda \int_\Omega \frac{b(\xi)}{\lambda - c(\xi)} m(d\xi),$$

or, after a little rearrangement,

$$\int_\Omega \frac{b(\xi)}{\lambda - c(\xi)} a(d\xi) \int_\Omega b(\eta)m(d\eta) = \int_\Omega b(\xi)m(d\xi).$$

The rest of the argument is identical to the one presented above.

Exercise 5.12 We have

$$(Km)(\omega) = \int_\Omega \Lambda(\eta)(\omega)m(d\eta) = \sum_{i=1}^{n} \alpha_i(\omega) \int_\Omega b_i(\eta)m(d\eta),$$

which shows that the range of K is spanned by $\{\alpha_i\}_{i=1}^{n}$. If $m = \sum_j c_j \alpha_j$ then $Km = \sum_i d_i \alpha_i$, with

$$d_i = \sum_{j=1}^{n} \int_\Omega b_i(\eta)\alpha_j(d\eta)c_j,$$

so the coefficients satisfy $d = Lc$, where L is the matrix with elements $l_{ij} = \int_\Omega b_i(\eta)\alpha_j(d\eta)$. Necessarily then, non-zero eigenvalues of K are in one-to-one correspondence with non-zero eigenvalues of L.

Exercise 5.13 We have

$$(Km)(i, \widetilde{\omega}) = \sum_{j=1}^{n} \int_{\widetilde{\Omega}} \Lambda_i(j, \varsigma)(\widetilde{\omega})m(j, d\varsigma)$$

$$= \alpha_i(\widetilde{\omega}) \sum_{j=1}^{n} \int_{\widetilde{\Omega}} b_{ij}(\varsigma)m(j, d\varsigma),$$

which shows that the range of K is spanned by the measures α_i on $\widetilde{\Omega}$. If $m(i, \widetilde{\omega}) = c_i \alpha_i(\widetilde{\omega})$ then $(Km)(i, \widetilde{\omega}) = d_i \alpha_i(\widetilde{\omega})$, with $d_i = \sum_{j=1}^{n} \int_{\widetilde{\Omega}} b_{ij}(\varsigma)\alpha_j(d\varsigma)c_j$, which shows that the coefficients transform according to a matrix L with elements $l_{ij} = \int_{\tilde{\Omega}} b_{ij}(\zeta)\alpha_j(d\zeta)$. Necessarily, R_0 is the dominant eigenvalue of this matrix.

Exercise 5.14 Let $\Omega = \{1, 2, ..., n\}$. We have

$$k_{ij} = \int_0^\infty h \sum_{l=1}^n c_{il} P(\tau, l, j) \, d\tau,$$

with

$$P(\tau, l, j) = (e^{\tau(\Sigma - D)})_{lj},$$

and hence

$$\int_0^\infty P(\tau, l, j) \, d\tau = -(\Sigma - D)_{lj}^{-1}.$$

Exercise 5.15

$$R_0^D = \frac{\beta}{\mu + \sigma} = \frac{\rho p}{\mu + \sigma}.$$

Exercise 5.16 The rates at which an individual has contacts with 0 and + individuals are respectively $\rho \frac{N_0}{N}$ and $\rho \frac{N_+}{N}$. If an individual is D-infected but not d-infected and it has contact with a D-susceptible that is not d-infected, the probability of transmission is p. If the D-infected individual is also d-infected, then this success ratio is increased by a factor w, and if the D-susceptible individual is also d-infected then there is an increase with a factor v. Hence

$$B = \frac{\rho p}{N} \begin{pmatrix} N_0 & N_0 w \\ N_+ v & N_+ v w \end{pmatrix}.$$

Exercise 5.17 Although not explicitly stated in the text, the idea is to describe demographic turnover as in the beginning of Section 3.2. This means that we have a population birth rate $B = \mu N$ and that all newborns are susceptible to d. Hence the differential equations read

$$\frac{dN_0}{dt} = \mu N - \mu N_0 - \zeta N_0 + \gamma N_+,$$

$$\frac{dN_+}{dt} = -\mu N_+ + \zeta N_0 - \gamma N_+,$$

and $N_0 + N_+ = N$. Using this last equation and putting the right-hand side of the differential equation for N_0 equal to zero, we arrive at $\mu(N - N_0) - \zeta N_0 + \gamma(N - N_0) = 0$, or $N_0 = \frac{\gamma + \mu}{\zeta + \gamma + \mu} N$. Hence $N_+ = N - N_0 = \frac{\zeta}{\zeta + \gamma + \mu} N$.

Exercise 5.18 To obtain G from the description in the previous exercise, we have to do two things: i) leave out the birth term, since we now concentrate on one particular individual, and ii) increase the death rate to $\mu + \sigma$, since we now consider a D-infected individual. Hence

$$G = \begin{pmatrix} -\zeta - \mu - \sigma & \gamma \\ \zeta & -\gamma - \mu - \sigma \end{pmatrix}$$

and $\det G = (\zeta + \mu + \sigma)(\gamma + \mu + \sigma) - \gamma\zeta = (\mu + \sigma)(\zeta + \mu + \sigma + \gamma)$, and so

$$-G^{-1} = \frac{1}{(\mu + \sigma)(\mu + \sigma + \gamma + \zeta)} \begin{pmatrix} \mu + \sigma + \gamma & \gamma \\ \zeta & \mu + \sigma + \zeta \end{pmatrix}.$$

That $K = -BG^{-1}$ follows from Exercise 5.14 and the above definitions of B and G.

Exercise 5.19 The determinant of B equals zero, so the columns are linearly dependent (indeed, the second is w times the first). So the range is spanned by the first column, i.e. by $\binom{N_0}{N_+ v}$. Stated differently: in the initial phase new D-cases arise in the ratio $N_0 : N_+ v$, with respect to being d-free or d-infected. The key assumption that underlies this result is that the success ratio is enlarged by the product vw when both individuals are d-infected. This product structure reflects the assumption that the states of the susceptible individual and the infectious individual have independent influence on the transmission probability.

Exercise 5.20 We have

$$Bx = \frac{\rho p}{N} \begin{pmatrix} N_0 & N_0 w \\ N_+ v & N_+ vw \end{pmatrix} \begin{pmatrix} x_1 \\ x_2 \end{pmatrix} = \frac{\rho p}{N}(x_1 + wx_2) \begin{pmatrix} N_0 \\ N_+ v \end{pmatrix}.$$

Since $K = -BG^{-1}$, the range of K is spanned by $\binom{N_0}{N_+ v}$ as well.

Exercise 5.21 We have

$$K\phi = -BG^{-1}\phi = -\frac{\rho p}{N}\left(\begin{pmatrix} 1 \\ w \end{pmatrix} \cdot G^{-1}\phi \right) \begin{pmatrix} N_0 \\ N_+ v \end{pmatrix},$$

which motivates us to consider $\phi = \frac{\rho p}{N}\binom{N_0}{N_+ v}$. For this choice of ϕ we have $K\phi = \lambda\phi$, with

$$\lambda = -\begin{pmatrix} 1 \\ w \end{pmatrix} \cdot G^{-1}\phi = -\frac{\rho p}{N}\begin{pmatrix} 1 \\ w \end{pmatrix} \cdot G^{-1}\begin{pmatrix} N_0 \\ N_+ v \end{pmatrix}.$$

Exercise 5.22 We identify R_0 with the only non-zero eigenvalue of K found in the preceding exercise. Before elaborating on the explicit expression, we note that for $v = w = 1$ we should retrieve the expression $R_0 = \frac{\rho p}{\mu + \sigma}$ that we found in Exercise 5.15, since in that case it is totally irrelevant for D-transmission whether or not an individual is d-infected.
   Since

$$-G^{-1}\begin{pmatrix} N_0 \\ N_+ v \end{pmatrix} = \frac{1}{(\mu + \sigma)(\mu + \sigma + \gamma + \zeta)} \begin{pmatrix} (\mu + \sigma + \gamma)N_0 + \gamma N_+ v \\ \zeta N_0 + (\mu + \sigma + \zeta)N_+ v \end{pmatrix},$$

we find that

$$\begin{aligned} R_0 &= -\frac{\rho p}{N}\begin{pmatrix} 1 \\ w \end{pmatrix} \cdot G^{-1}\begin{pmatrix} N_0 \\ N_+ v \end{pmatrix} \\ &= \frac{\rho p}{(\mu + \sigma)(\mu + \sigma + \gamma + \zeta)}\left\{ (\mu + \sigma + \gamma)\frac{N_0}{N} + \gamma v\frac{N_+}{N} \right. \\ &\quad \left. + w\zeta\frac{N_0}{N} + (\mu + \sigma + \zeta)vw\frac{N_+}{N} \right\}. \end{aligned}$$

If we now substitute the expressions for N_0 and N_+ derived in Exercise 5.17, take out the common factor $(\mu + \gamma + \zeta)^{-1}$ from the expression in $\{\cdots\}$ and rearrange terms, we obtain

$$R_0 = \rho p \frac{(\mu + \gamma)(\mu + \sigma + \gamma + \zeta w) + \zeta v w(\frac{\gamma}{w} + \mu + \sigma + \zeta)}{(\mu + \gamma + \zeta)(\mu + \sigma)(\mu + \sigma + \gamma + \zeta)}.$$

For $v = w = 1$ this does indeed reduce to $\frac{\rho p}{\mu + \sigma}$, as required.

Exercise 5.23 i) We rewrite F as

$$
\begin{aligned}
F &= \frac{\gamma + \mu}{\gamma + \mu + \zeta} \frac{\mu + \sigma + \gamma + \zeta w}{\mu + \sigma + \gamma + \zeta} + \frac{\zeta}{\gamma + \mu + \zeta} \frac{v\gamma + vw\mu + vw\sigma + vw}{\mu + \sigma + \gamma + \zeta} \\
&= \left(\frac{\gamma + \mu}{\gamma + \mu + \zeta} + \frac{\zeta}{\gamma + \mu + \zeta} \right) \frac{\mu + \sigma + \gamma + \zeta w}{\mu + \sigma + \gamma + \zeta} \\
&\quad + \frac{\zeta}{\gamma + \mu + \zeta} \frac{(v-1)\gamma + (vw-1)\mu + (vw-1)\sigma + (v-1)w\zeta}{\mu + \sigma + \gamma + \zeta},
\end{aligned}
$$

where we note that the first term in the sum is ≥ 1, and that the second term is ≥ 0 since all parameters are positive and since $v \geq 1$ and $w \geq 1$.

ii) Take the limit $\gamma \to \infty$ or $\zeta \to 0$ in the expression for F. We indeed expect a disease d of which infected individuals are cured very rapidly (i.e. have a very short infectious period $1/\gamma$), and with a very low infectious pressure on the population (i.e. a very small force of infection ζ), not to make much impact on the transmission of D.

iii) Note first that $F = 1$, if $v = w = 1$. Consider $F = F(\gamma, \zeta)$, for the case $v, w > 1$. To keep track of all parameters, we define (with apologies for the terrible notation—but you try and come up with something better for quick occasional analysis like this) $f(w) := \mu + \sigma + \gamma + \zeta w$, and $\phi := v(w\mu + w\sigma + \gamma + \zeta w)$ and note that $f(1) < f(w) < \phi < vwf(1)$ and $wf(1) > f(w)$ and that $df(w)/d\gamma = 1$, $df(w)/d\zeta = w$, $d\phi/d\gamma = v$, $d\phi/d\zeta = vw$. Then

$$F = \frac{(\gamma + \mu)f(w) + \zeta\phi}{(\gamma + \mu + \zeta)f(1)}.$$

Therefore

$$\frac{\partial F}{\partial \gamma} = \frac{(\gamma + \mu + \zeta)f(1)[f(w) + \gamma + \mu + \zeta v] - [(\gamma + \mu)f(w) + \zeta\phi][f(1) + \gamma + \mu + \zeta]}{(\gamma + \mu + \zeta)^2 f^2(1)},$$

and so $\partial F/\partial \gamma > 0$, using the inequalities that hold for f and ϕ.

Similarly,

$$\frac{\partial F}{\partial \zeta} = \frac{(\gamma + \mu + \zeta)f(1)[w\gamma + \mu + \phi + \zeta vw] - [(\gamma + \mu)f(w) + \zeta\phi][f(1) + \gamma + \mu + \zeta]}{(\gamma + \mu + \zeta)^2 f^2(1)},$$

and so $\partial F/\partial \zeta > 0$, using the same inequalities.

iv) We can rewrite F as

$$F = \frac{(\mu + \gamma)(\frac{\mu}{\zeta} + \frac{\sigma}{\zeta} + \frac{\gamma}{\zeta} + w)}{\zeta(\frac{\mu}{\zeta} + \frac{\gamma}{\zeta} + 1)(\frac{\mu}{\zeta} + \frac{\sigma}{\zeta} + \frac{\gamma}{\zeta} + 1)} + \frac{vw(\frac{\gamma}{w\zeta} + \frac{\mu}{\zeta} + \frac{\sigma}{\zeta} + 1)}{(\frac{\mu}{\zeta} + \frac{\gamma}{\zeta} + 1)(\frac{\mu}{\zeta} + \frac{\sigma}{\zeta} + \frac{\gamma}{\zeta} + 1)},$$

which leads to $F \to vw$ for $\zeta \to \infty$.

Exercise 5.24 i) We can rewrite F as

$$F = \frac{(\mu+\gamma)}{(\mu+\gamma+\zeta)} \frac{(\mu+\sigma+\gamma+\zeta v)}{(\mu+\sigma+\gamma+\zeta)} + \frac{\zeta v^2}{(\mu+\gamma+\zeta)} \frac{(\frac{\gamma}{v}+\mu+\sigma+\zeta)}{(\mu+\sigma+\gamma+\zeta)},$$

which gives the desired result by substituting the expression for N_+/N from Exercise 5.17.

ii) What we have in mind here is the following. The natural death rate μ is usually much smaller than 1 ($1/\mu$ is the life expectancy), but the cure rate γ is likely to be greater than one ($1/\gamma$ is the average time before the patient is cured of d) and therefore $\gamma \gg \mu$. For disease D we had HIV in mind, which usually leads to AIDS and is then fatal, where the time period from initial infection to death by the disease can be long and σ is therefore substantially smaller than one, but $\sigma > \mu$.

iii) Start from the expression obtained in i) and divide both numerators and denominators by $\gamma + \mu$:

$$F = \left(1 - \frac{N_+}{N}\right) \frac{(1 + \frac{\sigma}{\gamma+\mu} + \frac{\zeta v}{\gamma+\mu})}{(1 + \frac{\sigma}{\gamma+\mu}\frac{\zeta}{\gamma+\mu})} + \frac{N_+}{N}v^2 \frac{(\frac{\gamma}{v(\gamma+\mu)} + \frac{\mu}{\gamma+\mu} + \frac{\sigma}{\gamma+\mu} + \frac{\zeta}{\gamma+\mu})}{(1 + \frac{\sigma}{\gamma+\mu}\frac{\zeta}{\gamma+\mu})}.$$

Now use the approximations in ii) and take out the common factor $(1 + \frac{\zeta}{\gamma+\mu})^{-1} = 1 - \frac{N_+}{N}$ to find

$$F \approx \left(1 - \frac{N_+}{N}\right) \left\{ \left(1 - \frac{N_+}{N}\right)\left(1 + v\frac{\zeta}{\gamma+\mu}\right) + \frac{N_+}{N}v + \frac{N_+}{N}v^2\frac{\zeta}{\gamma+\mu} \right\}.$$

Now note that $(1 - \frac{N_+}{N})\frac{\zeta}{\gamma+\mu} = \frac{N_+}{N}$ and expand the parentheses:

$$F \approx (1 - \frac{N_+}{N})\left\{ 1 - \frac{N_+}{N} + 2\frac{N_+}{N}v \right\} + \left(\frac{N_+}{N}\right)^2 v^2$$

and

$$F \approx 1 + \left(\frac{N_+}{N}\right)^2 (1 - 2v + v^2) + 2v\frac{N_+}{N} - 2\frac{N_+}{N},$$

which finally leads to the desired approximation.

Exercise 5.25 Since we consider invasion of D, the steady-state values N_0 and N_+ pertaining to d do not change. Nor does the matrix B change (this matrix basically describes the transmission probabilities of D as a function of the status with respect to d of the two individuals, one D-infectious and the other D-susceptible, involved in a contact).

However, in the matrix G we should replace γ by γz, since this matrix pertains to a D-infected individual and describes its transitions with respect to d. So we get

$$R_0 = -\frac{pp}{N}\begin{pmatrix}1\\w\end{pmatrix} \cdot G^{-1}\begin{pmatrix}N_0\\N_+v\end{pmatrix}$$

$$= \frac{\rho p}{(\mu + \sigma)(\mu + \sigma + \gamma z + \zeta)} \left\{ (\mu + \sigma + \gamma z)\frac{N_0}{N} + \gamma z v \frac{N_+}{N} + w\zeta \frac{N_0}{N} \right.$$

$$\left. + (\mu + \sigma + \zeta)vw \frac{N_+}{N} \right\}$$

$$= \frac{\rho p}{(\mu + \sigma)(\mu + \sigma + \gamma z + \zeta)(\zeta + \gamma + \mu)} \{ (\mu + \sigma + \gamma z)(\gamma + \mu) + \gamma z v \zeta$$

$$+ w\zeta(\gamma + \mu) + (\mu + \sigma + \zeta)vw\zeta \}$$

$$= \rho p \frac{(\mu + \gamma)(\mu + \sigma + \gamma z + \zeta w) + \zeta vw(\frac{\gamma z}{w} + \mu + \sigma + \zeta)}{(\mu + \gamma + \zeta)(\mu + \sigma)(\mu + \sigma + \gamma z + \zeta)}.$$

Exercise 5.26 To understand the matrix, one has to keep in mind that the components of a four-vector correspond, in this order, to the states $(f, 0)$, $(f, +)$, $(m, 0)$ and $(m, +)$, and that of the two indices of a matrix element the second corresponds to a D-infectious individual and the first to a D-susceptible individual. With these bookkeeping conventions, the elements of B are a direct translation of the assumptions in the text. Concerning the range of B, simply observe that the two listed vectors are respectively the third and first columns of the matrix B and that the fourth column is w_m times the third while the second column is w_f times the first (and that the first and third columns are clearly linearly independent).

Exercise 5.27 We write a four-vector as a two-vector of two vectors. The result on the range of B then translates into the observation that an eigenvector of K corresponding to a non-zero eigenvalue is necessarily of the form $\binom{c_1\psi_1}{c_2\psi_2}$ with ψ_i as in the hint and c_i as arbitrary coefficients. The matrix K has, in self-explanatory notation, the structure

$$\begin{pmatrix} 0 & B_{fm} \\ B_{mf} & 0 \end{pmatrix} \begin{pmatrix} -G_f^{-1} & 0 \\ 0 & -G_m^{-1} \end{pmatrix},$$

and the range of B_{fm} is spanned by ψ_1 and the range of B_{mf} by ψ_2. The coefficient of ψ_1 is found by taking the inner product of $\binom{1}{w_m}$ with the two-vector to which B_{fm} is applied, and likewise in the case of B_{mf} we have to take the inner product with $\binom{1}{w_f}$ to find the coefficient of ψ_2. So K maps the vector $\binom{c_1\psi_1}{c_2\psi_2}$ onto a vector $\binom{b_1\psi_1}{b_2\psi_2}$, with $b = Lc$ for L as given in the hint. Consequently, $R_0 = \sqrt{l_{fm}l_{mf}}$, and all that remains to be done is to give explicit expressions for G_f^{-1} and G_m^{-1} to find a completely explicit expression for R_0.

Exercise 5.28 We have

$$N_{(i,0)} = \frac{\gamma + \mu}{\gamma + \mu + i\zeta} N_i, \qquad N_{(i,+)} = \frac{i\zeta}{\gamma + \mu + i\zeta} N_i,$$

$$\Sigma_i = \begin{pmatrix} -i\zeta & \gamma \\ i\zeta & -\gamma \end{pmatrix}, \quad G_i = \begin{pmatrix} -i\zeta - \mu - \sigma & \gamma \\ i\zeta & -\gamma - \mu - \sigma \end{pmatrix}.$$

Exercise 5.29 The number of contacts per unit of time between $(i, \cdot)$-individuals and $(j, \cdot)$-individuals is given by

$$\frac{\rho ij N_i N_j}{\sum_k k N_k}.$$

When we leave out the factor N_j, we get the per j-capita number of contacts per unit of time with $(i, \cdot)$-individuals.

Exercise 5.30 The operator K acts on a countable vector of two-vectors to produce again such a countable vector of two-vectors. The element K_{ij} of the matrix representation tells us how the jth two-vector of the original contributes to the ith two-vector of the image. We arrive at K_{ij} from the expression in Exercise 5.29 by realising that a fraction $N_{(i,0)}/N_i$ of contacts with $(i, \cdot)$-individuals is with $(i, 0)$-individuals etc., and by taking the probability of transmission and its various enhancement factors into account. From the expression for K_{ij}, one sees at once that the range of K has the ith component spanned by $\begin{pmatrix} N_{(i,0)} \\ vN_{(i,+)} \end{pmatrix}$, with the corresponding coefficient obtained from the inner product of $\begin{pmatrix} 1 \\ w \end{pmatrix}$ with $\frac{\rho p}{\sum_k k N_k} \sum_{j=0}^{\infty} j G_j^{-1} \phi_j$.

Exercise 5.31 The expression for R_0 now follows at once from the characterisation of the range and the expression for the coefficient as given at the end of the preceding exercise.

Exercise 5.33 i) $D = (\rho + \mu)(\sigma + 2\mu) - \rho(\sigma + \mu) = \mu(\sigma + 2\mu) + \rho\mu = \mu(\rho + \sigma + 2\mu)$.

The formula for $-G^{-1}$ is just the general formula $\begin{pmatrix} a & b \\ c & d \end{pmatrix}^{-1} = \frac{1}{D} \begin{pmatrix} d & -b \\ -c & a \end{pmatrix}$

applied to this particular case.

ii) The ijth element of $-G^{-1}$ is the expected time that an individual that currently has state j will spend in state i in the future. By summing over i, we get the expected future lifespan. Both columns of the matrix $\begin{pmatrix} \sigma + 2\mu & \sigma + \mu \\ \rho & \rho + \mu \end{pmatrix}$ sum to $\sigma + 2\mu + \rho$, so, taking the factor $1/D$ into account, we deduce that the expected future lifespan equals $1/\mu$, irrespective of the current state. Since being single or paired has, in this model, no influence on the survival probability, this outcome was predictable and serves more as a check that we did not make mistakes.

The expected time to be single is found in the first row and equals $\frac{1}{\mu} \frac{\sigma + 2\mu}{\rho + \sigma + 2\mu}$ if the current state is single, and $\frac{1}{\mu} \frac{\sigma + \mu}{\rho + \sigma + 2\mu}$ otherwise.

iii) We want to calculate the expected number Q of future partners of an individual that is single.

a) The expected duration of one partnership equals $\frac{1}{\sigma + 2\mu}$ and the expected total time of being paired for an individual that is currently single equals $\frac{\rho}{D}$. Since the durations of subsequent partnerships are independently identically distributed (i.i.d.), the expected number Q of partners equals the quotient $\frac{\rho(\sigma + 2\mu)}{D}$ of these two quantities.

b) A second derivation is based on the embedded discrete time process of state transitions (recall Exercise 1.33-vi). When an individual is single, its next state will be 'paired' with probability $\frac{\rho}{\rho + \mu}$, and in that case we have to add 1 to the number of partners. Return to the state 'single' is a conditio sine qua non for acquiring a new partner (in this model; admittedly, reality is less restrictive in its possibilities). With probability $\frac{\sigma + \mu}{\sigma + 2\mu}$, the next state will be 'single'. But then we are back to where we started, and the expected number of future partners is again Q. Hence

$$Q = \frac{\rho}{\rho + \mu} \left(1 + \frac{\sigma + \mu}{\sigma + 2\mu} Q \right),$$

and solving this equation once more leads to $Q = \frac{\rho(\sigma + 2\mu)}{D}$.

Exercise 5.34 i) If we are in state 2, the probability that the next state transition brings us to state 3 equals $\frac{h}{h + \sigma + 2\mu}$.

ii) If we are in state 3, the probability that the next state transition brings us to state 1 equals $\frac{\sigma + \mu}{\sigma + 2\mu}$.

iii) If we concentrate on the distinction between being paired or single and suppress all information related to the infection, we are exactly in the situation considered in Exercise 5.33. (A key point here is that infected individuals have exactly the same ρ, σ and μ as non-infected individuals. In particular, an additional death rate due to the infection would necessitate that we redo the earlier calculation.) To make this explicit we have to lump the current states 2 and 3 into one state, which is possible since the outgoing arrows of 2 and 3, including the corresponding rates, are identical.

iv) It follows from the results derived above that

$$R_0 = \frac{\sigma + \mu}{\sigma + 2\mu} \frac{\rho(\sigma + 2\mu)}{D} \frac{h}{h + \sigma + 2\mu} = \frac{\rho h(\sigma + \mu)}{\mu(\rho + \sigma + 2\mu)(h + \sigma + 2\mu)}.$$

Exercise 5.35 i) Given that the current state is 1, the probability that the next transition will bring the system to state 2 equals $\frac{\rho}{\rho + \mu}$. From state 2, we will go with probability $\frac{h}{h + \sigma + 2\mu}$ to state 3 (in which case we are sure to ever arrive at 3) and with probability $\frac{\sigma + \mu}{h + \sigma + 2\mu}$ back to state 1 (in which case the probability to ever arrive in 3 is again P_{13}). Hence

$$P_{13} = \frac{\rho}{\rho + \mu} \left\{ \frac{h}{h + \sigma + 2\mu} + \frac{\sigma + \mu}{h + \sigma + 2\mu} P_{13} \right\},$$

and by solving this equation we find

$$P_{13} = \frac{h\rho}{(h + \mu)\rho + (h + \sigma + 2\mu)\mu}.$$

ii) In order to return to 3 from 3, one has to go through 1. The probability to go from 3 to 1 equals $\frac{\sigma + \mu}{\sigma + 2\mu}$. Hence

$$P_{33} = \frac{\sigma + \mu}{\sigma + 2\mu} P_{13}.$$

iii) A newly infected individual is in state 3 and is, by definition, expected to produce R_0 secondary cases. When it returns to state 3, it has produced one secondary case and is, by the Markov property, expected to produce R_0 more. Hence $R_0 = P_{33}(1 + R_0)$. So

$$\begin{aligned}
R_0 &= \frac{P_{33}}{1 - P_{33}} = \frac{(\sigma + \mu)P_{13}}{\sigma + 2\mu - (\sigma + \mu)P_{13}} \\
&= \frac{(\sigma + \mu)h\rho}{(\sigma + 2\mu)\{(h + \mu)\rho + (h + \sigma + 2\mu)\mu\} - (\sigma + \mu)h\rho}
\end{aligned}$$

$$= \frac{(\sigma + \mu)h\rho}{\mu\{(h + \mu)\rho + (h + \sigma + 2\mu)\mu\} + (\sigma + \mu)\mu(\rho + h + \sigma + 2\mu)}$$

$$= \frac{(\sigma + \mu)h\rho}{\mu\{(h + \mu)\rho + (h + \sigma + 2\mu)(\sigma + 2\mu) + \rho(\sigma + \mu)\}}$$

$$= \frac{(\sigma + \mu)h\rho}{\mu(h + \sigma + 2\mu)(\sigma + 2\mu + \rho)}.$$

iv) It follows from $R_0 = \frac{P_{33}}{1 - P_{33}}$ that $R_0 > 1 \Leftrightarrow P_{33} > \frac{1}{2}$. When returning to state 3, an infectious individual has 'cloned' itself once or, in other words, the number of infected individuals in state 3 has doubled from 1 to 2. So if this event has probability $P_{33} > \frac{1}{2}$, we have a net increase, and exponential growth of the subpopulation of infecteds is bound to happen within a deterministic context.

Exercise 5.36 Individuals in state 3 will enter state 1 with probability $\frac{\sigma + \mu}{\sigma + 2\mu}$. When we arrive in state 3 from state 1, there are two infected individuals to consider. Hence

$$\tilde{R}_0 = 2\frac{\sigma + \mu}{\sigma + 2\mu}P_{13} = 2P_{33}.$$

Exercise 5.37 i) By dividing both the numerator and the denominator by σ^2, we arrive at

$$R_0 = \frac{(1 + \frac{\mu}{\sigma})\frac{h}{\sigma}\rho}{\mu(\frac{h}{\sigma} + 1 + \frac{2\mu}{\sigma})(1 + \frac{2\mu}{\sigma} + \frac{\rho}{\sigma})},$$

which, for $\sigma, h \to \infty$ with $\frac{h}{\sigma} \to p$, has the limit

$$R_0 = \frac{\rho p}{\mu(p + 1)} = \frac{\rho q}{\mu},$$

with $q = \frac{p}{p+1}$.

ii) When $h = p\sigma$ for fixed p, we have

$$R_0 = \frac{(\sigma + \mu)p\rho}{\mu(p + 1 + \frac{2\mu}{\sigma})(\sigma + 2\mu + \rho)}.$$

Now note that both $\frac{\sigma + \mu}{\sigma + 2\mu + \rho}$ and $\frac{1}{p + 1 + \frac{2\mu}{\sigma}}$ are increasing functions of σ.

iii) When $\sigma \gg \mu$, the expected duration of a partnership is approximately $1/\sigma$. Consequently, the probability that transmission occurs during a partnership of an infectious and a susceptible individual is approximately h/σ. In ii) we have taken h/σ to be constant while σ varied. In that case the effect of increasing σ is primarily that an individual is expected to have more partners and hence, since the success ratio per partner is approximately constant, is expected to produce more secondary cases. If, however, h is taken to be fixed (i.e. independent of σ), an increase of σ has the effect that transmission is less likely during the shorter period that a partnership lasts. So then it is not clear how R_0 depends on σ.

In the present class of models we think of transmission opportunities on two levels: on the one hand formation and dissociation of pairs and on the other hand (sexual) contacts within partnerships. When analysing the dependence on parameters, one

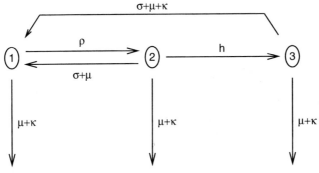

Figure 12.1

has to realise that there may be opposing effects on the different levels. This can create results that at first sight are counterintuitive (such as a decrease of R_0 when σ is increased). When making comparative statements, therefore, the choice and implications of the gauging should be clearly stated and analysed. This is a non-trivial issue that all too often does not get the attention it deserves.

Exercise 5.38 i) When the disease causes an additive death rate κ, the scheme of transitions is as depicted in Figure 12.1.

To compute R_0, we follow the method of Exercise 5.35. This yields

$$P_{13} = \frac{\rho}{\rho + \mu + \kappa} \left(\frac{h}{h + \sigma + 2\mu + \kappa} + \frac{\sigma + \mu}{h + \sigma + 2\mu + \kappa} P_{13} \right),$$

which gives

$$P_{13} = \frac{\rho h}{(\rho + \mu + \kappa)(h + \sigma + 2\mu + \kappa) - \rho(\sigma + \mu)}$$

$$= \frac{\rho h}{(\mu + \kappa)(h + \sigma + 2\mu + \kappa) + \rho(h + \mu + \kappa)}$$

and

$$P_{33} = \frac{\sigma + \mu + \kappa}{\sigma + 2\mu + 2\kappa} P_{13}.$$

Now recall that

$$R_0 = P_{33}(1 + R_0) \Rightarrow R_0 = \frac{P_{33}}{1 - P_{33}},$$

which leads to

$$R_0 = \frac{(\sigma + \mu + \kappa)P_{13}}{\sigma + 2\mu + 2\kappa - (\sigma + \mu + \kappa)P_{13}}$$

$$= \frac{(\sigma + \mu + \kappa)\rho h}{(\sigma + 2\mu + 2\kappa)\{(\mu + \kappa)(h + \sigma + 2\mu + \kappa) + \rho(h + \sigma + \mu + \kappa)\} - (\mu + \sigma + \kappa)\rho h}$$

$$= \frac{(\sigma + \mu + \kappa)\rho h}{(\mu + \kappa)\{(\sigma + 2\mu + 2\kappa)(h + \sigma + 2\mu + \kappa) + \rho(h + \sigma + 2\mu + 2\kappa)\}}.$$

ii) In the scheme in Figure 12.2, the states 1-3 have the same meaning as before and 4 is the state of mourning.

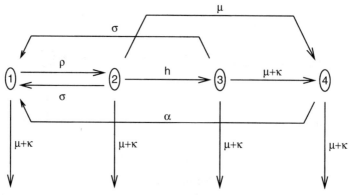

Figure 12.2

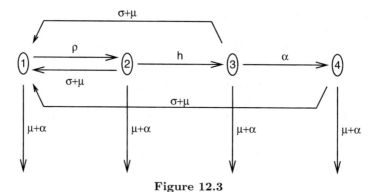

Figure 12.3

Here we have assumed that this state results whenever the partner dies, irrespective of the cause of death (a different model would be to assume that the state 4 is only visited after the death of the partner due to the disease). To compute R_0, one can use, as before,

$$R_0 = P_{33}(1 + R_0),$$

but now with

$$P_{33} = \left(\frac{\sigma}{\sigma + 2\mu + 2\kappa} + \frac{\mu + \kappa}{\sigma + 2\mu + 2\kappa} \frac{\alpha}{\alpha + \mu + \kappa} \right) P_{13}$$

where P_{13} is calculated from

$$
\begin{aligned}
P_{13} = {} & \frac{\rho}{\rho + \mu + \kappa} \left[\frac{h}{h + \sigma + 2\mu + \kappa} + \left(\frac{\sigma}{h + \sigma + 2\mu + \kappa} \right. \right. \\
& \left. \left. + \frac{\mu}{h + \sigma + 2\mu + \kappa} \frac{\alpha}{\alpha + \mu + \kappa} \right) P_{13} \right]
\end{aligned}
$$

(note that for $\alpha \to \infty$ we recover the earlier set of identities). We refrain from the derivation of the explicit expression for R_0.

Exercise 5.39 The scheme is now as in Figure 12.3.

Here 4 means 'paired to an immune individual'. Since the arrows going out of 3 and 4 are exactly the same, we may actually lump states 3 and 4 without any loss

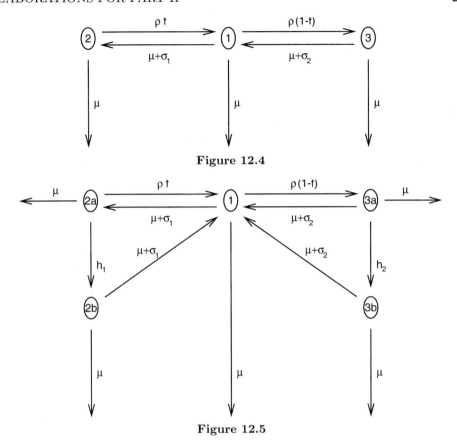

Figure 12.4

Figure 12.5

of information. Applying the computational scheme as before, we find the explicit expression

$$R_0 = \frac{h\rho(\mu + \sigma)}{(\mu + \alpha)(\alpha + h + 2\mu + \sigma)(\alpha + 2\mu + \rho + \sigma)}$$

We should like to draw your attention to a subtle difference between the setting of this exercise and that of Exercise 5.38-i. In the present situation an increase of α leads to a decrease of R_0, as one can deduce immediately from the explicit expression. In the situation of Exercise 5.38-i, the additional death rate has two effects: it shortens the period of infectiousness of the individual we consider (which is just the same as loss of infectiousness), but it may also bring about an early death of an infected partner, thus creating the potential for an increased number of partners. This is reflected in the fact that κ appears in the numerator of the expression for R_0 in Exercise 5.38-i. A more refined analysis establishes that R_0 is still a decreasing function of κ (in other words, the first effect is the more dominant one).

Exercise 5.40 The demographic part of the model is described by the scheme in Figure 12.4.

When we superimpose the transmission of an infective agent, we may split both 2 and 3 into two states; see Figure 12.5.

In Figure 12.5, the individuals in states marked with a have a partner that is susceptible and those in a state marked with b have a partner that is infectious. So there are now two different states, 2b and 3b, in which an individual can begin its life as an infected and we have to set up a 2×2 next-generation matrix and compute R_0 as its dominant eigenvalue. We will see that the matrix has one-dimensional range, the underlying reason being that a newly infected individual necessarily has to pass through state 1 in order to infect other individuals.

An individual in state 2b will with probability $\pi_1 = \frac{\mu+\sigma_1}{2\mu+\sigma_1}$ go to state 1 before dying. For an individual in state 3b, the corresponding probability is $\pi_2 = \frac{\mu+\sigma_2}{2\mu+\sigma_2}$. Consider an infected individual in state 1. Let R_s be the expected number of serious partners that it will infect and R_c the expected number casual partners that it will infect. Then the 2×2 next-generation matrix is

$$M = \begin{pmatrix} \pi_1 R_s & \pi_2 R_s \\ \pi_1 R_c & \pi_2 R_c \end{pmatrix}.$$

Clearly $Mx = (\pi_1 x_1 + \pi_2 x_2)\begin{pmatrix} R_s \\ R_c \end{pmatrix}$, that is, the range of M is spanned by $\begin{pmatrix} R_s \\ R_c \end{pmatrix}$. Hence we have

$$R_0 = \pi_1 R_s + \pi_2 R_c.$$

It remains to calculate R_s and R_c.

To an individual in state 1, three things can happen. It can die, in which case it will make no further 'victims'. With probability $\frac{\rho f}{\rho+\mu}$ it forms a serious partnership and with probability $\frac{\rho(1-f)}{\rho+\mu}$ a casual partnership. In the first case it will transmit the infectious agent to its partner with probability $\frac{h_1}{h_1+2\mu+\sigma_1}$, and it will return to state 1 with probability $\frac{\mu+\sigma_1}{2\mu+\sigma_1}$ (note that the distinction 2a/2b is irrelevant for the probabilities of dying or returning to 1, so for computing the probability of return to state 1 we can lump 2a and 2b). In the second case we have the same expressions, but with index 1 replaced by index 2.

We now combine this information into the identities

$$R_s = \frac{\rho f}{\rho+\mu}\left(\frac{h_1}{h_1+2\mu+\sigma_1} + \frac{\mu+\sigma_1}{2\mu+\sigma_1}R_s\right) + \frac{\rho(1-f)}{\rho+\mu}\frac{\mu+\sigma_2}{2\mu+\sigma_2}R_s,$$

$$R_c = \frac{\rho f}{\rho+\mu}\frac{\mu+\sigma_1}{2\mu+\sigma_1}R_c + \frac{\rho(1-f)}{\rho+\mu}\left(\frac{h_2}{h_2+2\mu+\sigma_2} + \frac{\mu+\sigma_2}{2\mu+\sigma_2}R_c\right).$$

After solving these equations and inserting the result into the expression for R_0, we find

$$R_0 = \frac{\rho F}{(\rho+\mu)(2\mu+\sigma_1)(2\mu+\sigma_2) - \rho f(\mu+\sigma_1)(2\mu+\sigma_2) - \rho(1-f)(2\mu+\sigma_1)(\mu+\sigma_2)}.$$

with

$$F = f\frac{h_1(\mu+\sigma_1)(2\mu+\sigma_2)}{h_1+2\mu+\sigma_1} + (1-f)\frac{h_2(2\mu+\sigma_1)(\mu+\sigma_2)}{h_2+2\mu+\sigma_2}.$$

The present example is ideal for illustrating the fact that one can sometimes meaningfully define another threshold quantity $\tilde{R}_0$, which has a similar interpretation but yet is different from R_0. Consider an infected individual in state 1. Let $\tilde{R}_0$ be the expected number of infected individuals (including the originally considered individual itself) entering state 1 after one potential pair formation event. Then

$$\tilde{R}_0 = \frac{\rho f}{\rho + \mu} \left(1 + \frac{h_1}{h_1 + 2\mu + \sigma_1} \right) \frac{\mu + \sigma_1}{2\mu + \sigma_1} + \frac{\rho(1 - f)}{\rho + \mu} \left(1 + \frac{h_2}{h_2 + 2\mu + \sigma_2} \right) \frac{\mu + \sigma_2}{2\mu + \sigma_2}$$

where the first term is the contribution due to a serious pair being formed and the second term is that due to a casual pair being formed. One easily checks that $R_0 = 1 \Leftrightarrow \tilde{R}_0 = 1$.

To check the calculations, we may take $\sigma_1 = \sigma_2$ and $h_1 = h_2$ and recover the expression for R_0 derived in Exercise 5.34. As a small research programme, one might study the effect of neglecting variations in types of pairs by comparing R_0 as computed in the present exercise with the expression of Exercise 5.34 with $\frac{1}{\sigma} = \frac{f}{\sigma_1} + \frac{1-f}{\sigma_2}$ and h equal to h_1(expected time in pair of type 2)/(expected time in pair) $+h_2$(expected time in pair of type 3)/(expected time in pair).

Exercise 5.41 The point is the phenomenological parameter ρ, which is the probability per unit of time with which a single acquires a partner. When we consider males and females, we have to impose consistency. If the sex ratio is 1:1, this is automatically guaranteed. If we have a different sex ratio, however, we should also have different ρs for males and females.

Exercise 5.42 There are two types at birth: a plant can either become infected while standing in the field, or when being planted or standing in the nursery. Since disease-induced mortality rates and transmission rate constants differ for these types, we have to take both into account. Since these are the only types at birth, R_0 will be the dominant eigenvalue of a 2×2 matrix. For the computation of the elements of this next-generation matrix, we follow Section 5.4, since the transitions in type can, according to the assumptions given in the text, be described by a Markov process. The types at birth are numbered 1 and 2, for plants that are in the field or in the nursery at the moment of infection, respectively. To determine the upper-left element of K, we have to specify the expected number of new cases born with type 1, caused by an individual that was just born with type 1 itself, during its entire infectious life. The three other elements have similar interpretations. We first describe type change, then infectivity. The transition matrix $G = \Sigma - D$ is given by

$$\begin{pmatrix} -(\mu_1 + \rho_1) & \zeta \\ 0 & -(\mu_2 + \rho_2 + \zeta) \end{pmatrix}.$$

We then know that the first column of $-G^{-1}$ gives the fraction of the time period that an infected born in the field is expected to spent in the field (first element) and in the nursery (second element, which will be zero, since field plants do not move to the nursery) respectively. The second column of $-G^{-1}$ has a similar interpretation for individuals whose type at birth is the nursery. Inverting G and multiplying the result by -1 gives the first ingredient for K:

$$-G^{-1} = \begin{pmatrix} \frac{1}{\mu_1 + \rho_1} & \frac{\zeta}{(\mu_1 + \rho_1)(\mu_2 + \rho_2 + \zeta)} \\ 0 & \frac{1}{\mu_2 + \rho_2 + \zeta} \end{pmatrix}.$$

From Section 5.4, we know that if we multiply these expected sojourn times with the type-dependent expected rate of making new victims, we obtain K.

For the infectivity matrix the assumptions lead to

$$\begin{pmatrix} \beta_1 N_1 & 0 \\ p\gamma & \beta_2 N_2 \end{pmatrix},$$

since field plants are expected to produce $\beta_1 N_1$ new field infections per unit of time, and are expected to produce $p\gamma$ infected cuttings (i.e. individuals that are born with type 2, by vertical transmission); infected nursery plants do not have contact with field plants, but are expected to produce $\beta_2 N_2$ new nursery infections per unit of time.

Multiplying the two ingredients in the right order, we find

$$K = \begin{pmatrix} \frac{\beta_1 N_1}{\mu_1+\rho_1} & \frac{\zeta\beta_1 N_1}{(\mu_1+\rho_1)(\mu_2+\rho_2+\zeta)} \\ \frac{p\gamma}{\mu_1+\rho_1} & \frac{\beta_2 N_2}{\mu_2+\rho_2+\zeta} + \frac{p\gamma\zeta}{\mu_2+\rho_2+\zeta} \end{pmatrix}.$$

Finally,

$$R_0 = \frac{1}{2}(k_{11} + k_{22}) + \frac{1}{2}\sqrt{k_{11}^2 - 2k_{11}k_{22} + k_{22}^2 + 4k_{12}k_{21}}.$$

Exercise 5.43 i) There are two types at birth: transiently infected animals (entering class E; type 1) and persistently infected newborns (entering class P; type 2). Transiently infected animals can arise in two ways: by contact either with another transiently infected animal or with a persistently infected animal. Persistently infected animals only arise in one way: through birth from a transiently infected mother in immune class Z_2. Since persistently infected animals have an increased death rate, reduced fertility and a much higher horizontal (mass-action) transmission rate constant (for causing transient infections), compared with transiently infected animals, it makes good sense to explicitly distinguish the two types at birth.

ii) Suppose $\beta_1 = 0$. For the next-generation matrix K, we have to model four elements. The element $k_{11} = 0$ by assumption, since transiently infected animals are, for the moment, not assumed to horizontally infect other animals. Element k_{12} is also straightforward, since it describes the expected number of infections of type 1, caused by a persistently infected individual (type 2) during its entire life. Life expectancy of a persistently infected animal is $\frac{1}{\mu+b}$, and the mass-action transmission rate is $\beta_2 N$. So, $k_{12} = \frac{\beta_2 N}{\mu+b}$. The element k_{22} describes the expected number of persistent infections caused by a persistently infected animal during its entire life. Since persistent infections directly caused by a given individual can only arise vertically through giving birth, we have that $k_{22} = \frac{\mu-a}{\mu+b}$, since all calves will be infected (by assumption). It remains to characterise the element k_{21}. This describes the expected number of persistently infected animals arising from a transiently infected animal during its entire infectious life. We note first that from the modelling assumptions in the text, it follows immediately that this number will be at most one (assuming one calf per pregnancy). The reason is that persistently infected calves can only be born if the mother enters immune state Z_2 after which she becomes permanently immune and so cannot produce an infected calf again. Consider an animal that has just become transiently infected.

With probability $\frac{\alpha}{\mu+\alpha}$ the animal will survive the latency period, and with probability $\frac{\gamma}{\mu+\gamma}$ it will survive the subsequent infectious period and recover. With probability π_2 it will recover into class Z_2. Finally, the animal will survive in the recovered class long enough to give birth with probability $\frac{\phi_2}{\mu+\phi_2}$, and with probability θ the calf will be persistently infected. Combining all terms gives

$$k_{21} = \frac{\alpha}{\mu+\alpha}\frac{\gamma\pi_2}{\mu+\gamma}\frac{\theta\phi_2}{\mu+\phi_2}.$$

The next-generation matrix K has now been completely specified, and R_0 can be calculated as its dominant eigenvalue:

$$R_0 = \frac{1}{2}B + \frac{1}{2}\sqrt{B^2 + 4A},$$

with $B := k_{22}$ and $A := k_{21}k_{12}$.

iii) We have the following series of implications:

$$
\begin{aligned}
R_0 &= \frac{1}{2}B + \frac{1}{2}\sqrt{B^2 + 4A} > 1 \\
&\Leftrightarrow \quad \sqrt{B^2 + 4A} > 2 - B \\
&\Leftrightarrow \quad B^2 + 4A > 4 - 4B + B^2 \\
&\Leftrightarrow \quad A + B > 1.
\end{aligned}
$$

iv) In this case the element k_{11} changes, since now transiently infected animals can cause transient infections. For this, the animal has to survive the latency period, which happens with probability $\frac{\alpha}{\mu+\alpha}$. Subsequently, it is expected to cause $\beta_1 N$ new transient cases per unit of time, during the infectious period of average length $\frac{1}{\mu+\gamma}$. So,

$$k_{11} = \frac{\alpha}{\mu+\alpha}\frac{\beta_1 N}{\mu+\gamma},$$

and the other elements of K are as before.

Exercise 5.44 i) There are three relevant h-states for the rival agent. The population of individuals that can become infected by the rival agent has been structured into susceptible (index 1), infected (index 2) and removed (index 3) individuals (all with respect to the resident agent). Note that the type of an individual is dynamic, since even though the steady state is fixed as $N_1 = \overline{S}$, $N_2 = \overline{I}$ and $N_3 = \overline{R}$, individual hosts will turn from susceptible (to resident) to infected (by resident) and immune (for the resident) (note also that these are the only type changes allowed). Depending on its status with respect to the resident agent, a host individual might well have different susceptibility and infectivity with respect to the rival agent. It therefore makes good sense to distinguish the three states as different types at birth for the rival agent.

ii) Let q_i and θ_i be the susceptibility and infectivity factors with respect to the rival agent. By this we mean that an individual of type i carrying the rival transmits the rival agent at rate $\theta_i q_j N_j$ to type j individuals. Then the next-generation matrix $K = (k_{ij})$ has a product structure, i.e. $k_{ij} = q_i N_i \xi_i$ (with ξ_i still to be specified; see below) and $R_0 = \sum_{i=1}^{3} q_i N_i \xi_i$.

For $j = 3$ we take $\xi_3 = \sigma_3 \theta_3$, with σ_3 the expected period of time that a rival infected, resident immune, individual will be rival infectious. A possible choice for σ_3 with this interpretation is $\sigma_3 = 1/(\tilde{\alpha} + \mu)$, corresponding to an exponentially distributed (with parameter $\tilde{\alpha}$) rival infectious period.

For $j = 2$ we have to take into account that the host may experience the resident infectious $\rightarrow$ resident immune transition while being rival infectious. Therefore we put $\xi_2 = \sigma_2 \theta_2 + \pi_2 \sigma_3 \theta_3$, where π_2 denotes the probability that such a transition does indeed occur. We might take $\sigma_2 = 1/(\tilde{\alpha} + \mu + \alpha)$ and $\pi_2 = \alpha/(\tilde{\alpha} + \mu + \alpha)$ (those readers who cannot interpret the expression for π_2 should consult Exercise 1.33-vi).

Likewise, we take $\xi_1 = \sigma_1 \theta_1 + + \pi_1 \sigma_2 \theta_2 + \pi_1 \pi_2 \sigma_3 \theta_3$, with, for instance, $\sigma_1 = 1/(\tilde{\alpha} + \mu + \beta \overline{S})$ and $\pi_1 = \beta \overline{S}/(\tilde{\alpha} + \mu + \beta \overline{S})$.

Note that we have implicitly assumed, when giving such expressions for σ_i and π_i, that the various parameters governing the dynamics of the host's status with respect to the resident agent are not changed by the presence of the rival agent.

Obedient readers remembering Section 5.4 would probably have written

$$k_{ij} = \int_0^\infty h(\tau) \sum_{\ell=1}^3 c(i, \ell) P(\tau, \ell, j) \, d\tau$$

and then have assumed product structure for c. Essentially this is what we did as well, except that we allowed the infectivity h to depend on ℓ and that we immediately introduced the integrals with respect to τ in the form of the σs. Readers are invited to check for themselves that working with the expression above gives the same result for R_0.

iii) Since k_{ij} is a separable kernel, we have the expression

$$R_0 = \sum_{i=1}^3 q_i N_i \xi_i = q_1(\sigma_1 \theta_1 + + \pi_1 \sigma_2 \theta_2 + \pi_1 \pi_2 \sigma_3 \theta_3) \overline{S} + q_2(\sigma_2 \theta_2 + \pi_2 \sigma_3 \theta_3) \overline{I} + q_3 \sigma_3 \theta_3 \overline{R}$$

for the basic reproduction ratio of the rival agent in the environmental conditions as set by the resident agent. From the steady-state conditions for the resident, we have that

$$\frac{\beta}{\alpha + \mu} \overline{S} = 1.$$

So whenever $q_1 \sigma_1 \theta_1 = \frac{\beta}{\alpha + \mu}$ (which means that the rival does equally well as the resident on the reservoir of susceptible (to both agents) individuals) and any of the remaining terms in the expression for R_0 are strictly positive, the rival agent can grow when still rare. This is due to a certain asymmetry in the assumptions concerning the resident and the mutant. Indeed, for the resident, superinfection has no effect, and the R category consists of immunes. But for the mutant, superinfection yields a positive contribution to overall transmission (when $q_2 \sigma_2 \theta_2 > 0$) and/or cross-immunity is only partial (when $q_3 \sigma_3 \theta_3 > 0$). This asymmetry persists under reversal of the role of resident and mutant, and consequently we have mutual invadability resulting in coexistence, but due to a debatable assumption. (From an adaptive dynamics point of view, the next issue is to investigate the invasibility of a coalition of two agents by a third rival; see e.g. Andreasen & Pugliese (1995)[1] for additional remarks).

To make an investigation of the above type truly useful, one would need to specify a submodel for the interaction of the agents with each other via the immune system of the host, and try to derive the q, θ, σ and π vectors from that.

12.2   Elaborations for Chapter 6

Exercise 6.1 As the epidemic systematically affects more unvaccinated than vaccinated individuals (compared with the ratio in which they occur), the proportion of vaccinated individuals among the susceptibles will gradually increase. As we ignore this effect for the time being, we have to add the adjective 'early'.

Exercise 6.2 Because of the one-dimensional range for a two-dimensional mapping, the eigenvalues are 0 and R_v, and hence their sum equals R_v. Since the trace of a matrix (the sum of the diagonal terms) always equals the sum of the eigenvalues (counting multiplicity), we find the stated expression.

The alternative is to apply the matrix to the vector $(1 - p, pf)^\top$ that spans the range. This yields the vector

$$(1 - p + pf\phi)R_0 \begin{pmatrix} 1 - p \\ pf \end{pmatrix},$$

and hence $R_v = (1 - p + pf\phi)R_0$.

Exercise 6.3 The incidence $i(t)$ is now a vector with two components, the first corresponding to unvaccinated individuals and the second to vaccinated individuals. Exactly as (1.9) was derived in Section 1.2.3, we deduce that

$$i(t) = \int_0^\infty cA(\tau) \begin{pmatrix} 1 - p & (1 - p)\phi \\ pf & pf\phi \end{pmatrix} i(t - \tau) \, d\tau.$$

Upon making the Ansatz that $i(t) = e^{rt}i^s$ for the (as-yet unknown) stable distribution vector i^s, it follows that the matrix

$$\int_0^\infty cA(\tau)e^{-r\tau} \, d\tau \begin{pmatrix} 1 - p & (1 - p)\phi \\ pf & pf\phi \end{pmatrix}$$

should have dominant eigenvalue one.

Referring back to Exercise 2.12, we see that the time dependence enters in the same way for any susceptible-infective combination (i.e. any position in the matrix) and that consequently i^s equals the stable distribution in a generation perspective. The one-dimensional range property allowed us to calculate the latter explicitly as $\begin{pmatrix} 1-p \\ pf \end{pmatrix}$, modulo a normalisation factor. Inserting this expression for i^s (or using the trace argument of Exercise 6.2), we find that r_v should satisfy the equation

$$1 = \frac{R_v \int_0^\infty e^{-r_v\tau} A(\tau) \, d\tau}{\int_0^\infty A(\tau) \, d\tau}.$$

---

[1] V. Andreasen & A. Pugliese: Pathogen coexistence induced by density dependent host mortality. *J. theor. Biol.*, **177** (1995), 159-165.

Exercise 6.4 As a warm-up, do Exercise 1.6 (or read its elaboration). We introduce the symbol $\widetilde{q}^{1}(k_1, k_2)$ to denote the probability that a type 1 infected individual infects in total k_1 type 1 individuals and k_2 type 2 individuals, and the symbol $\widetilde{q}^{2}(k_1, k_2)$ to denote the corresponding probability for an infected individual of type 2. By so-called first-step analysis (as it is used more generally in the theory of Markov processes; see e.g. Taylor and Karlin (1984)), we arrive at the consistency conditions

$$\pi_i = \sum_{k_1,k_2=0}^{\infty} \widetilde{q}^{i}(k_1, k_2)\pi_1^{k_1}\pi_2^{k_2}, \quad i = 1, 2.$$

To proceed, we have to determine the $\widetilde{q}^{i}(k_1, k_2)$ from the assumptions about the contact process and the probability of transmission. To do so, we make a short cut based on the ideas presented in (the elaboration of) Exercise 1.9. That is, we note that the numbers of infected individuals of the two types are independent random variables with a Poisson distribution, with parameters that are easily expressed in terms of R_0, p, f and ϕ. In fact, we have

$$\widetilde{q}^{i}(k_1, k_2) = q^{i,1}(k_1)q^{i,2}(k_2),$$

with

$$q^{1,1}(k_1) = \frac{((1-p)R_0)^{k_1}e^{-(1-p)R_0}}{k_1!},$$

$$q^{1,2}(k_2) = \frac{(pfR_0)^{k_2}e^{-pfR_0}}{k_2!},$$

$$q^{2,1}(k_1) = \frac{((1-p)\phi R_0)^{k_1}e^{-(1-p)\phi R_0}}{k_1!},$$

$$q^{2,2}(k_2) = \frac{(pf\phi R_0)^{k_2}e^{-pf\phi R_0}}{k_2!}.$$

Evaluation of the sums now leads to the equations (6.1).

Exercise 6.5 If we write $\pi_i = e^{A_i}$ then $A_2 = \phi A_1$, so we have $\pi_2 = \pi_1^{\phi}$, which reflects that the infectivity of type 2 individuals is reduced by a factor ϕ.

Exercise 6.6 We start with the last question: if we exposed the population to infectious material from outside and one individual were to become infected, this would with probability $\frac{1-p}{1-p+pf}$ be a type 1 individual and with probability $\frac{pf}{1-p+pf}$ a type 2 individual. In the first case, we would get a major outbreak with probability $1-\pi_1$ and in the second case one with probability $1-\pi_2$. So, modulo the normalising factor $(1-p+pf)^{-1}$, ξ is the probability of a major outbreak, given that one individual is infected as a result of uniform exposure to infectious material coming from outside the population.

The definition of ξ and the equations (6.1) imply that $\pi_1 = e^{-R_0\xi}$ and $\pi_2 = e^{-\phi R_0\xi}$, and upon substitution of these relations into the defining relation for ξ, equation (6.2) for ξ is obtained.

Exercise 6.7 The probability to escape is the zero term of a Poisson distribution with an intensity parameter involving the total force of infection exerted during the entire epidemic, which equals $R_0((1-p)(1-\sigma_1) + \phi p(1-\sigma_2))$ in the case of type 1 individuals. This quantity is reduced by a factor f in the case of type 2 individuals. Thus we arrive at the equations (6.3).

Exercise 6.8 We first note that $\sigma_2 = \sigma_1^f$, which reflects that the susceptibility of type 2 individuals is reduced by a factor f. In terms of θ, the equations (6.3) can be rewritten as $\sigma_1 = e^{-R_0\theta}$ and $\sigma_2 = e^{-fR_0\theta}$, and upon substitution of these relations into the defining relation for θ, equation (6.4) for θ is obtained. The quantity θ is (proportional to) the size of a major outbreak when we measure 'size' in terms of the output of infectious material (indeed, note the factor ϕ in the second term).

Exercise 6.9 The equations (6.2) and (6.4) for ξ and θ respectively are identical except for an interchange of f and ϕ. To express this more precisely, we first incorporate the dependence on f and ϕ into the notation by writing $\xi = \xi(f, \phi)$ and $\theta = \theta(f, \phi)$, and then note that $\xi(f, \phi) = \theta(\phi, f)$.
  Let k with $0 < k < 1$ be given and define

$$J(f) := \xi(f, \frac{k}{f}) \, \theta(f, \frac{k}{f})$$

for $k \le f \le 1$. The goal is to minimise J.
  Our first observation is that

$$J(1) = \xi(1, k) \, \theta(1, k) = \theta(k, 1) \, \xi(k, 1) = J(k),$$

or, in words, that the boundary values are equal. In fact, we have the more general symmetry relation

$$J(f) = \xi(f, \frac{k}{f}) \, \theta(f, \frac{k}{f}) = \theta(\frac{k}{f}, f) \, \xi(\frac{k}{f}, f) = J(\frac{k}{f}),$$

which implies that J has an extreme in the 'midpoint' $f = \sqrt{k}$. Is this a maximum or a minimum?
  The first attempt is to find an intuitive argument that yields the answer. Alas, this is in vain. The second attempt is to calculate and investigate the derivative of J (as well as the second derivative). Alas, this is also in vain. The third attempt is to look at extreme situations: very large R_0 on the one hand, and $R_0(1-p+kp)$ only slightly greater than 1 on the other hand.
  For $R_0 \to \infty$, we have $\xi \to 1-p+fp$ (since the exponentials in equation (6.2) go to zero) and $\theta \to 1-p+\phi p$ (for an analogous reason). So, $\xi\theta \to (1-p)^2 + kp^2 + (f + \frac{k}{f})p(1-p)$. As one easily verifies, the function $f \mapsto f + \frac{k}{f}$ has a *minimum* at $f = \sqrt{k}$. We conclude that for large values of R_0 vaccines that put all their reduction potential into either susceptibility or infectivity (irrelevant which of the two) are better than vaccines that have a 'mixed' strategy.
  Next, let us examine the critical situation characterised by $R_0(1-p+kp) = 1$ (see Exercise 6.2). In that situation both θ and ξ are small and we may approximate the exponentials by the first few terms in their Taylor expansion. This leads to

$$\xi = (1-p)\left(R_0\xi - \frac{1}{2}R_0^2\xi^2\right) + fp\left(\phi R_0\xi - \frac{1}{2}R_0^2\phi^2\xi^2\right) + \text{h.o.t.},$$

and, after dividing by ξ, neglecting the 'h.o.t' and solving for ξ, to

$$\xi = \frac{(1-p+kp)R_0 - 1}{1-p+kp\phi} \frac{2}{R_0^2} + \cdots.$$

Likewise we have (by replacing ϕ by f)

$$\theta = \frac{(1-p+kp)R_0 - 1}{1-p+kpf} \frac{2}{R_0^2} + \cdots.$$

The product $\theta\xi$ therefore has, in leading order, all f dependence in the factor

$$(1-p+kp\phi)(1-p+kpf) = (1-p)^2 + k^3 p^2 + k\left(f+\frac{k}{f}\right)p(1-p)$$

in the denominator. Since $f \mapsto f + \frac{k}{f}$ has a minimum at $f = \sqrt{k}$, the product $\theta\xi$ has a maximum at $f = \sqrt{k}$. We conclude that, near criticality, the best vaccines are those that divide their reduction potential equally over susceptibility and infectivity.

The final conclusion is that, within the setting considered in this section, it depends on quantitative details what type of vaccine performs best.

As an encore (but who asked for it?), we look briefly at the special case $p = 1$ (everybody vaccinated). Then

$$\xi = fp(1 - e^{-\phi R_0 \xi}) \Rightarrow \phi\xi = k(1 - e^{-\phi R_0 \xi})$$

and

$$\theta = \phi p(1 - e^{-\phi R_0 \theta}) \Rightarrow f\theta = k(1 - e^{-\phi R_0 f}),$$

which shows that both $\phi\xi$ and $f\theta$ depend only on k and R_0 (and not on f and ϕ separately) and are in fact equal to each other. Hence $\xi\theta = \frac{1}{k}(\phi\xi)(f\theta)$ is independent of f when $f\theta = k$. That is, if all individuals are vaccinated, it makes no difference whether the vaccine reduces infectivity or susceptibility.

Exercise 6.11 The operator

$$(K_r\phi)(\xi) = N(\xi)a(\xi) \int_\Omega \int_0^\infty b(\tau, \eta)e^{-r\tau}\, d\tau\, \phi(\eta)\, d\eta$$

has a one-dimensional range spanned by $N(\cdot)a(\cdot)$, and hence the one and only non-zero eigenvector equals

$$\int_\Omega \int_0^\infty b(\tau, \eta)e^{-r\tau}\, d\tau\, N(\eta)a(\eta)\, d\eta,$$

and the equation for r is obtained by requiring that this expression equals 1.

Exercise 6.14 Define

$$I(t, \xi) := \int_{-\infty}^t e^{-\alpha(\xi)(t-\tau)} i(\tau, \xi)\, d\tau,$$

which is the total size of the infective subpopulation with h-state ξ at time t. Differentiation of this expression with respect to t gives

$$\frac{dI}{dt} = i(t, \xi) - \alpha(\xi)I(t, \xi), \tag{12.1}$$

where $i(t, \xi)$ is given by (6.6), in the nonlinear version. If we substitute the assumption on A into the nonlinear version of (6.6), we obtain

$$i(t, \xi) = S(t, \xi) \int_0^\infty \int_\Omega \beta(\xi, \eta) e^{-\alpha(\eta)\tau} i(t - \tau, \eta) \, d\eta \, d\tau.$$

Exchange of the order of integration using Fubini's Theorem (see e.g. Rudin (1974)) and substitution of $\tau \leftrightarrow t - \tau$ leads to

$$i(t, \xi) = S(t, \xi) \int_\Omega \beta(\xi, \eta) \int_{-\infty}^t e^{-\alpha(\eta)(t-\tau)} i(\tau, \eta) \, d\tau \, d\eta.$$

So, together with the definition of $I(t, \eta)$, we have

$$i(t, \xi) = S(t, \xi) \int_\Omega \beta(\xi, \eta) I(t, \eta) \, d\eta. \tag{12.2}$$

Finally, substitution of (12.1) into (12.2) gives the differential equation for $I(t, \xi)$ we were looking for:

$$\frac{dI}{dt}(t, \xi) = S(t, \xi) \int_\Omega \beta(\xi, \eta) I(t, \eta) \, d\eta - \alpha(\xi) I(t, \xi).$$

Exercise 6.15 Define

$$I(t, a) := \int_0^a i(t - \tau, a - \tau) e^{-\int_{a-\tau}^a \alpha(\sigma) \, d\sigma} \, d\tau,$$

and find by differentiation

$$\frac{\partial I}{\partial t} + \frac{\partial I}{\partial a} = i(t, a) - \alpha(a) I(t, a). \tag{12.3}$$

If we substitute the assumption on A into the nonlinear version of (6.6), we obtain

$$i(t, a) = S(t, a) \int_0^\infty \int_0^\infty \beta(a, \eta + \tau) e^{-\int_\eta^{\eta+\tau} \alpha(\sigma) \, d\sigma} i(t - \tau, \eta) \, d\eta \, d\tau.$$

Substitution of $a' = \eta + \tau$ gives

$$i(t, a) = S(t, a) \int_0^\infty \int_\tau^\infty \beta(a, a') e^{-\int_{a'-\tau}^{a'} \alpha(\sigma) d\sigma} i(t - \tau, a' - \tau) \, da' \, d\tau,$$

and then, using Fubini's Theorem,

$$i(t, a) = S(t, a) \int_0^\infty \beta(a, a') \int_0^{a'} e^{-\int_{a'-\tau}^{a'} \alpha(s) \, ds} i(t - \tau, a' - \tau) \, d\tau \, da'.$$

We can rewrite this as

$$i(t, a) = S(t, a) \int_0^\infty \beta(a, a') I(t, a') \, da'$$

and substitute the result into (12.3) to obtain the partial differential equation we were looking for.

Exercise 6.16 $\frac{\partial S}{\partial t}(t, x)$ is the number of new cases per unit time per unit space (if S is a spatial density) at time t, with h-state x at the moment of becoming infected. The right-hand side is the product of two factors, the first being the density $S(t, x)$ of susceptibles at time t with h-state x. So the second factor is, by definition, the force of infection, i.e. the probability for a susceptible with h-state x to become infected at time t.

The modelling assumption is that the force of infection is the sum of contributions of individuals that themselves became infected τ units of time earlier (so at time $t - \tau$), while having h-state ξ. The size of that contribution is measured by the modelling ingredient $A(\tau, x, \xi)$, which has to be specified in more detail (and eventually even quantitatively, perhaps) in order to make the model concrete and applicable to a specific infectious agent in a specific 'host' population. It is the expected force of infection exerted on individuals with h-state x by individuals that had h-state ξ at the moment they became infected, τ units of time ago.

The equation may serve to describe, for example, a fungal disease in an agricultural crop, by simply interpreting x and ξ as referring to spatial position (and to let S indeed be a spatial density). A special feature in that case is that the h-state is static (plants do not move), and that consequently it may make sense to let $A(\tau, x, \xi)$ only depend on the spatial distance $|x - \xi|$ and not on x and ξ separately (see Chapter 8 for elaboration).

Inflow of new susceptibles is *not* included, since we equate the incidence to $\frac{\partial S}{\partial t}(t, x)$, requiring that there are no changes in S due to causes other than infection. So, for the same reason, return to the class of susceptibles by loss of immunity after having experienced infection is also *not* included.

Exercise 6.17 Divide both sides of (6.20) by $S(t, x)$ and integrate over time from $-\infty$ to t. The left-hand side then becomes

$$\int_{-\infty}^{t} \frac{1}{S(\sigma, x)} \frac{\partial S}{\partial t}(\sigma, x) \, d\sigma = \int_{-\infty}^{t} \left[\frac{d}{dt} \ln S(t, x) \right]_{t=\sigma} d\sigma$$

$$= \ln S(t, x) - \ln S(-\infty, x) = \ln \frac{S(t, x)}{S(-\infty, x)},$$

while for the right-hand side we have

$$\int_{-\infty}^{t} \int_{\Omega} \int_{0}^{\infty} A(\tau, x, \xi) \frac{\partial S}{\partial t}(\sigma - \tau, \xi) \, d\tau \, d\xi \, d\sigma$$

$$= \int_{-\infty}^{t} \int_{\Omega} \int_{0}^{\infty} A(\tau, x, \xi) \frac{d}{d\sigma} S(\sigma - \tau, \xi) \, d\tau \, d\xi \, d\sigma$$

$$= \int_{\Omega} \int_{0}^{\infty} A(\tau, x, \xi) \int_{-\infty}^{t} \frac{d}{d\sigma} S(\sigma - \tau, \xi) \, d\sigma \, d\tau \, d\xi$$

$$= \int_{\Omega} \int_{0}^{\infty} A(\tau, x, \xi) \{S(t - \tau, \xi) - S(-\infty, \xi)\} \, d\tau \, d\xi.$$

Together this yields equation (6.22), which, by taking the limit $t \to \infty$, implies (6.21).

Exercise 6.18 Take exponentials of both sides of (6.21) after having slightly rewritten the right-hand side, and you arrive at (6.23). The point of this exercise, however, is to show that one may derive (6.23) directly by employing the arguments introduced in Section 1.3.1.

The left-hand side of (6.23) is the fraction of individuals with h-state x that escape infection, which we equate to the probability to escape infection. The right-hand side is the zero term of a Poisson distribution with intensity

$$\int_\Omega \int_0^\infty A(\tau, x, \xi) \, d\tau \, S(-\infty, \xi) \left\{ 1 - \frac{S(\infty, \xi)}{S(-\infty, \xi)} \right\} d\xi,$$

which is indeed the total force of infection exerted on individuals with h-state x, given that for any ξ a fraction $S(\infty, \xi)/S(-\infty, \xi)$ fell victim. Thus (6.23) is indeed shown to be a consistency condition.

Exercise 6.19 R_0 was defined in Chapter 5; see in particular Section 5.2. The question should be reformulated as: how does the kernel $k(x, \xi)$ of the integral operator K relate to the present modelling ingredients $S(-\infty, \xi)$ and $A(\tau, x, \xi)$? The answer is

$$k(x, \xi) = S(-\infty, \xi) \int_0^\infty A(\tau, x, \xi) \, d\tau.$$

The linear integral operator with this kernel has R_0 as its dominant positive eigenvalue (assuming the spectral radius is indeed an eigenvalue).

If we introduce

$$u(x) := -\ln \frac{S(\infty, x)}{S(-\infty, x)},$$

we can rewrite (6.23) as

$$u(x) = \int_\Omega k(x, \xi) \left(1 - e^{-u(\xi)} \right) d\xi. \tag{12.4}$$

On the basis of the interpretation we expect that (6.23) has a solution $S(\infty, \xi) < S(-\infty, \xi)$ when $R_0 > 1$ (note that we, of course, consider $S(-\infty, \xi)$ as given and $S(\infty, \xi)$ as unknown). This translates into (12.4) having a positive solution for $R_0 > 1$. What kind of arguments enable us to check our expectations?

In the rest of this elaboration we shall sketch the essence of such arguments, without going into the technical details that are involved in complete and rigorous mathematical proofs.

Two key observations are as follows:

1. $u \mapsto 1 - e^{-u}$ is an increasing (and bounded) function.
2. $1 - e^{-u} \geq (1 - \varepsilon)u$ for ε positive and small and $0 \leq u \leq \bar{u}$, where $\bar{u} = \bar{u}(\varepsilon)$ is the positive root of $1 - e^{-u} = (1 - \varepsilon)u$. (Draw a picture for yourself; use the fact that the graph of $u \mapsto 1 - e^{-u}$ has tangent line $u \mapsto u$ in the origin.)

We can rewrite (12.4) as

$$u = Tu,$$

where T is the nonlinear integral operator defined by

$$(Tu)(x) = \int_\Omega k(x, \xi) \left(1 - e^{-u(\xi)}\right) d\xi.$$

For functions defined on Ω we shall use the order relation $\phi \geq \psi$ (defined by $\phi \geq \psi$ if and only if $\phi(x) \geq \psi(x)$ for all $x \in \Omega$, or, for almost all $x \in \Omega$ in the case that we work with integrable functions that are not necessarily defined for all $x \in \Omega$). From now on we restrict attention to positive (i.e. non-negative) functions.

The operator T is monotone, i.e. $\phi \geq \psi \Rightarrow T\phi \geq T\psi$ (this is a direct consequence of the positivity of k and the monotonicity of $u \mapsto 1 - e^{-u}$, as observed under (1) above). Bounded monotone sequences converge (this can be extended to sequences of functions under an additional technical condition (compactness) related to the precise form of convergence that one wants to consider. (On the one hand, in the present situation $L_1(\Omega)$ convergence is most natural, but, on the other hand, uniform convergence, i.e. $C(\Omega)$ convergence, is more easy to work with; we shall not dwell on this, but confine ourselves to the remark that one needs a minor technical condition on the kernel k that guarantees that the integral operator T is compact in the function space considered.) We want to generate a monotone sequence by applying the operator T repeatedly, i.e. by considering the iteration scheme

$$u_{n+1} = Tu_n.$$

Since T is (assumed to be) continuous, a convergent sequence u_n yields a fixed point of T (just pass to the limit in $u_{n+1} = Tu_n$), i.e. a solution of (12.4). To generate a monotone sequence, we only have to get started, i.e. to find u_0 such that

$$u_1 := Tu_0 \geq u_0,$$

since the monotonicity of T then guarantees, by induction, that $u_{k+1} \geq u_k$ for all k. We now concentrate on finding u_0. It is here that the assumption $R_0 > 1$ and the key observation (2) above are crucial.

Choose ε positive but small enough to guarantee that $(1 - \varepsilon)R_0 > 1$. Let ϕ be the eigenfunction of the linear integral operator K corresponding to the eigenvalue R_0. Assume that ϕ is bounded (this is another minor assumption on the kernel k, as a rule subsumed under earlier ones). Now choose $u_0 = \delta\phi$ with δ such that $u_0(x) \leq \bar{u} = \bar{u}(\varepsilon)$. Then

$$(Tu_0)(x) = \int_\Omega k(x, \xi) \left(1 - e^{-u_0(\xi)}\right) d\xi \geq (1 - \varepsilon) \int_\Omega k(x, \xi)u_0(\xi) \, d\xi,$$

which leads to

$$(Tu_0)(x) \geq (1 - \varepsilon)\delta \int_\Omega k(x, \xi)\phi(\xi) \, d\xi = (1 - \varepsilon)R_0\delta\phi(x) = (1 - \varepsilon)R_0u_0(x) \geq u_0(x).$$

For $u_0 = \delta\phi$, therefore, the iteration scheme $u_{n+1} = Tu_n$ yields a non-decreasing sequence. Since $1 - e^{-u} \leq 1$ for $u \geq 0$, boundedness is immediate:

$$u_n(x) \leq \int_\Omega k(x, \xi)d\xi.$$

This ends our sketch of the proof of the implication $R_0 > 1 \Rightarrow$ '(6.21) has a non-trivial solution'.

What about the uniqueness of the non-trivial solution of (6.21)? To study uniqueness, the notion of *sublinearity* as introduced by Krasnoselskiĭ is very helpful.[2] Let $\widetilde{u}$ be any positive number. On the interval $(0, \widetilde{u})$ the graph of $u \mapsto 1 - e^{-u}$ lies above the straight line through the origin and $(\widetilde{u}, 1 - e^{-\widetilde{u}})$. Analytically this is expressed by

$$1 - e^{-\theta \widetilde{u}} > \theta(1 - e^{-\widetilde{u}}) \quad \text{for } 0 < \theta < 1.$$

Now let u and w be two strictly positive solutions of (12.1). Let $\widetilde{u}$ be a common upper bound for u and w. Define $\theta = \inf_{x \in \Omega} \frac{u(x)}{w(x)}$. Assume that the infimum is actually attained at some point $x = \overline{x}$ (alternatively, assume that Ω is compact and that u and w are continuous). Suppose that $\theta < 1$. Then we arrive at a contradiction by noting that

$$u(\overline{x}) = \int_\Omega k(\overline{x}, \xi) \left(1 - e^{-u(\xi)}\right) d\xi \geq \int_\Omega k(\overline{x}, \xi) \left(1 - e^{-\theta w(\xi)}\right) d\xi,$$

so

$$u(\overline{x}) > \theta \int_\Omega k(\overline{x}, \xi) \left(1 - e^{-w(\xi)}\right) d\xi = \theta w(\overline{x}).$$

Therefore $\theta = 1$ or $u(x) \geq w(x)$. But, as we can consider $\frac{w(x)}{u(x)}$ in exactly the same way, we also find $u(x) \leq w(x)$, and hence can conclude that $u(x) = w(x)$.

Exercise 6.20 i) $B(x)$ is the rate at which individuals with h-state x are born. We are supposing here that the h-state is static! Individuals with h-state x die from 'natural' causes at a per capita rate $\mu(x)$. The incidence of individuals with h-state x at time t is $i(t, x)$.

ii) The infection-free steady state is $\widetilde{S}(x) = \frac{B(x)}{\mu(x)}$.

iii) R_0 is the dominant eigenvalue of the linear integral operator with kernel

$$k(x, \xi) = \widetilde{S}(x) \int_0^\infty A(\tau, x, \xi) \, d\tau.$$

iv) An endemic steady state is characterised by the two equations

$$
\begin{aligned}
B(x) - \mu(x)\overline{S}(x) &= \overline{i}(x), \\
\overline{i}(x) &= \frac{\overline{S}(x)}{\widetilde{S}(x)} \int_\Omega k(x, \xi)\overline{i}(\xi) \, d\xi,
\end{aligned}
$$

which, by elimination of $\overline{S}(x)$ and use of the expression for $\widetilde{S}(x)$, combine into the equation

$$\overline{i}(x) = \left(1 - \frac{\overline{i}(x)}{B(x)}\right) \int_\Omega k(x, \xi)\overline{i}(\xi) \, d\xi$$

for the unknown $\overline{i}$ with $0 \leq \overline{i}(x) \leq B(x)$.

---

[2] M.A. Krasnoselskiĭ: *Positive Solutions of Operator Equations.* Noordhoff, Groningen, 1964, (Sections 6.1.3 and 6.1.7).

v) Writing the previous equation as $\bar{i} = T\bar{i}$, we see that the linearisation of T at the trivial (infection-free) fixed point $\bar{i} = 0$ gives the linear integral operator K, which has R_0 as its dominant eigenvalue. So the situation is reminiscent of the situation considered in the preceding exercise. A major difference, however, is that T is *not* monotone in the present case. One has to resort to more complicated arguments.[3]

Exercise 6.21 What is the probability that the 'line' originating with a 'child' of an individual of type x goes extinct? If that child has type ξ, this probability equals, by definition, $\pi(\xi)$. We do not know the type of the child, however; we only know the probability distribution $m(x, \cdot)$. Accordingly, the probability that the line goes extinct is

$$\int_\Omega \pi(\xi) m(x, d\xi).$$

If there are k children, the probability that all k lines will go extinct is simply the kth power of this quantity, by (assumed) independence. The identity (6.26) is now self-evident.

Exercise 6.22 The aim is to relate p_k and m to the modelling ingredients $S(-\infty, \cdot)$ and A. The quantity $f(y, x)$ is the expected number of 'children' with type y produced by an individual of type x (in the initial phase, where the reduction of susceptibles by infection can still be ignored). If we integrate this quantity with respect to y, we get the total expected number of offspring. Although perhaps not stated very explicitly, an assumption made throughout is that contacts (and hence transmissions) are made according to a Poisson process. As the Poisson process is completely specified by one parameter, the mean, we can determine $p_k(x)$ from this mean, and the result is (6.27).

The interpretation of $f(\cdot, x)$ and $m(x, \cdot)$ implies at once that f is the density of m, apart from the normalisation that is incorporated in m, but not in f. Hence (6.28) should hold.

If we substitute (6.27) and (6.28) into (6.26), a factor $\left(\int_\Omega f(y, x)\, dy\right)^k$ in $p_k(x)$ cancels the same factor in the denominator due to the normalisation of m, and (6.29) results.

Exercise 6.23 The assumption (6.30) says that the expected force of infection acting on individuals of type x exerted by individuals that had state ξ at the moment of becoming infected factorises as a product, with one factor depending only on the characteristic x of the potential victim and one factor depending only on the characteristics ξ and τ, the time elapsed since infection, of the infective considered. Recalling the function f and the measure m as introduced in Exercise 6.24, we find that the distribution of the h-state of newly infected individuals has density $y \mapsto a(y)S(-\infty, y)/\int_\Omega a(\eta)S(-\infty, \eta)\, d\eta$, for all x (the function $y \mapsto f(y, x) = a(y)S(-\infty, y)/\int_0^\infty b(\tau, x)\, d\tau$ has an x-dependent factor, but this is eliminated when we normalise).

Exercise 6.24 The idea behind the Ansatz is that the force of infection acting on individuals with h-state x factorises into an x-specific factor $a(x)$ and a function of time. Somewhat loosely, we may think of this function of time as the amount

---

[3] R.D. Nussbaum: *The Fixed Point Index and Some Applications.* Seminaire de Mathématiques Superieures. Les Presses de l'Université Montreal, 1985.

of infectious material 'in the air' at time t. The function $w(t)$ then measures the cumulative amount of infectious material to which individuals have been exposed. By integration (after dividing both sides by $S(t, x)$), we find

$$S(t, x) = S(-\infty, x)e^{-a(x)w(t)}.$$

If we substitute this into (6.22) and use (6.30), we find (6.32). The key point is that the unknown $w(t)$ in (6.32) is a *scalar* quantity. In fact (6.32) is a so-called nonlinear renewal equation (which is a Volterra integral equation of convolution type), written in time-translation-invariant form. To obtain the more traditional form, split the interval of τ integration into $[0, t)$ and $[t, \infty)$ and pretend that the contribution of $[t, \infty)$ yields a 'known' function $f(t)$ (since, indeed, it gives the contribution to the cumulative quantity $w(t)$ of those individuals that were infected before time zero).

Exercise 6.25 The kernel k also factorises:

$$k(x, \xi) = a(x)S(-\infty, x)\int_0^\infty b(\tau, \xi)\, d\xi.$$

Hence the integral operator K has one-dimensional range and we obtain for R_0 the explicit expression

$$R_0 = \int_\Omega \left(\int_0^\infty b(\tau, \xi)d\tau\right) a(\xi)S(-\infty, \xi)\, d\xi.$$

Consider the right-hand side of the final-size equation (6.34) as a function $g = g(w(\infty))$ of $w(\infty)$. For small values of w we have $g(w) \sim R_0 w$ (since $1 - e^{-aw} \sim aw$ for $w \to 0$). So, when $R_0 > 1$, the graph of g starts out above the 45° line. For $w \to \infty$ we find that g tends to a constant (since $e^{-aw} \to 0$ for $w \to \infty$) and therefore, for large enough values of w, the graph of g lies below the 45° line. It follows that the graph of g must cross the 45° line. This geometrical observation translates into the conclusion that (6.34) must have a positive solution $w(\infty)$.

Exercise 6.26 The appropriate Ansatz is that

$$\frac{S(\infty, x)}{S(-\infty, x)} = e^{-a(x)w(\infty)}.$$

Once we subsitute this into (6.23) and use, in addition, (6.30), we obtain (6.34).

Exercise 6.27 The first part is a gap exercise. If we put $\frac{\partial S}{\partial t} = 0$ in (6.36), we find

$$\bar{S}(x) = \frac{B(x)}{\mu(x) + a(x)\bar{v}}.$$

In the steady-state version of (6.37) we can factor out $\bar{v}$. Combining these observations, we obtain (6.38). Call the right-hand side of (6.38) a function $g = g(\bar{v})$ of $\bar{v}$. Then $g(0) = R_0$, where R_0 is given by the expression derived in Exercise 6.27 with $S(-\infty, \xi)$ replaced by $\bar{S}(\xi)$. Moreover, g is a strictly decreasing function that tends to zero at infinity. Hence (6.38) has a unique positive solution when $R_0 < 1$.

Exercise 6.28 With the assumption (6.39), we can write the consistency condition (6.26) as

$$\pi(x) = \sum_{k=0}^{\infty} p_k(x) \left(\int_{\Omega} \pi(\xi) v(d\xi) \right)^k = \sum_{k=0}^{\infty} p_k(x) z^k,$$

which is (6.42). So z should equal the integral of the function of x on the right-hand side against the measure v, that is

$$z = \sum_{k=0}^{\infty} z^k \int_{\Omega} p_k(x) v(dx),$$

which is (6.41).

Exercise 6.29 Since constant (i.e. independent of x and ξ) factors can be incorporated in either a or b, we can still add a normalisation condition on either a and b to (6.30). In the present case we chose (6.43). Next, recall that, under the assumption (6.30), we have for f as introduced in Exercise 6.22 that

$$f(y, x) = S(-\infty, y) a(y) \int_{0}^{\infty} b(\tau, x) \, d\tau$$

and hence we have

$$\int_{\Omega} f(y, x) dy = \int_{0}^{\infty} b(\tau, x) \, d\tau. \tag{12.5}$$

Using this information in (6.28), we find

$$m(x, \omega) = \int_{\omega} S(-\infty, y) a(y) \, dy$$

which is indeed independent of x and so can be identified with $v(\omega)$ of the preceding exercise, giving (6.44). Finally, inserting (12.5) into the expression (6.27) for $p_k(x)$, we find (6.45).

12.3   Elaborations for Chapter 7

Exercise 7.1 i) The distribution function of the length of life is $1 - \mathcal{F}_d(a)$ and the corresponding probability density function is therefore $-\frac{d}{da}\mathcal{F}_d(a) = +\mu(a)\mathcal{F}_d(a)$. Hence the expected length of life equals $\int_0^{\infty} a\mu(a)\mathcal{F}_d(a) \, da$. The expected length of life is also the mean age at death.

ii) The stable age distribution is (cf. (7.1)) $Ce^{-ra}\mathcal{F}_d(a)$; hence the mean age of those dying at any particular moment in time equals

$$\frac{C \int_0^{\infty} a\mu(a)e^{-ra}\mathcal{F}_d(a) \, da}{C \int_0^{\infty} \mu(a)e^{-ra}\mathcal{F}_d(a) \, da},$$

and the factors C cancel.

iii) The growth of a population positively affects the relative frequency of the younger age groups and this perseveres in samples of individuals at the point of death.

iv) On the one hand, $\int_0^\infty a\mu e^{-\mu a} \, da = -ae^{-\mu a}|_0^\infty + \int_0^\infty e^{-\mu a} \, da = \frac{1}{\mu}$, the life expectancy. On the other hand, we have that $\int_0^\infty \mu e^{-(r+\mu)a} \, da = \frac{\mu}{r+\mu}$ and $\int_0^\infty a\mu e^{-(r+\mu)a} \, da = -a\frac{\mu}{r+\mu}e^{-(r+\mu)a}|_0^\infty + \frac{\mu}{r+\mu}\int_0^\infty e^{-(r+\mu)a} \, da = \frac{\mu}{(r+\mu)^2}$, so the mean age of those dying at any particular moment in time equals $\frac{1}{r+\mu}$.

Exercise 7.2 When r is large, $N(a)$ given by (7.1) is decreasing rapidly with age. When a population census is represented graphically, one often uses two adjacent histograms, one for males and one for females. Hence the description 'pyramid'.

Exercise 7.3 R_0 is the number of children that a newborn individual is expected to beget in its entire life:

$$R_0 = \int_0^\infty \beta(a)\mathcal{F}_d(a) \, da.$$

Exercise 7.4 i) For sexual contacts and kissing, shaking hands, etc., yes. But, for example, for parents (and day-care centre employees) cleaning children's noses and changing diapers, no. And for blood transfusion certainly no.

ii) A contact may be 'defined' as any event generating a transmission opportunity, so this has, by definition, a non-zero transmission probability. Some types of contact, like sexual intercourse or shaking hands, may be defined in other terms, unrelated to any agent, and then it makes sense to introduce the 'probability of transmission, given a contact' as a separate factor. But 'hugging' occurs with variable intensity and duration, and one would expect the probability of transmission of influenza to be a monotone function of both.

In any case, an asymmetric probability of transmission (such as for STDs, where male $\rightarrow$ female transmission is sometimes easier than female $\rightarrow$ male transmission) can spoil the strict symmetry requirement that stems from bookkeeping considerations.

Exercise 7.5 i) An individual infected at age α will be alive τ units of time later with probability $\mathcal{F}_d(\alpha + \tau)/\mathcal{F}_d(\alpha)$ and then has infectivity $h(\tau, \alpha)$. At that time it is aged $\alpha + \tau$ and therefore makes $c(a, \alpha + \tau)N(a)$ contacts per unit of time with individuals of age a. Since 'infectivity' is used here in the sense of 'probability of transmission, given a contact with a susceptible', we only have to integrate the product with respect to τ to obtain the expected number of secondary cases with age a at the moment of becoming infected.

ii) If $c(a, \alpha) = f(a)g(\alpha)$ then $k(a, \alpha) = f(a)N(a)\psi(\alpha)$, with (compare (7.3))

$$\psi(\alpha) = \int_0^\infty h(\tau, \alpha)g(\alpha + \tau)\frac{\mathcal{F}_d(\alpha + \tau)}{\mathcal{F}_d(\alpha)} \, d\tau.$$

We see that K has a one-dimensional range spanned by fN, and consequently R_0 is given by

$$R_0 = \int_0^\infty \psi(\alpha)f(\alpha)N(\alpha) \, d\alpha,$$

which is (7.2).

When $f = g$, we can interpret this quantity as age-specific social activity, and then the assumption on c means that the contact intensity of an individual is proportional to its social activity and to the social activity of the entire population and that contacts are made according to supply, without any selection of age. When $f \neq g$, one could try a similar line of reasoning, while distinguishing between active and passive contact (or, in any case, a role difference between the two individuals involved in a contact). But the authors admit that they did not know any convincing concrete example, until Don Klinkenberg suggested blood transfusion.

iii) Let now $c(a, \alpha) = \sum_{k=1}^{n} f_k(a)g_k(\alpha)$. Define

$$\psi_k(\alpha) = \int_0^\infty h(\tau, \alpha)g_k(\alpha + \tau)\frac{\mathcal{F}_d(\alpha + \tau)}{\mathcal{F}_d(\alpha)} \, d\tau;$$

then

$$(K\phi)(a) = \sum_{k=1}^{n} f_k(a)N(a) \int_0^\infty \psi_k(\alpha)\phi(\alpha) \, d\alpha,$$

from which we see that the range of K is spanned by $\{f_k N\}_{k=1}^{n}$. So let $\phi(a) = \sum_{j=1}^{n} c_j f_j(a)N(a)$; then $(K\phi)(a) = \sum_{k=1}^{n} d_k f_k(a)N(a)$, with $d_k = \sum_{j=1}^{n} c_j \int_0^\infty \psi_k(\alpha)f_j(\alpha)N(\alpha) \, d\alpha$. So, in other words, the vector c is transformed into the vector $d = Mc$ where M is the matrix with elements

$$m_{ij} = \int_0^\infty \psi_i(\alpha)f_j(\alpha)N(\alpha) \, d\alpha.$$

Necessarily then, R_0 is the dominant eigenvalue of M.

Exercise 7.6 The 'short-disease approximation' amounts to putting $\psi(\alpha) = H(\alpha)g(\alpha)$ and $\psi_k(\alpha) = H(\alpha)g_k(\alpha)$. Otherwise, the expressions for R_0 in Exercise 7.5-ii and for the matrix element m_{ij} in Exercise 7.5-iii remain identical.

Exercise 7.7 i)

$$\mathcal{F}_d(a) = \begin{cases} 1, & a \leq M, \\ 0, & a > M. \end{cases}$$

ii) The uniform mixing implies that $c(a, \alpha) = c_m$, a constant.

iii) This third assumption allows us to compute $h(\tau, \alpha)$. In fact, h is independent of α and given explicitly by

$$h(\tau, \alpha) = \beta e^{-\gamma\tau},$$

where β is the constant infectivity during the infectious period. Applying the formula (7.2) we find

$$R_0 = \int_0^\infty \int_0^\infty \beta e^{-\gamma\tau} c_m \frac{\mathcal{F}_d(\alpha + \tau)}{\mathcal{F}_d(\alpha)} \, d\tau \, N(\alpha) \, d\alpha,$$

which, upon substitution of the formula for $\mathcal{F}_d$ and of $N(a) = k$ for $0 \leq a \leq M$, while $N(a) = 0$ for $a > M$, simplifies to

$$R_0 = \int_0^M \int_0^{M-\alpha} \beta v e^{-\gamma\tau} c_m \, d\tau \, d\alpha = \beta c_m v \int_0^M \frac{1 - e^{-\gamma(M-\alpha)}}{\gamma} \, d\alpha$$

$$= \frac{\beta c_m v}{\gamma}\left(M - \frac{1 - e^{-\gamma M}}{\gamma}\right).$$

Exercise 7.8 The difficulty is: what does 'same age' mean? Born in the same year, month, week, day, hour, minute, second, ...? In the continuous-time formulation, we idealise and pretend that we can even go beyond milliseconds. But clearly such a distinction is meaningless when we aim to describe social behaviour. So, biologically speaking, this does not make any sense. This answer provides a good motivation for the next section.

Exercise 7.9 In the short-disease approximation we have for this special case that

$$\psi_k(\alpha) = H(\alpha) \sum_{l=1}^{n} c_{kl} \chi_{I_l}(\alpha),$$

and hence

$$
\begin{aligned}
m_{ij} &= \int_0^\infty H(\alpha) \sum_{l=1}^{n} c_{il} \chi_{I_l}(\alpha) \chi_{I_j}(\alpha) N(\alpha) \, d\alpha \\
&= c_{ij} \int_{I_j} H(\alpha) N(\alpha) \, d\alpha.
\end{aligned}
$$

Exercise 7.10 The probability that an individual is both alive and uninfected at age a is $\mathcal{F}_i(a)\mathcal{F}_d(a)$, and such individuals have probability $\lambda(a)$ of becoming infected. Hence we want to compute the expected value of the stochastic variable 'age at infection', which has probability density function

$$\frac{\lambda(a)\mathcal{F}_i(a)\mathcal{F}_d(a)}{\int_0^\infty \lambda(\alpha)\mathcal{F}_i(\alpha)\mathcal{F}_d(\alpha) \, d\alpha},$$

where the denominator serves to achieve the right normalisation. This is then exactly the quantity $\bar{a}$.

When $\mathcal{F}_d(a) = 1$ for $0 \le a < M$, and $\mathcal{F}_d(a) = 0$ for $a \ge M$, we have

$$\int_0^M a\lambda(a)\mathcal{F}_i(a) \, da = -a\mathcal{F}_i(a)\big|_0^M + \int_0^M \mathcal{F}_i(a) \, da = \int_0^M \mathcal{F}_i(a) \, da,$$

which leads to the second expression for $\bar{a}$.

Exercise 7.11 For small λ we have

$$\lambda(a)e^{-\int_0^a \lambda(\alpha) \, d\alpha} \approx \lambda(a),$$

and hence the right-hand side of (7.5) can be approximated by its linearisation (as an operator transforming $\lambda = \lambda(a)$ into another function of a)

$$(\widetilde{K}\lambda)(a) = \int_0^\infty \int_0^\infty h(\tau, \alpha)c(a, \alpha + \tau)\lambda(\alpha)N(\alpha)\frac{\mathcal{F}_d(\alpha + \tau)}{\mathcal{F}_d(\alpha)} \, d\tau \, d\alpha.$$

Putting $\phi(a) = \lambda(a)N(a)$ and $(K\phi)(a) = N(a)(\widetilde{K}\lambda)(a)$, we see that K thus defined is indeed the next-generation operator K with the kernel from Exercise 7.5-i. (Note that the relation between λ and ϕ makes sense, since λ is a force of infection and ϕ a number density.) Equality of left- and right-hand sides for the linearisation would imply $R_0 = 1$. We refer to (the elaboration of) Exercise 6.22 for a sketch of the proof that $R_0 > 1$ guarantees that a non-trivial solution $\lambda > 0$ exists.

Exercise 7.12 i) Under the assumption on c, the right-hand side of (7.5) is of the form

$$\sum_{k=1}^{n} f_k(a) \int_0^{\infty} \int_0^{\infty} h(\tau, \alpha) g_k(\alpha + \tau) \lambda(\alpha) S(\alpha) \frac{\mathcal{F}_d(\alpha + \tau)}{\mathcal{F}_d(\alpha)} \, d\tau \, d\alpha,$$

which shows that any solution λ must be of the form

$$\lambda(a) = \sum_{j=1}^{n} d_j f_j(a).$$

Substituting this expression into the equation we find that the coefficients d_k should satisfy the nonlinear system of equations

$$d_k = \sum_{j=1}^{n} d_j \int_0^{\infty} \int_0^{\infty} h(\tau, \alpha) g_k(\alpha + \tau) f_j(\alpha) \frac{\mathcal{F}_d(\alpha + \tau)}{\mathcal{F}_d(\alpha)} N(\alpha) e^{-\sum_{i=1}^{n} d_i \int_0^{\alpha} f_i(\sigma) d\sigma} \, d\tau \, d\alpha.$$

In the special case $n = 1$ we can divide both sides by $d_1 = d$ (thus cancelling the trivial solution $d = 0$) and arrive at the simple form

$$1 = \int_0^{\infty} \int_0^{\infty} h(\tau, \alpha) g(\alpha + \tau) f(\alpha) \frac{\mathcal{F}_d(\alpha + \tau)}{\mathcal{F}_d(\alpha)} N(\alpha) e^{-d \int_0^{\alpha} f(\sigma) d\sigma} \, d\tau \, d\alpha$$

which has, because of the monotonicity of the right-hand side as a function of d, a unique positive solution provided that $R_0 > 1$.

ii) In the short-disease approximation, the system simplifies to

$$d_k = \sum_{j=1}^{n} d_j \int_0^{\infty} H(\alpha) g_k(\alpha) f_j(\alpha) N(\alpha) e^{-\sum_{i=1}^{n} d_i \int_0^{\alpha} f_i(\sigma) d\sigma} \, d\alpha$$

and, for $n = 1$, this reduces to

$$1 = \int_0^{\infty} H(\alpha) g(\alpha) f(\alpha) N(\alpha) e^{-d \int_0^{\alpha} f(\sigma) d\sigma} \, d\alpha$$

while with interval decomposition we find

$$d_k = \sum_{j=1}^{n} c_{kj} \int_{I_j} H(\alpha) N(\alpha) e^{-\sum_{i=1}^{n} d_i \int_0^{\alpha} \chi_{I_i}(\sigma) d\sigma} \, d\alpha \, d_j$$

Note that effectively the summation in the exponent is only up to j, which corresponds to $\mathcal{F}_i(a)$ depending only on λ up to a.

iii) The question is: suppose one can measure the d_i, can one determine from these the c_{ij}? This would amount to determining n^2 unknowns from n data, which is impossible for $n > 1$. We can improve a bit if we require symmetry, i.e. $c_{ij} = c_{ji}$, since that assumption reduces the number of unknowns to $\frac{1}{2}n(n+1)$, but this still exceeds n even for $n = 2$.

iv) According to Anderson & May (1991), the ijth element of this matrix is the probability per unit time that an infective in age class j infects a susceptible in age class i. Apart from an infectivity factor, this corresponds to the matrix element c_{ij}.

v) As discussed in iii), symmetry is the first assumption that suggests itself. An obvious possibility is to choose $c_{ij} = b_i b_j$ for some $b_1, ..., b_n$. We refer to the cited literature for other possible choices.

Exercise 7.13 i) Suppose one can, from serological data, determine the fraction $S(a)/N(a)$ of susceptibles for various values of a. According to the assumptions, we have $S(a) = N(a)e^{-Qa}$, so $\frac{1}{a}\ln\frac{N(a)}{S(a)}$ should be approximately constant, and its value yields an estimate for Q.

An alternative is to determine the mean age at infection $\bar{a}$ from data and to put $Q = \bar{a}^{-1}$, which is exact if every individual lives exactly L units of time (this is an easy calculation from Exercise 7.10, the second formula for $\bar{a}$). If, however, $\mathcal{F}_d(a) = e^{-\mu a}$ then we already found in Exercise 3.6 (formula (3.10)), that $\bar{a} = \frac{1}{\mu+Q}$ and hence $Q = \frac{1}{\bar{a}} - \mu$.

ii) Assume that f is identically one, g is constant, h does not depend on α and $\mathcal{F}_d(\alpha+\tau)/\mathcal{F}_d(\alpha) = e^{-\mu\tau}$. The equation derived in Exercise 7.12-i then simplifies to (note that d is replaced by Q now)

$$1 = D\int_0^\infty e^{-(r+\mu+Q)\alpha}\,d\alpha = \frac{D}{r+\mu+Q},$$

where $D = C\int_0^\infty h(\tau)ge^{-\mu\tau}\,d\tau$, with C from (7.1). Similarly, the expression (7.2) for R_0 simplifies to

$$R_0 = D\int_0^\infty e^{-(r+\mu)\alpha}\,d\alpha = \frac{D}{r+\mu}.$$

Hence $R_0 = \frac{r+\mu+Q}{r+\mu} = 1 + \frac{Q}{r+\mu}$, a formula that allows one to estimate R_0 from the observable quantities r, μ and Q. Inserting $Q = \frac{1}{\bar{a}} - \mu$ and putting $r = 0$, we recover the formula

$$R_0 = \frac{1}{\mu\bar{a}} = \frac{L}{\bar{a}} = \frac{\text{life expectancy}}{\text{mean age at infection}}$$

derived in Exercise 3.6. By now we are much better aware of all assumptions that are required to arrive at this identity.

iii) We have $1 = \int_0^\infty \psi(\alpha)N(\alpha)e^{-Q\alpha}\,d\alpha$ and $R_0 = \int_0^\infty \psi(\alpha)N(\alpha)\,d\alpha$. If ψ is (approximately) constant we can eliminate ψ from these equations and find the desired result.

iv) The second formula of Exercise 7.10 implies that, under the assumptions, $\bar{a} = 1/Q$. The result from iii) can be rewritten, using (7.1), as

$$R_0 = \frac{\int_0^\infty e^{-ra}\mathcal{F}_d(a)\,da}{\int_0^\infty e^{-(r+Q)a}\mathcal{F}_d(a)\,da},$$

which, given the choice of survival function, can be written as

$$R_0 = \frac{\int_0^L e^{-ra}\,da}{\int_0^L e^{-(r+Q)a}\,da} = \frac{\frac{1}{r} - \frac{1}{r}(1 - rL + \cdots)}{\frac{1}{r+Q} - \frac{1}{r+Q}e^{-(r+Q)L}}.$$

For $r \approx 0$, this leads to

$$R_0 \approx \frac{QL}{1 - e^{-QL}},$$

which, for $QL \gg 1$, finally gives

$$R_0 \approx QL = \frac{L}{a}.$$

The condition $QL \gg 1$ is not too difficult to meet for human hosts, since life expectancy is usually of the order of 50 years or more.

Exercise 7.14 i) Let $\mathcal{F}(a)$ denote the probability to be alive at age a, then

$$\mathcal{F}(a) = \mathcal{F}_d(a)\left\{\mathcal{F}_i(a) + \theta(1 - \mathcal{F}_i(a))\right\}.$$

ii) Let $P(a)$ denote the probability to be alive at age a; given that one does not die of disease-unrelated causes. Then $P(a)$ equals the conditional probability

$$P(a) = \frac{\mathcal{F}(a)}{\mathcal{F}_d(a)} = \mathcal{F}_i(a) + \theta(1 - \mathcal{F}_i(a)).$$

iii) Conversely, if we consider the probability to be alive at age a, given that one does not die of the disease (or, equivalently, in a situation in which the infectious agent is eliminated), then this is just $\mathcal{F}_d(a) = \frac{\mathcal{F}(a)}{P(a)}$ (which is also obtained by putting $\theta = 0$, of course).

iv) For constant force of infection Q we have $\mathcal{F}_i(a) = e^{-Qa}$, and hence $\frac{1}{P(a)} \to \frac{1}{\theta}$ for $a \to \infty$. In fact

$$\frac{1}{P(a)} = \frac{1}{e^{-Qa} + \theta(1 - e^{-Qa})}.$$

Now recall that the solution of the logistic differential equation $\frac{dx}{dt} = rx(1 - \frac{x}{K})$ with initial condition $x(0) = x_0$ is given by

$$x(t) = \frac{1}{e^{-rt}\frac{1}{x_0} + \frac{1}{K}(1 - e^{-rt})},$$

and observe that $\frac{1}{P(a)}$ is of this form with $x_0 = 1$, $K = \frac{1}{\theta}$ and $r = Q$.

Exercise 7.15 i) We assumed that the feed is randomly distributed over the herd, and so i) follows directly from the demographic steady state (imposed by the farmer) by taking the age-dependent susceptibility $\beta(a)$ into account. Note that this factor describes not only the actual susceptibility, but also the age-dependent exposure to infection (i.e. the share of potentially infected feed in an animal's diet).

ii) The important parameters are age-dependent, which would call for an infinite number of types at birth. The fact that the individuals infected by feed have an age distribution (at 'birth') that is independent of the age of the infector, allows us to consider them as having one fixed (but stochastic) state at birth (cf. Sections 5.3.1 and 5.3.3). So it is possible to work with just two types at birth: animals that become infected by feed (type 1; distributed over all ages as described by the density in i)), and maternally infected newborns (type 2).

iii) We first look at element k_{21}, i.e. we take an animal that was just infected through feed and want to estimate the number of infected calfs it will produce during the remainder of its life. We know that its age at 'birth' is distributed according to the expression in i). The probability that the animal will survive (natural mortality) from infection to infection-age τ is $\mathcal{F}_d(a + \tau)/\mathcal{F}_d(a)$. The probability that it will still be infectious at infection-age τ is $\mathcal{F}_i(\tau)$, and the infectivity at that infection-age is $\gamma(\tau)$. Finally, the animal has probability per unit of time $b(a + \tau)$ of producing offspring at infection-age τ, and with probability m newborns will be infected. If we integrate over all infection-ages, and integrate over all ages at birth, according to the distribution in i), we obtain

$$k_{21} = \frac{\int_0^\infty \beta(a) \int_0^\infty mb(a + \tau)\mathcal{F}_d(a + \tau)\mathcal{F}_i(\tau)\gamma(\tau)\, d\tau\, da}{\int_0^\infty \beta(a)\mathcal{F}_d(a)\, da}.$$

For element k_{12}, we have to count the expected number of feed infections that are caused by one maternally infected animal. For these animals, true age is equal to infection-age. The probability to survive and remain infectious until age τ is $\mathcal{F}_d(\tau)\mathcal{F}_i(\tau)$ and the rate at which animals are culled is $\mu(\tau) + \nu(\tau)$. The animal, once rendered into feed, will come into 'contact' with animals whose age distribution is given by $\overline{S}(a)$, with susceptibility $\beta(a)$. So, we obtain

$$k_{12} = \int_0^\infty \beta(a)\overline{S}(a) \int_0^\infty (\mu(\tau) + \nu(\tau))\mathcal{F}_d(\tau)\mathcal{F}_i(\tau)\gamma(\tau)\, d\tau.$$

The other two elements of K are derived with completely analogous reasoning. We find

$$k_{22} = \int_0^\infty mb(\tau)\mathcal{F}_d(\tau)\mathcal{F}_i(\tau)\gamma(\tau)\, d\tau$$

for the expected number of maternally infected animals produced by a maternally infected animal. And finally,

$$k_{11} = \int_0^\infty \beta(a)\overline{S}(a) \int_0^\infty (\mu(a + \tau) + \nu(\tau))\frac{\mathcal{F}_d(a + \tau)}{\mathcal{F}_d(a)}\mathcal{F}_i(\tau)\gamma(\tau)\, d\tau\, da.$$

Of course, many of the convenient assumptions in this exercise need to be relaxed if one wants to use R_0 to study, for example, the effects of changes in the rendering process on the spread of the agent. Think about how you could include in the above differences in infectivity between the types and differences in success of the rendering process.

iv) For convenience, we assume that $\beta(a)$ now only describes susceptibility and not diet. In the expressions for k_{11} we add

$$\int_0^\infty w\beta(a)\overline{S}(a) \int_0^\infty \frac{\mathcal{F}_d(a + \tau)}{\mathcal{F}_d(a)}\mathcal{F}_i(\tau)\gamma(\tau)\, d\tau\, da,$$

which can be summarised by substituting $\mu(a) + \nu(a) + w$ in the appropriate place in the expression for k_{11} in iii). This substitution also takes care of the necessary change in k_{12}. The other elements of K remain unchanged.

Exercise 7.16 i) By looking at k_{11}, we find

$$A(\alpha, \tau) = (\mu(\alpha + \tau) + \nu(\tau)) \frac{\mathcal{F}_d(\alpha + \tau)}{\mathcal{F}_d(\alpha)} \mathcal{F}_i(\tau)\gamma(\tau).$$

ii) Substitution of the Ansatz into the integral equation, and cancellation of the factor e^{rt} on both sides, gives

$$f(a) = \overline{S}(a)\beta(a) \int_0^\infty \int_0^\infty A(\alpha, \tau)f(\alpha)e^{-r\tau}\, d\tau\, d\alpha.$$

Now define an operator M, acting on a function f, by the right-hand side of this equation, and we see that a solution $f(\cdot)$ satisfying the Ansatz will be an eigenfunction of M corresponding to eigenvalue 1. Now note that the operator satisfies the criterion of separable mixing, and therefore has a one-dimensional range. From Section 5.3.1, we then know that the (only) eigenfunction is given by $\beta(\cdot)\overline{S}(\cdot)$. If we substitute this into the relation $f = Mf$ above, the terms $\beta(a)\overline{S}(a)$ cancel on both sides and we find the condition (7.8). In this condition only r is an unknown quantity, and r can therefore be defined by this implicit relation.

iii) If we define a function $g(r)$ by the left-hand side of (7.8), we note that $g(0) = R_0$ (for feed transmission only, i.e. $R_0 = k_{11}$), that $g' < 0$ and that $\lim_{r\to\infty} g(r) = 0$. Therefore there exists a unique value r^* such that $g(r^*) = 1$, and we see that $r^* > 0 \Leftrightarrow R_0 > 1$ (recall Figure 1.2). Since the value is unique, we can use, for example, Newton iteration to find the unique zero of the function $g(r) - 1$ to calculate r^* numerically.

Exercise 7.17 In the formula (7.2) we replace $N(a)$ by $N(a)\mathcal{F}_v(a)$, to incorporate that only non-vaccinated individuals are susceptible.

Exercise 7.18 If $\mathcal{F}_v(a) = 1 - q$ for $a > 0$ then $R_v = (1 - q)R_0$ and so $R_v < 1$ corresponds to $q > 1 - \frac{1}{R_0}$.

Exercise 7.19 Next let $\mathcal{F}_v(a) = 1$ for $0 < a < a_v$; and $\mathcal{F}_v(a) = 1 - q$ for $a > a_v$, then

$$R_v = R_0 - q \int_{a_v}^\infty \psi(\alpha)f(\alpha)N(\alpha)\, d\alpha,$$

where $\psi(\alpha)$ is defined by (7.3), and so $R_v < 1$ corresponds to

$$q > \frac{R_0 - 1}{\int_{a_v}^\infty \psi(\alpha)f(\alpha)N(\alpha)\, d\alpha}.$$

Exercise 7.20 i) This is Exercise 7.12-i, the special case $n = 1$, but now with $N(a)$ replaced by $N(a)\mathcal{F}_v(a)$.

ii) The short-disease approximation amounts to replacing $\psi(\alpha)$, given by (7.3), by $H(\alpha)g(\alpha)$.

Exercise 7.21 When a fraction of the population is vaccinated, the force of infection will decrease. We expect that, as a result, the mean age at infection $\overline{a}$ will increase, since with a lowered infection pressure in the population, it would take longer for a susceptible to meet an infectious individual. The formula (3.10) shows that this relation indeed holds when both the force of infection and the force of mortality are constant. Naively, the authors set out to prove a general result, starting from (7.7). But soon they discovered that the normalisation factor (the denominator) has subtle effects and that, in fact, one can construct a counterexample in which the mean age at infection decreases, despite the fact that the force of infection decreases. The key feature is that the force of infection at high ages decreases much more strongly than the force of infection at low ages. We still think that $\bar{a}$ should increase when $\lambda(a)$ is multiplied by some factor less than one, but could not prove so within a short time.

The special case of no mortality (i.e. $\mathcal{F}_d(a) \equiv 1$) is much simpler, since then the denominator equals 1. By partial integration, the numerator can be rewritten as

$$\int_0^\infty e^{-\int_0^a \lambda(\alpha)\, d\alpha}\, da$$

(provided $\exp(-\int_0^a \lambda(\alpha)\, d\alpha) \to 0$ for $a \to \infty$, i.e. provided no individual escapes for ever from being infected), which clearly shows a monotone decreasing dependence on λ.

After having filled and erased the blackboard many times, we started to think: what do we actually want to show? Well, we want to show that a decrease in the force of infection leads to an increase of the risk to become infected at a relatively advanced age. Let $a_2 > a_1 > 0$. The risk of becoming infected at an age between a_1 and a_2, given that one does not die before a_2, is given by

$$\mathcal{F}_i(a_1) - \mathcal{F}_i(a_2) = \left(1 - \frac{\mathcal{F}_i(a_2)}{\mathcal{F}_i(a_1)}\right) \mathcal{F}_i(a_1)$$
$$= \left(1 - e^{-\int_{a_1}^{a_2} \lambda(\alpha)\, d\alpha}\right) e^{-\int_0^{a_1} \lambda(\alpha)\, d\alpha}.$$

Clearly, the first factor decreases when λ increases, but the second factor *increases*, and that may very well be the dominant effect. Thus vaccination may, in the case of rubella and for those females that were not vaccinated (or for whom vaccination failed), lead to an increase of the risk of becoming infected while being pregnant.

12.4   Elaborations for Chapter 8

Exercise 8.1 Define $u(t, x) = w(t, x)e^{kt}$; then $u_t = w_t e^{kt} + ku$, and consequently (8.1) can be rewritten in the form

$$w_t = D\triangle w$$

(the terms ku cancel, and then e^{kt} can be factored out). As you can find in any text book on partial differential equations (PDE; see e.g. Courant & Hilbert (1962)), the so-called fundamental solution of this equation is

$$w(t, x) = \frac{1}{4\pi Dt} \exp\left(-\frac{|x|^2}{4Dt}\right).$$

For completeness, however, we verify that w thus defined satisfies the PDE. To do so, we first compute $\frac{\partial}{\partial x_1} w(t, x)$:

$$\frac{\partial}{\partial x_1} \frac{1}{4\pi Dt} \exp\left(-\frac{x_1^2 + x_2^2}{4Dt}\right) = -\frac{2x_1}{4Dt} w(t, x).$$

Hence

$$\frac{\partial^2}{\partial x_1^2} \frac{1}{4\pi Dt} \exp\left(-\frac{x_1^2 + x_2^2}{4Dt}\right) = -\frac{2}{4Dt} w(t, x) + \frac{4x_1^2}{(4Dt)^2} w(t, x).$$

By symmetry, we have

$$\frac{\partial^2}{\partial x_2^2} \frac{1}{4\pi Dt} \exp\left(-\frac{x_1^2 + x_2^2}{4Dt}\right) = -\frac{2}{4Dt} w(t, x) + \frac{4x_2^2}{(4Dt)^2} w(t, x),$$

and consequently

$$D\triangle w = -\frac{1}{t} w + \frac{|x|^2}{4Dt^2} w.$$

Taking the derivative with respect to t we obtain

$$\frac{\partial}{\partial t} \frac{1}{4\pi Dt} \exp\left(-\frac{|x|^2}{4Dt}\right) = -\frac{1}{4\pi Dt^2} \exp\left(-\frac{|x|^2}{4Dt}\right) + \frac{|x|^2}{4Dt^2} \frac{1}{4\pi Dt} \exp\left(-\frac{|x|^2}{4Dt}\right)$$

$$= -\frac{1}{t} w + \frac{|x|^2}{4Dt^2} w,$$

which upon combination with the expression for $D\triangle w$ establishes that w satisfies the PDE.

Exercise 8.2 This is most easily achieved by applying a transformation to polar coordinates:

$$\int_{-\infty}^{\infty} \int_{-\infty}^{\infty} e^{-a(x_1^2 + x_2^2)} \, dx_1 \, dx_2 = \int_0^{\infty} \int_0^{2\pi} e^{-ar^2} r \, d\varphi \, dr = 2\pi \frac{-1}{2a} e^{-ar^2} \Big|_0^{\infty} = \frac{\pi}{a}.$$

So when $a = \frac{1}{4Dt}$ this yields $4\pi Dt$. But w has an additional factor $\frac{1}{4\pi Dt}$ and we conclude that $\int_{\mathbb{R}^2} w(t, x) \, dx = 1$. Hence $\int_{\mathbb{R}^2} u(t, x) \, dx = e^{kt}$.

Exercise 8.3 For $t \downarrow 0$ we have $\frac{1}{4\pi Dt} \uparrow \infty$. However, when $|x| \geq \varepsilon$, the factor $\exp(-\frac{|x|^2}{4Dt})$ goes to zero much faster than $\frac{1}{4\pi Dt}$ goes to infinity, and hence $\lim_{t \downarrow 0} u(t, x) = 0$ for $|x| \geq \varepsilon$.

Exercise 8.4 Define $\xi = \xi(t, x) = x \cdot \nu - ct = x_1 \nu_1 + x_2 \nu_2 - ct$; then $\frac{\partial \xi}{\partial t} = -c$ and $\frac{\partial \xi}{\partial x_1} = \nu_1$, $\frac{\partial \xi}{\partial x_2} = \nu_2$. Hence, by the chain rule,

$$\frac{\partial}{\partial t} w(\xi) = -cw'(\xi), \quad \frac{\partial}{\partial x_1} w(\xi) = \nu_1 w'(\xi),$$

$$\frac{\partial^2}{\partial x_1^2} w(\xi) = \nu_1^2 w''(\xi), \quad \frac{\partial^2}{\partial x_2^2} w(\xi) = \nu_2^2 w''(\xi).$$

Inserting (8.3) into (8.1) and using these identities, we find $-cw' = Dw'' + kw$ (since, ν being a unit vector, $\nu_1^2 + \nu_2^2 = 1$), which is (8.4). Equation (8.4) is a second-order, linear, homogeneous differential equation. Standard theory (see e.g. Hale (1969)) tells us that we should look for two (linearly independent) solutions. Substitution of $w(\xi) = e^{\lambda \xi}$ into (8.4) yields (8.5), as the factor $e^{\lambda \xi}$ cancels. Hence (8.6) holds.

Exercise 8.5 For uniform (i.e. x-independent) solutions, (8.1) predicts exponential growth at rate k If we consider travelling wave with minimal velocity, however, and measure the growth rate at an arbitrary position, we find that it is twice as large. The mechanism is spillover: population density is substantially larger to the left than it is to the right, and so diffusion contributes to net growth rate at any position.

Exercise 8.6 For w we find the equation

$$Dw'' + (c - \theta)w' + kw = 0,$$

and so we should have $c - \theta \geq c_0$, which amounts to $c \geq c_0 + \theta$. If we replace σ by $-\sigma$, that is, if we look for solutions of the form

$$u(t, x) = w(-\sigma \cdot x - ct),$$

we find for w the equation

$$Dw'' + (c + \theta)w' + kw = 0,$$

and so we should have that $c + \theta \geq c_0$, which amounts to $c \geq c_0 - \theta$. If $\theta > c_0$, the minimal wave speed in the direction $-\sigma$ is *negative*! This has to be interpreted as lack of spread in this direction. In other words, the growth of the species is confined to a region that is blown off towards infinity in the σdirection.

Exercise 8.7 We first perform the transformation $\eta = x - \xi$ of the integration variable in (8.8):

$$u(t, x) = \int_0^\infty \int_{\mathbb{R}^2} B(\tau, |\eta|) \, u(t - \tau, x - \eta) \, d\eta \, d\tau$$

(here we have used that $(-1)^2 = 1$). Now we substitute (8.9) and obtain

$$w(x \cdot \nu - ct) = \int_0^\infty \int_{\mathbb{R}^2} B(\tau, |\eta|) \, w(x \cdot \nu - \eta \cdot \nu - ct + c\tau) \, d\eta \, d\tau.$$

Next, call $x \cdot \nu - ct = \theta$ and write

$$w(\theta) = \int_0^\infty \int_{\mathbb{R}^2} B(\tau, |\eta|) \, w(\theta - \eta \cdot \nu + c\tau) d\eta d\tau.$$

The final step consists of introducing a tailor-made coordinate system. We supplement the unit vector ν by the orthogonal unit vector $\nu^\perp$, which is chosen such that the determinant of the matrix $(\nu \nu^\perp)$ equals one. Let α and σ be coordinates relative to the basis defined by ν and $\nu^\perp$, i.e. let

$$\eta = \alpha \nu + \sigma \nu^\perp;$$

then

$$w(\theta) = \int_0^\infty \int_{-\infty}^\infty \int_{-\infty}^\infty B(\tau, \sqrt{\alpha^2 + \sigma^2}) \, w(\theta - \alpha + c\tau) \, d\alpha \, d\sigma \, d\tau.$$

With the change of variable $\zeta = \alpha - c\tau$, we obtain from this

$$w(\theta) = \int_0^\infty \int_{-\infty}^\infty \int_{-\infty}^\infty B(\tau, \sqrt{(\zeta + c\tau)^2 + \sigma^2}) \, w(\theta - \zeta) \, d\zeta \, d\sigma \, d\tau,$$

which leads to

$$w(\theta) = \int_{-\infty}^\infty V_c(\zeta) \, w(\theta - \zeta) \, d\zeta$$

after interchanging the order of integration.

Exercise 8.8 All you have to do is to cancel the common factor $e^{\lambda\theta}$ on both sides of the equality.

Exercise 8.9 Substituting (8.11) into (8.13), we find

$$L_c(\lambda) = \int_{-\infty}^\infty e^{-\lambda\zeta} \int_0^\infty \int_{-\infty}^\infty B(\tau, \sqrt{(\zeta + c\tau)^2 + \sigma^2}) \, d\sigma \, d\tau \, d\zeta,$$

which, upon reversing the substitution $\alpha = \zeta + c\tau$ made earlier, can be written as

$$L_c(\lambda) = \int_0^\infty \int_{-\infty}^\infty \int_{-\infty}^\infty e^{-\lambda\alpha} e^{\lambda c\tau} B(\tau, \sqrt{\alpha^2 + \sigma^2}) \, d\sigma \, d\alpha \, d\tau,$$

which, essentially, is (8.14). By putting $\lambda = 0$ and reversing the coordinate transformation $\eta = \alpha\nu + \sigma\nu^\perp$, we arrive at the first identity of (8.15), while the second is a direct consequence of the interpretation of B and R_0. If we differentiate with respect to λ, we get two terms, one involving an additional factor $-\alpha$ and the other involving an additional factor $c\tau$. The first of these terms vanishes by symmetry considerations (whatever the function f, $\int_{-\infty}^\infty \alpha f(\alpha^2) \, d\alpha = 0$, since $\int_{-\infty}^0 \alpha f(\alpha^2) \, d\alpha = \int_\infty^0 -\sigma f(\sigma^2) \, d(-\sigma) = -\int_0^\infty \sigma f(\sigma^2) \, d\sigma$). These observations, together with those leading to (8.15), give the equality in (8.16), and the inequality (8.17) is then a direct consequence of the non-negativity of B and the (implicitly!) assumed non-negativity of c.

If we differentiate twice with respect to λ, we obtain three terms involving additional factors α^2, $-2\alpha c\tau$ and $c^2\tau^2$ respectively. For reasons of symmetry, the middle one of these factors vanishes, and we arrive at the conclusion that this second derivative is positive.

The assertion (8.18) is a straightforward consequence of the fact that all the c dependence is concentrated in the factor $\exp(\lambda c\tau)$.

For $c = 0$, the convex function $\lambda \mapsto L_c(\lambda)$ achieves its minimum at $\lambda = 0$ and, when $R_0 > 1$, therefore $L_c(\lambda) > 1$ for all λ. By the property (8.19) the set $\{c : \text{'there exists } \lambda < 0 \text{ such that } L_c(\lambda) < 1'\}$ is non-empty and, moreover, contains $[\bar{c}, \infty)$ whenever it contains $\bar{c}$. Hence this set is either the whole line or a half-line. Since zero does not belong to the set, it must be a half-line. In line with earlier notation, we call the boundary point c_0.

Exercise 8.10 L_c is a convex function of λ that, for $c = c_0$, 'touches' the level one, but does not dip below this level. Equation (8.12) reads $L_c(\lambda) = 1$ and equation (8.18) says we should have a minimum when varying λ. Together therefore, the two equations state that the minimum should be one. To make more precise assertions, we have to distinguish the case where the minimum is actually attained for a finite value of λ from the case where L_c has, as a function of λ, an infimum at minus infinity. This involves the behaviour of $L_c(\lambda)$ for $\lambda \to -\infty$. This behaviour is, in turn, determined by the competition between the factor $\exp(-\lambda\alpha)$, tending to $+\infty$ for $\alpha > 0$, and the factor $\exp(\lambda c\tau)$, tending to zero. Hence it is determined by the support of the function B (i.e. the interval(s) where $B > 0$). We think that it is helpful to be aware of the possibility that c_0 is characterised by $\lim_{\lambda \to -\infty} L_{c_0}(\lambda) = 1$, but that this is kind of exceptional and that it makes good pragmatic sense to try and determine c_0 and λ_0 from $L_c(\lambda) = 1$ and (8.18).

Exercise 8.11 Ignoring the reduction of the density of susceptibles, the incidence is the product of the force of infection A and the host population density S_0. So, with the right interpretation of the word 'offspring' in the definition of B, we arrive at (8.21). In order for B thus defined to only depend on $|x - \xi|$, it is necessary that, among other things, S_0 is constant as a function of x.

12.5   Elaborations for Chapter 9

Exercise 9.1 An age representation is not applicable to macroparasitic infections since infected individuals can receive additional doses of the infectious agent after the start-up dose. In the microparasite case the fast reproduction after the start-up dose will cause further incoming doses of the same agent to go 'unnoticed' in the large amount already built up inside the host. Consequently, we can neglect this influence. In the macroparasite case the additional doses make the crucial difference between various infected individuals, and both the number and size of the doses received and the timing of these re-infections determines the impact of the parasite on the host and the contribution of the host to the spread of infection in the population.

Exercise 9.2 i) Invasion is characterised as a 'best case' event (from the point of view of the parasite). In other words, we are in the situation where the transmission process can proceed optimally. For microparasites we mean for that no infectious material enters hosts that are already infected. For macroparasites one should interpret 'adverse conditions' as anything that hinders development of parasites. This means, for example, that parasites do not compete with each other for resources within the host, or interact with the immune system of the host, but can each optimally produce eggs to be shed into the environment. In other words, there are no density-dependent feedback effects acting on the parasite in any of the stages of its life cycle, i.e. there is no nonlinear interaction between parasites, or between parasites and the environment they inhabit (including hosts).

   ii) In the invasion phase there are no density-dependent constraints, and parasite densities in all stages of the life cycle will be very low. Also there will be few infected hosts, and the chances of infecting the same host more than once will be relatively small in the invasion phase. This can be a problem if the parasite species has obligatory sexual reproduction in the host, since in that case at least one male and one female parasite need to be present within the same individual at the same time. In the very beginning this is obviously not the case, which implies that the epidemic can never take off. In practice, parasites are both distributed (as eggs) and picked up (as larvae) in clumps, thus increasing the probability that in new infections both sexes are present.

iii) The steady states are given by the intersections of the isocline $I = \frac{\mu N}{A} m$ with the isocline $I = \frac{C(m)N}{\delta + C(m)N}$, i.e. by the intersections of a straight line with a sigmoidal curve (by assumption) in the (I, m) plane that both pass through the origin (draw a picture for yourself). For small A, the straight line increases too steeply to intersect the sigmoid isocline for positive values of m. When we increase A, the straight line turns clockwise. It will first touch the sigmoid isocline (for some critical value of A) and then intersect it in exactly two positions, if A is increased further. So, taking the zero steady state into account, the number of steady states is one for A small and three for A above the critical level.

R_0 equals zero (for all values of A), since $C'(0) = 0$, and so in the linearisation around the zero steady state no new infected snails are produced.

The critical value of A is the value for which the isoclines touch. It corresponds to a saddle-node bifurcation (see e.g. Kuznetsov (1998)) of endemic steady states. The lower (in terms of I) steady state is a saddle point, and its stable manifold separates the domains of attraction of the infection-free steady state and the stable endemic steady state.

Exercise 9.3 $\triangle t$ should be small. We assume that the probability is, to a good approximation, equal to a probability per unit time (prescribed by the modelling assumptions) multiplied by the length of the time interval. Does the hint make sense? The assumption is really that these 'rates' are waiting-time-independent.

Exercise 9.4 The key events are death and 'birth' of a parasite within the host (where 'birth' is interpreted as 'entering the host'). Death of a parasite has probability $\mu \triangle t + o(\triangle t)$. Hence the probability that one of the parasites dies is $i \mu \triangle t + o(\triangle t)$. The parasite load increases by one with probability $\beta \triangle t + o(\triangle t)$.

Exercise 9.5 The probability $p_i(t)$ decreases with rate $\beta p_i(t)$ due to a transition $i \to i+1$ corresponding to a larva entering the host and turning into an adult parasite. Likewise, $p_i(t)$ increases with rate $\beta p_{i-1}(t)$ due to the same event happening to a host carrying up to that moment $i - 1$ parasites.

The probability $p_i(t)$ decreases with rate $i \mu p_i(t)$ due to one of the i inhabitant parasites dying. Likewise, $p_i(t)$ increases with rate $(i + 1)\mu p_{i+1}(t)$ due to the same event happening within a host carrying up to that moment $i + 1$ parasites.

Combining the above statements, one arrives at (9.1), if $i \geq 1$. In the case of $p_0(t)$, i.e. the hosts carrying no parasites, no parasite can die and there are no hosts with one less parasite. Thus (9.2) is derived from (9.1) by putting $i = 0$ and adopting the convention that $p_{-1}(t) = 0$ for all t.

Exercise 9.6 Each adult parasite, by assumption, produces larvae with rate λ. The total number of parasites in the population is given by $\sum_{i=1}^{\infty} i p_i(t) =: P(t)$, and therefore the larval population increases with rate $\lambda P(t)$. Larvae die in the environment with per capita rate ν, and larvae have contact with hosts according to the law of mass action, i.e. with a rate proportional to $NL(t)$. Since no other things are assumed to happen to larvae, we obtain equation (9.3) to describe the changes in the larval population. The term $-\theta NL(t)$ should correspond to $\sum_{i=0}^{\infty}\beta p_i(t) = \beta N$ so we should have $\beta = \theta N$.

Exercise 9.7 The number $p_i(t)$ decreases with rate $(d + i\kappa)p_i(t)$ due to death of hosts. The number $p_i(0)$ increases with rate $b\sum_{i=0}^{\infty}p_i(t)$ due to reproduction of hosts. Otherwise the equations remain unchanged.

Exercise 9.8 If you throw a dice, the probability that the outcome is 1 is $\frac{1}{6}$. The law-of-large numbers now states that throwing a dice many times in succession will indeed give the outcome 1 in about one-sixth of the attempts, with 'about' becoming more and more 'precise' as the number of throws increases. Imagine now a large population from which we draw individuals at random and check if their parasite load happens to be i. If there are sufficiently many individuals, the probability to find an individual with i parasites will be 'about' the same as the fraction of the population with i parasites.

Exercise 9.9 If we differentiate the defining relation of $g(t, z)$ with respect to t, we find

$$\frac{\partial g}{\partial t}(t, z) = \sum_{i=0}^{\infty}\frac{dp_i}{dt}(t)z^i$$

$$= -\beta g(t, z) - \mu\sum_{i=0}^{\infty}ip_i(t)z^i + \mu\sum_{i=0}^{\infty}(i + 1)p_{i+1}(t)z^i + \beta\sum_{i=0}^{\infty}p_{i-1}(t)z^i$$

$$= \beta(z - 1)g(t, z) - \mu(z - 1)\frac{\partial g}{\partial z}(t, z).$$

Exercise 9.10 If we differentiate (9.7) with respect to z, we obtain

$$\frac{\partial^2 g}{\partial t \partial z} = \beta(z - 1)\frac{\partial g}{\partial z} + \beta g - \mu(z - 1)\frac{\partial^2 g}{\partial z^2} - \mu\frac{\partial g}{\partial z}.$$

Now note that $\frac{\partial g}{\partial z}(t, z)|_{z=1} = P(t)$ and $g(t, 1) = N(t)$. If we evaluate the above PDE in $z = 1$ and substitute these relations, we find the desired result.

Exercise 9.11 If we combine (9.4) and (9.5), we obtain

$$\frac{dN}{dt} = \sum_{i=0}^{\infty}\frac{dp_i}{dt}$$

$$= b\sum_{i=0}^{\infty}p_i(t) - (\beta + d)\sum_{i=0}^{\infty}p_i(t) - (\mu + \kappa)\sum_{i=1}^{\infty}ip_i(t)$$

$$+ \mu\sum_{i=0}^{\infty}(i + 1)p_{i+1}(t) + \beta\sum_{i=1}^{\infty}p_{i-1}(t)$$

$$= (b - d)N - \beta\sum_{i=0}^{\infty}p_i(t) - (\mu + \kappa)\sum_{i=1}^{\infty}ip_i(t)$$

$$+\mu \sum_{j=1}^{\infty} j p_j(t) + \beta \sum_{j=0}^{\infty} p_j(t)$$

$$= \quad (b-d)N - \kappa \sum_{i=1}^{\infty} i p_i(t) = (b-d)N - \kappa P.$$

In the same way, we can derive a differential equation for P:

$$\frac{dP}{dt} \quad = \quad \sum_{i=0}^{\infty} i \frac{dp_i}{dt} = \sum_{i=1}^{\infty} i \frac{dp_i}{dt}$$

$$= \quad -(\beta+d) \sum_{i=1}^{\infty} i p_i - (\mu+\kappa) \sum_{i=1}^{\infty} i^2 p_i + \mu \sum_{i=1}^{\infty} i(i+1) p_{i+1} + \beta \sum_{i=1}^{\infty} i p_{i-1}$$

$$= \quad -(\beta+d) \sum_{i=1}^{\infty} i p_i - (\mu+\kappa) \sum_{i=1}^{\infty} i^2 p_i + \mu \sum_{j=2}^{\infty} (j-1) j p_j + \beta \sum_{k=0}^{\infty} (k+1) p_k$$

$$= \quad -dP - \beta P + \beta P + \beta N - \mu p_1 - \mu \sum_{i=2}^{\infty} i^2 p_i + \mu \sum_{j=2}^{\infty} j^2 p_j$$

$$-\mu \sum_{j=2}^{\infty} j p_j - \kappa \sum_{i=1}^{\infty} i^2 p_i$$

$$= \quad \beta N - dP - \mu P - \kappa \sum_{i=1}^{\infty} i^2 p_i.$$

Exercise 9.12 For details on the method of characteristics for solving quasilinear first-order differential equations see a textbook on PDE (e.g. Courant & Hilbert (1962)).

In the present case we have the equation (9.7) and find that the characteristic equations are

$$\frac{dt}{ds} \quad = \quad 1,$$

$$\frac{dz}{ds} \quad = \quad \mu(z-1),$$

$$\frac{dg}{ds} \quad = \quad \beta(z-1)g.$$

Assume $t(0) = 0$ and $z(0) = z_0$; then the first equation gives $t = s$, and the second equation can be solved by the variation-of-constants formula. (This gives the solution of the ODE $x'(t) = ax(t) + b(t)$ as $x(t) = e^{at} x_0 + \int_0^t e^{a(t-s)} b(s)\, ds$.) We therefore find

$$z(t) = e^{\mu t}(z_0 - 1) + 1.$$

Define an operator

$$T(t, x) := e^{\mu t}(x - 1) + 1;$$

then we can interpret the solution $z(t)$ for a given initial value z_0 as the result of the operator $T(t, \cdot)$ acting on z_0. By the uniqueness of solutions, $T(\cdot, x)$ satisfies the semigroup property (see Section 4.1): $T(t_2, T(t_1, x)) = T(t_2 + t_1, x)$. Calculating back in time from $z(t)$, we then find $T(-t, z) = T(-t, T(t, z_0)) = T(0, z_0) = z_0$. In this way, we can express z_0 in terms of z (rather than the other way around).

Finally, we can now 'solve' the third equation in terms of z:

$$
\begin{aligned}
g(t, z) &= g(0, z_0)e^{\beta \int_0^t (z(a)-1)\, da} \\
&= g_0(z_0) \exp\left(\beta \int_0^t T(a, z_0)\, da - \beta t \right) \\
&= g_0(T(-t, z)) \exp\left(\beta \int_0^t T(a, T(-t, z))\, da - \beta t \right) \\
&= g_0(T(-t, z)) \exp\left(\beta \int_0^t T(a - t, z)\, da - \beta t \right).
\end{aligned}
$$

Therefore we find that the solution to (9.7) is given by

$$
g(t, z) = g_0(T(-t, z)) \exp\left(\frac{\beta}{\mu}(z - 1)(1 - e^{-\mu t}) \right),
$$

which gives the desired result if we re-substitute the expression for $T(-t, z)$.

For $t \to \infty$ we have convergence of g towards

$$
g(\infty, \tau) = g_0(1) \exp\left(\frac{\beta}{\mu}(z - 1) \right),
$$

which is the generating function of the Poisson distribution with parameter β/μ (see e.g. Chung (1979)). This is not at all surprising: we consider hosts with an eternal life and immigrating parasites at rate β that die at per capita rate μ.

Exercise 9.13 i) We have $g'(t, 1) := \frac{\partial}{\partial z} g(t, z)|_{z=1} = \frac{\partial}{\partial z} \sum_{i=0}^{\infty} r_i(t) z^i|_{z=1} = \sum_{i=0}^{\infty} i r_i(t) z^{i-1}|_{z=1} = \sum_{i=0}^{\infty} i r_i(t) = \frac{1}{N(t)} \sum_{i=0}^{\infty} i p_i(t) = \frac{P(t)}{N(t)}$.

ii) On the one hand, we have $\sum_{i=0}^{\infty} i^2 p_i(t) = N(t) \sum_{i=0}^{\infty} i^2 r_i(t)$. On the other hand, we have $N(t)(g''(t, 1) + g'(t, 1)) = N(t)(\sum_{i=0}^{\infty} i(i - 1) r_i(t) + \sum_{i=0}^{\infty} i r_i(t)) = N(t) \sum_{i=0}^{\infty} i^2 r_i(t)$. The variance is calculated as $E(i^2) - E(i)^2$, where E denotes 'expectation'. Therefore, the variance is given by $\sum_{i=0}^{\infty} i^2 r_i(t) - (\sum_{i=0}^{\infty} i r_i(t))^2 = g''(t, 1) + g'(t, 1) - (g'(t, 1))^2$. Finally, we find that $\sum_{i=0}^{\infty} i^2 r_i(t) = g''(t, 1) + g'(t, 1) - (g'(t, 1))^2 + (g'(t, 1))^2 = $ variance + mean^2 = mean(mean + variance/mean).

iii) From ii), we know that $\sum_{i=0}^{\infty} i^2 p_i = N(g''(1) + g'(1))$ (where we have dropped t since now g does not depend on time), so it suffices to calculate $g''(1)$ and $g'(1)$. First calculate the derivatives

$$
\begin{aligned}
g'(z) &= -k\left(1 + \frac{m}{k}(1 - z)\right)^{-(k+1)} \left(-\frac{m}{k}\right) = m\left(1 + \frac{m}{k}(1 - z)\right)^{-(k+1)} \\
g''(z) &= -(k+1)m\left(1 + \frac{m}{k}(1 - z)\right)^{-(k+2)} \left(-\frac{m}{k}\right) \\
&= \frac{k+1}{k} m^2 \left(1 + \frac{m}{k}(1 - z)\right)^{-(k+2)},
\end{aligned}
$$

and then substitute $z = 1$ to find $g'(1) = m$ and $g''(1) = \frac{k+1}{k}m^2$. From $m = P/N$ and the relation derived in ii), we now immediately find

$$\sum_{i=0}^{\infty} i^2 p_i = N\left(\left(\frac{P}{N}\right)^2 \frac{k+1}{k} + \frac{P}{N}\right).$$

Exercise 9.14 i) We can arrive at the expression in two ways. A heuristic argument would go as follows: in a steady state the number of larvae entering hosts per unit time, βN, should equal the number of larvae produced per unit time, λP, multiplied by the fraction $\frac{\theta N}{\nu + \theta N}$ of these that ever make their way into a host. A more formal argument calculates the quasi-steady state by putting $dL/dt = 0$ in (9.3) and using the definition $\beta N = \theta N L$.

ii) See text.

Exercise 9.15 i) One adult parasite produces λ larvae per unit time and will, on average and in a density-independent situation, survive for $\frac{1}{\mu + d + \kappa}$ time units. Each larva that is produced has a probability $\frac{\theta N}{\nu + \theta N}$ of becoming an adult. This gives the expression for R_0. When the density of hosts N becomes large (for example when the host population grows exponentially, i.e. when $b > d$), larvae will increasingly more likely be removed from the environment by ingestion by a host (θN) than by death (ν). Therefore for large N most larvae are expected to make it to adulthood.

ii) Yes, since for R_0 it does not matter what the duration of the larval stage is. The only important thing is the fraction of larvae that eventually reach the adult stage.

Exercise 9.16 We first restate the system of ODE

$$\frac{dN}{dt} = (b - d)N - \kappa P, \qquad (12.6)$$

$$\frac{dP}{dt} = P\left(\frac{\lambda N}{c + N} - (\mu + d + \kappa) - \kappa \frac{k+1}{k}\frac{P}{N}\right),$$

with $b - d > 0$, by assumption, and where we have put $c := \nu/\theta$.

i) From Exercise 9.15-i, we know that $R_0 = \frac{\lambda}{\mu + d + \kappa}$ in this situation and that therefore the parasite cannot be sustained in the host population if $\lambda < \mu + d + \kappa$. And indeed, in that case, dP/dt equals P times a negative quantity, so the parasite population will go extinct, $P \to 0$, and (12.6) shows that N will eventually increase exponentially with rate $b - d$.

ii) First note that, if we assume as an Ansatz that N eventually grows exponentially, we can put asymptotically $\frac{N}{c+N} = 1$. We make this assumption and check afterwards that N indeed grows exponentially for large t. From i), we see that if $\lambda > \mu + d + \kappa$, the invasion will be successful and P will start to grow away from 'zero'. Initially, N will grow exponentially, but this is not guaranteed if P increases fast enough. To check whether that happens, we derive a differential equation for the mean $m(t) := \frac{P(t)}{N(t)}$. We have

$$\frac{dm}{dt} = \frac{\lambda PN - (\mu + d + \kappa)PN - \kappa\frac{k+1}{k}P^2 - (b - d)PN + \kappa P^2}{N^2}$$

$$= (\lambda - (b + \mu + \kappa))m - \frac{\kappa}{k}m^2.$$

This differential equation has two steady states:

$$\overline{m} = 0,$$
$$\overline{m} = \frac{k}{\kappa}(\lambda - (b + \mu + \kappa)).$$

One sees immediately that for $\lambda < b + \mu + \kappa$ there really is only one biologically feasible steady state, $\overline{m} = 0$, and that this steady state is stable, since it is clear from the differential equation that small perturbations from $\overline{m} = 0$ decay to zero. Therefore for $\mu + d + \kappa < \lambda < b + \mu + \kappa$ the parasite population will grow, but not fast enough to keep up with the growing host population N, and the parasites will be diluted as expressed by $\frac{P}{N} \to 0$. The host will asymptotically grow exponentially with rate $b - d$, and therefore the assumption $\frac{N}{c+N} = 1$ was justified.

iii) For $\lambda > b + \mu + \kappa$ the differential equation for m has two steady states, and the picture of $\frac{dm}{dt}$ is qualitatively like that of Figure 8.3. For $\lambda > b + \mu + \kappa$, therefore, the steady state $\overline{m} = \frac{k}{\kappa}(\lambda - (b + \mu + \kappa))$ is stable and we conclude that now P does grow fast enough to keep up with a growing N. Since P will behave asymptotically as $\overline{m}N$ we have that, for large t,

$$\frac{dN}{dt} \sim (b - d - \kappa\overline{m})N.$$

We see that again the assumption that $\frac{N}{c+N} = 1$, asymptotically, is justified as long as

$$b - d - \kappa\overline{m} > 0.$$

If we substitute the expression for $\overline{m}$ into this inequality, we can rewrite it as an inequality for λ, and we find that N will grow exponentially as long as

$$-\lambda k + bk + \mu k + \kappa k + b - d > 0,$$
$$-\lambda + b + \mu + \kappa + \frac{b-d}{k} > 0,$$
$$-\lambda + \mu + d + \kappa + b - d + \frac{b-d}{k} > 0,$$
$$\mu + d + \kappa + (b-d)\frac{k+1}{k} > \lambda.$$

For the growth rate of P we have asymptotically

$$\frac{dP}{dt} \sim (\lambda - (\mu + d + \kappa) - \kappa\frac{k+1}{k}\overline{m})P$$
$$= (\lambda - (\mu + d + \kappa) - \kappa\overline{m} - \frac{\kappa}{k}\overline{m})P$$
$$= (b - d - \kappa\overline{m})P,$$

where we have substituted the expression for $\overline{m}$ in the final term. We conclude that P and N grow with the same exponential rate.

iv) Putting $\frac{dN}{dt} = 0$ and $\frac{dP}{dt} = 0$, we find

$$(b-d)\overline{N} = \kappa\overline{P},$$
$$\frac{\lambda\overline{N}}{c+\overline{N}} = (\mu + d + \kappa) - \kappa\frac{k+1}{k}\frac{\overline{P}}{\overline{N}}$$

for the steady-state values $\overline{N}$ and $\overline{P}$, with $\overline{P} > 0$. After substitution of $\kappa \overline{P}$ from the first equation into the second one and rearranging the terms, we find

$$\overline{N} = \frac{c\lambda_2}{\lambda - \lambda_2}, \quad \overline{P} = \frac{b-d}{\kappa}\overline{N},$$

where

$$\lambda_2 := \mu + d + \kappa - (b-d)\frac{k+1}{k},$$

and we conclude that the endemic steady state is biologically feasible only if $\lambda > \lambda_2$, i.e. in the region where both P and N cease to grow exponentially for large t. We will skip the stability analysis of this endemic steady state.

Exercise 9.17 The preferred choice is to multiply the contributions. The reason is that the probability per unit of time (of producing offspring) is reduced and that the assumption of independent influence of each parasite would lead to inconsistency when we add contributions. Indeed, if we would add the reductions, there is a possibility that the birth rate could become negative, which would make the model badly posed from a biological point of view.

Exercise 9.18 i) Stage 2 refers to the adult stage where reproduction takes place, stage 1 individuals are then the larvae (see the definition of the life cycle preceding the exercise). The rate d_1 is then the loss rate of larvae: $d_1 = \theta N + \nu$ (loss of larvae through ingestion by hosts and loss by death); the rate of loss of adults is $d_2 = \mu + d + \kappa$ (loss of adult parasites by death and loss due to death of the host, both natural and parasite-induced). The rate m_1 is the maturation rate of larvae, i.e. the rate at which larvae become adults. In the present set-up this is equal to the rate of becoming ingested by a host: $m_1 = \theta N$; the rate m_2 is the rate of production of larvae: $m_2 = \lambda$. We see from Exercise 9.15 that

$$R_0 = \frac{m_1 m_2}{d_1 d_2}.$$

ii) First look at i) again. We see that the quotient $\frac{m_1}{d_1}$ is to be interpreted as the fraction of larvae that reach the adult stage (the rate of ingestion m_1 multiplied by the average life span $\frac{1}{d_1}$). The quotient $\frac{m_2}{d_2}$ is to be interpreted as the number of larvae produced per adult (the number of larvae produced per adult per unit time multiplied by the average life span of an adult). If there are n stages in the life cycle, we obtain n of these quotients, each giving either the fraction of the population in the current stage that reaches the next stage or the number of individuals of the next stage produced per individual of the present stage. Summarising both in one statement, one can say that one parasite of stage i is expected to produce $\frac{m_i}{d_i}$ individuals of stage $i+1$. The product of these quotients then gives the expected number of individuals of any stage i produced per individual of stage i, i.e. the product equals R_0.

Exercise 9.19 The transition matrix of the Markov process is given by

$$G = \begin{pmatrix} -d_1 & 0 & \cdots & 0 & 0 \\ m_1 & -d_2 & \ddots & 0 & 0 \\ 0 & m_2 & \ddots & 0 & 0 \\ \vdots & \ddots & \ddots & \ddots & \vdots \\ 0 & \cdots & 0 & m_{n-1} & -d_n \end{pmatrix},$$

since the rate of leaving stage i is d_i and the rate of entering stage $i+1$ is m_i. In stage n reproduction is assumed to occur, and the state of an individual that has arrived in state n does not change anymore; the state is only left by death. We have seen in Section 5.4 that the element $(-G^{-1})_{ij}$ of $-G^{-1}$ has the interpretation of the expected time that an individual born in state j will spend in state i. We are only interested in the time that an individual born in state 1 ultimately will spend in state n, since that is the time period in which it can produce offspring. The expected number of offspring produced by one newborn state 1 individual is then given by

$$m_n(-G^{-1})_{n1} = R_0.$$

We therefore wish to calculate $(-G^{-1})_{n1}$. We can extract the element $n1$ from the matrix $-G^{-1}$ by applying $-G^{-1}$ to the vector $(1, 0, ..., 0)^\top$, which gives the first column of $-G^{-1}$, and multiplying this from the right by the row-vector $(0, ..., 0, 1)$, to give the last element of that column, i.e.

$$(-G^{-1})_{n1} = -(0, ..., 0, 1)G^{-1} \begin{pmatrix} 1 \\ 0 \\ \vdots \\ 0 \end{pmatrix}.$$

We find that

$$(-G^{-1})_{n1} = \frac{m_1 m_2 \cdots m_{n-1}}{d_1 d_2 \cdots d_n},$$

and therefore finally

$$R_0 = \frac{m_1 m_2 \cdots m_n}{d_1 d_2 \cdots d_n}.$$

Exercise 9.20 i) One possibility is to indeed note that the product of n copies of K, with

$$K = \begin{pmatrix} 0 & 0 & \cdots & 0 & \frac{m_n}{d_1} \\ \frac{m_1}{d_2} & 0 & \ddots & 0 & 0 \\ 0 & \frac{m_2}{d_3} & \ddots & 0 & 0 \\ \vdots & \ddots & \ddots & \ddots & \vdots \\ 0 & \cdots & 0 & \frac{m_{n-1}}{d_n} & 0 \end{pmatrix},$$

is given by

$$K^n = \frac{m_1 m_2 \cdots m_n}{d_1 d_2 \cdots d_n} I,$$

where I is again the identity matrix. Since the spectral radius of the matrix K^n is equal to $\rho(K)^n$, and since K^n clearly has only one eigenvalue, we find

$$\rho(K)^n = \frac{m_1 m_2 \cdots m_n}{d_1 d_2 \cdots d_n}.$$

Another approach is to write out the characteristic equation for K to obtain

$$\det(K - \lambda I) = \lambda^n - \frac{m_1 m_2 \cdots m_n}{d_1 d_2 \cdots d_n} = 0,$$

for both even and odd values of n. Therefore the positive real eigenvalue of K is equal to the spectral radius and is given by

$$\lambda = \rho(K) = \left(\frac{m_1 m_2 \cdots m_n}{d_1 d_2 \cdots d_n} \right)^{\frac{1}{n}}$$

(the other eigenvalues are nth roots of unity multiplied by $\rho(K)$). From the previous exercises, we already know that

$$R_0 = \frac{m_1 m_2 \cdots m_n}{d_1 d_2 \cdots d_n}$$

has the right interpretation and threshold. We therefore have to define here

$$R_0 = \rho(K)^n$$

to arrive at the same result.

ii) From a biological point of view, only the nth power (with n equal to the length of the life cycle) makes sense if we insist on a generation interpretation, since going full circle involves n 'transformations' (of course, if we only care about the threshold property, any power of $\rho(K)$ is OK, and so are other monotone functions that have the value one at one). Remember that in Exercise 9.15.ii we argued that it did not matter for R_0 whether we explicitly took the larval stage into account or whether we applied a time-scale argument to decrease the dimension of the system. If we did not take the nth power, this would mean that adding an explicit additional larval stage to the life cycle would give a different value for R_0 compared with the situation where we collapse the additional stage by using a time-scale argument (thereby taking the effect of the stage in the cycle into account, without explicitly modelling the stage itself). As we have seen, this is not the case, since the R_0 values are the same.

Exercise 9.21 Parasite individuals of each type pass through the life cycle with type specific rates. We can collect these rates into matrices D_i and M_i for stage i individuals. The matrix giving the expected contribution to stage $i+1$ of an individual in stage i is then $M_i D_i^{-1}$ (this matrix has elements $m_i(j)/d_i(j)$ on the diagonal, for each possible type $j = 1, ..., k$). To describe the entire life cycle for all types simultaneously we have to take the product of the n matrices $M_1 D_1^{-1}, ..., M_n D_n^{-1}$, where the expected number of adults of type j produced per adult of type j is the jjth element.

Exercise 9.22 We are as a first step interested in characterising $z(t, k)$, the incidence at time t of infected hosts born with k parasites, as a fraction of the population. The probability that an individual becomes infected with k parasites by an infected individual who currently has i parasites is given by cq_{ki}. This infecting individual was itself born with some number of parasites $j \geq i$, τ units of time ago. Therefore

$$c \sum_{i=1}^{j} x(i, j, \tau) q_{ki} =: A(\tau, k, j)$$

is the probability that an individual who was born with j parasites infects an individual with k parasites, τ units of time later. Taking into account all possible birth states j

and infection-ages τ, and realising that the fraction of the population that was born with j parasites and that currently has infection-age τ is given by $z(t-\tau,j)$, we arrive at

$$z(t,k) = \int_0^\infty \sum_{j\geq 1}(c\sum_{i=1}^{j} x(i,j,\tau)q_{ki})z(t-k,j)d\tau.$$

The fraction of the population that at time t carries i parasites is found by summing over the fraction of the population that was born with $j \geq i$ parasites at any time τ before t and that has lost $j-i$ parasites in the time interval τ. Again, the fraction of the population that was born with j parasites a time τ ago is given by $z(t-\tau,j)$. In symbols, we find

$$r_i(t) = \int_0^\infty \sum_{j\geq i} x(i,j,\tau)z(t-\tau,j)\,d\tau.$$

If parasites die independently of each other, we have that each parasite has probability $e^{-\mu\tau}$ to still be alive after a time τ, and a probability $1-e^{-\mu\tau}$ to have died in the meantime. The probability to have i parasites left out of j initial parasites after a time τ is then given by

$$x(i,j,\tau) = \binom{j}{i}e^{-i\mu\tau}(1-e^{-\mu\tau})^{j-i},$$

i.e., of the j parasites, i must have survived and $j-i$ must have died before τ, and these i remaining parasites can be chosen from the original j parasites in $\binom{j}{i}$ ways.

Now rewrite $r_i(t)$ as

$$\begin{aligned}
r_i(t) &= \int_{-\infty}^t \sum_{j\geq i} x(i,j,t-\sigma)z(\sigma,j)\,d\sigma \\
&= \sum_{j\geq i}\int_{-\infty}^t x(i,j,t-\sigma)z(\sigma,j)\,d\sigma,
\end{aligned}$$

and therefore

$$\begin{aligned}
\frac{dr_i}{dt}(t) &= \sum_{j\geq i}x(i,j,0)z(t,j) - \sum_{j\geq i}x(i,j,\infty)z(-\infty,j) \\
&\quad + \sum_{j\geq i}\int_{-\infty}^t \frac{d}{dt}x(i,j,t-\sigma)z(\sigma,j)\,d\sigma \\
&= z(t,i) + \sum_{j\geq i}\int_{-\infty}^t \frac{d}{dt}x(i,j,t-\sigma)z(\sigma,j)\,d\sigma.
\end{aligned}$$

We consider the second term first, and obtain (using the chain rule)

$$\sum_{j\geq i}\int_{-\infty}^t \left(-i\mu\binom{j}{i}e^{-i\mu(t-\sigma)}(1-e^{-\mu(t-\sigma)})^{j-i}\right.$$
$$\left. +(j-i)\mu e^{-\mu(t-\sigma)}\binom{j}{i}e^{-i\mu(t-\sigma)}(1-e^{-\mu(t-\sigma)})^{j-i-1}\right)d\sigma$$

$$
\begin{aligned}
&= -i\mu r_i(t) + (i+1)\mu \int_{-\infty}^{t} \frac{j-i}{i+1} \binom{j}{i} e^{-(i+1)\mu(t-\sigma)} (1 - e^{-\mu(t-\sigma)})^{j-(i+1)} \, d\sigma \\
&= -i\mu r_i(t) + (i+1)\mu \int_{-\infty}^{t} \binom{j}{i+1} e^{-(i+1)\mu(t-\sigma)} (1 - e^{-\mu(t-\sigma)})^{j-(i+1)} \, d\sigma \\
&= -i\mu r_i(t) + (i+1)\mu r_{i+1}(t).
\end{aligned}
$$

From the integral equation for z, for the first term we obtain

$$
\begin{aligned}
z(t,i) &= c \int_0^\infty \sum_{j\geq 1} \left(\sum_{k=1}^{j} x(k,j,\tau) q_{ik} \right) z(t-\tau, j) \, d\tau \\
&= c \int_0^\infty \sum_{k\geq 1} \sum_{j=k}^{\infty} x(k,j,\tau) z(t-\tau, j) \, d\tau \, q_{ik} \\
&= c \sum_{k\geq 1} r_k(t) q_{ik}.
\end{aligned}
$$

Putting everything together, we find that

$$
\frac{dr_i}{dt}(t) = -i\mu r_i(t) + (i+1)\mu r_{i+1}(t) + c \sum_{j\geq 1} r_j(t) q_{ij}.
$$

Exercise 9.23 i) Each parasite is assumed to have c (indirect) contacts with susceptibles per unit time and at each contact on average η parasites become established in the contacted host. So, each parasite is expected to produce $c\eta$ new parasites per unit time. The life expectancy of a parasite is $\frac{1}{\mu}$, so that we find

$$
R_0 = \frac{c\eta}{\mu}.
$$

ii) We have from (9.14) that the elements of U are given by

$$
U_{ij} = cq_{ij} - i\mu\delta_{ij} + (i+1)\mu\delta_{i+1,j},
$$

where δ_{ij} is the Kronecker delta. The elements of the transpose $U^\top$ are then given by exchanging the indices: $U_{ij}^\top = U_{ji}$. We now look at the vector $\phi(i) = i$ and calculate

$$
\begin{aligned}
(U^\top \phi)(i) &= \sum_{j=0}^{\infty} U_{ij}^\top j = \sum_{j=0}^{\infty} j(cq_{ji} - j\mu\delta_{ji} + (j+1)\mu\delta_{j+1,i}) \\
&= c \sum_{j=0}^{\infty} j q_{ji} - i^2\mu + (i-1)i\mu = ic\eta - i\mu = i(c\eta - \mu),
\end{aligned}
$$

and we see that the vector $\phi(i) = i$ is a formal eigenvector of $U^\top$ and therefore $c\eta - \mu$ also belongs to the spectrum of U. We cannot conclude that it is also an *eigenvalue* of U, since in the infinite-dimensional case there is no direct correspondence of eigenvalues.

iii) Using the $x(i, j, \tau)$ given in Exercise 9.22, we obtain

$$
A_{kj} = \int_0^\infty A(\tau, k, j) \, d\tau = c \int_0^\infty \sum_{i=1}^j x(i, j, \tau) q_{ki} \, d\tau
$$

$$
= c \sum_{i=1}^j q_{ki} \int_0^\infty \binom{j}{i} e^{-i\mu\tau} (1 - e^{-\mu\tau})^{j-i} \, d\tau.
$$

Now the integral is the expected amount of time that an infected individual carries i parasites, when parasites die independently of each other with rate $\frac{1}{\mu}$. The expectation is therefore $\frac{1}{i\mu}$ (as can also be checked by working out the integral, carrying out repeated partial integrations). Therefore

$$
A_{kj} = c \sum_{i=1}^j q_{ki} \frac{1}{i\mu} = \frac{c}{\mu} \sum_{i=1}^j \frac{q_{ki}}{i}
$$

and

$$
(K\phi)(k) = \sum_{j \geq 1} A_{kj} \phi(j) = \frac{c}{\mu} \sum_{j \geq 1} \sum_{i=1}^j \frac{q_{ki}}{i} \phi(j).
$$

iv) The elements of $K^\top$ are given by $(K^\top)_{kj} = A_{jk}$. We check that the vector $\phi(j) = j$ is indeed an eigenvector of $K^\top$:

$$
(K^\top \phi)(k) = \sum_{j \geq 1} j A_{jk}
$$

$$
= \frac{c}{\mu} \sum_{j \geq 1} j \sum_{i=1}^k \frac{q_{ji}}{i}
$$

$$
= \frac{c}{\mu} \sum_{j \geq 1} j \left(q_{j1} + \frac{1}{2} q_{j2} + \cdots + \frac{1}{k} q_{jk} \right)
$$

$$
= \frac{c}{\mu} \left(\sum_{j \geq 1} j q_{j1} + \frac{1}{2} \sum_{j \geq 1} j q_{j2} + \cdots + \frac{1}{k} \sum_{j \geq 1} j q_{jk} \right)
$$

$$
= \frac{c}{\mu} \left(\eta + \frac{1}{2} 2\eta + \cdots + \frac{1}{k} k\eta \right) = k \frac{c\eta}{\mu}.
$$

We conclude that $\frac{c\eta}{\mu}$ is indeed the dominant eigenvalue, since it corresponds to a strictly positive eigenvector. This shows again that

$$
R_0 = \frac{c\eta}{\mu}.
$$

12.6 Elaborations for Chapter 10

Exercise 10.1 For mass action, the contacts per unit time per individual rise linearly with the population density. For low densities this might be a reasonable description of the contact process. For higher densities, however, one can hardly expect that a population twice as dense will lead to twice as many contacts for a given infective, except possibly when transmission is via aerosols. One reason is surely that contacts take time, whereas mass action assumes them to be instantaneous, and therefore only a limited number of contacts can occur per unit of time. We expect the contacts to saturate at high population density. Another effect that certainly causes saturation is satiation, which plays a role in sexual contacts, but also in blood meals taken by mosquitoes.

Exercise 10.2 i) This fraction is $C(N)S$, and the probability that a contact is with an infective equals I/N.

ii) The general idea is that both the fraction of the individuals and the fraction of time for one individual are equal to the probability for an individual to be engaged in a contact. Let $F(N)$ be the average fraction of the time that an individual spends having contact in a population of size N. We show that $F(N) = C(N)$.

Consider the time interval $[0, T]$. Consider Figure 12.6, where the contact patterns for a few individuals are drawn. Here a thin line indicates that the individual is single in that time interval and a thick line indicates that the individual is engaged in a contact during that period.

The idea now is to count contacts in two ways: by integrating 'horizontally' over time and 'vertically' over individuals. Let $\chi_i(t, N)$ be the indicator function of individual i, taking the value 1 for all t that the individual is involved in a contact, and the value 0 for all t that the individual is single. First we calculate what the fraction of time is that this individual i is engaged in contacts:

$$\frac{1}{T} \int_0^T \chi_i(t, N) \, dt$$

To calculate $F(N)$, we now have to average this over all individuals

$$F(N) = \frac{1}{N} \sum_{i=1}^{N} \frac{1}{T} \int_0^T \chi_i(t, N) \, dt.$$

For $C(N)$ we first count vertically in Figure 12.6 by fixing a time t and calculating first the fraction of the population that is engaged in a contact at that time,

$$\frac{1}{N} \sum_{i=1}^{N} \chi_i(t, N),$$

and proceed to average this fraction over all t:

$$C(N) = \frac{1}{T} \int_0^T \frac{1}{N} \sum_{i=1}^{N} \chi_i(t, N) \, dt.$$

By interchanging summation and integration, we see that $F(N) = C(N)$.

Exercise 10.3 Some reasonable properties are i) $C(N) > 0$, for $N > 0$; ii) $C(\cdot)$ non-decreasing; iii) $C(N)$ linear for small N; iv) $C(N)$ constant for large N. For

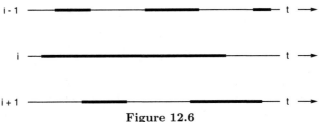

Figure 12.6

$C(N) = N$, (10.1) describes interaction governed by mass action. Examples where C would be approximately constant as a function of N probably include sexual contacts and blood meals taken by mosquitoes.

Exercise 10.4 Denote the search time by T_s. Then $T_s = T - T_h Z$. Also, we can express Z as $Z = aNT_s$ by Holling's assumption, where a is the proportionality constant, interpreted as the effective search rate (also called 'search efficiency'). So, $T - T_h Z = \frac{Z}{aN}$, which we can rewrite as

$$Z = \frac{aNT}{1 + aNT_h}.$$

The functional response, i.e. the number of prey caught per predator per unit of time, is then given by

$$\frac{Z}{T} = \frac{aN}{1 + aNT_h}.$$

Exercise 10.5 The point is that we pretend that N is constant when deriving $\bar{p}_0$. The idea is that N varies slowly, that p_0 adapts quickly to the current value of N, and that therefore we obtain a good approximation by putting $p_0 = \bar{p}_0$, while subsequently studying the slow dynamics of N. The adjectives 'quasi' and 'pseudo' reflect that we cheated and that, actually, N is not constant and $\bar{p}_0$ not a true steady state.

We find from $dp_0/dt = 0 = -aN\bar{p}_0 + e\bar{p}_1$ and $\bar{p}_0 + \bar{p}_1 = 1$ (predators are assumed to be either searching for prey or busy handling prey) that

$$\bar{p}_0 = \frac{e}{e + aN} = \frac{1}{1 + aNT_h}.$$

Now note (you might recall the basic idea of Exercise 10.2) that the number of prey caught per predator per unit of time is given by $aN\bar{p}_0$. We therefore again find that the functional response is given by

$$aN\bar{p}_0 = \frac{aN}{1 + aNT_h}.$$

Exercise 10.6 In the predator-prey formulation, there are two types of individuals that are treated differently: prey and predator (and the predator can be in two states). The prey is considered to be abundant relative to the predator, and, on the short time scale, the predation process consequently does not significantly affect the prey abundance. In the infective-susceptible formulation, there is only one type of individual, which can be in two states: susceptible and infective. The susceptibles therefore do not play the role of the prey, since they are time-limited in the same way as the infectives and they are not much more abundant. The infectives therefore do not encounter susceptibles at their total density, since a fraction of the susceptibles will be engaged in contacts themselves, and only contacts with 'free' susceptibles can be initiated. The consequences are that the Holling argument in Exercises 10.4 and 10.5 does not apply and that the formula (10.2) will need a correction to take into account that the available susceptibles do not have density N.

Exercise 10.7 The following is the basis for the equations given: if one pair dissolves, with probability per unit of time σ, it gives rise to two singles (hence the factor 2); singles meet other singles to form pairs according to mass-action kinetics with reaction rate constant ρ. There are two singles needed to make one pair (hence the factor $\frac{1}{2}$).

Exercise 10.8 We have

$$\frac{dX}{dt} = \frac{dS_1}{dt} + \frac{dI_1}{dt} = -\rho(S_1 + I_1)(S_1 + I_1) + 2\sigma(S_2 + M + I_2)$$
$$= -\rho X^2 + 2\sigma P$$

and

$$\frac{dP}{dt} = \frac{dS_2}{dt} + \frac{dM}{dt} + \frac{dI_2}{dt} = \frac{1}{2}\rho S_1^2 - \sigma S_2 + \rho S_1 I_1 - \sigma M - \beta M + \frac{1}{2}\rho I_2^2 - \sigma I_2 + \beta M$$
$$= \frac{1}{2}\rho(S_1^2 + 2S_1 I_1 + I_1^2) - \sigma(S_2 + M + I_2)$$
$$= \frac{1}{2}\rho X^2 - \sigma P.$$

Exercise 10.9 Putting the right-hand side of the differential equations equal to zero, we both times find the same equation, viz. $\rho X^2 - 2\sigma P = 0$. This reflects the conservation of individuals, and so we have to supplement this equation by the normalisation condition $X + 2P = N$. The resulting quadratic equation in X reads (with $\nu := \rho/\sigma$)

$$\nu X^2 + X - N = 0.$$

It has a unique positive solution

$$\overline{X} = \frac{-1 + \sqrt{1 + 4\nu N}}{2\nu}.$$

If we substitute the result into $\overline{P} = \frac{1}{2}(N - \overline{X})$ we find

$$\overline{P} = \frac{1 + 2\nu N - \sqrt{1 + 4\nu N}}{4\nu}.$$

Using $X + 2P = N$, we see that the system (10.4) is fully described by the one-dimensional ODE $\frac{dX}{dt} = -\rho X^2 + \sigma(N - X)$, and a graphical argument (draw a graph of the right-hand side) now shows that all relevant solutions (i.e. those starting with $0 \leq X \leq N$) converge to $\overline{X}$ for $t \to \infty$. So the steady state is a global attractor.

Exercise 10.10

$$\frac{dS}{dt} = \frac{dS_1}{dt} + \frac{dM}{dt} + 2\frac{dS_2}{dt} = -\rho S_1^2 + 2\sigma S_2 - \rho S_1 I_1 + \sigma M + \rho S_1^2 - 2\sigma S_2$$
$$+\rho S_1 I_1 - \sigma M - \beta M = -\beta M,$$

$$\frac{dI}{dt} = \frac{dI_1}{dt} + \frac{dM}{dt} + 2\frac{dI_2}{dt} = -\rho S_1 I_1 - \rho I_1^2 + 2\sigma I_2 + \sigma M + \rho I_1^2 - 2\sigma I_2 2\beta M$$
$$+\rho S_1 I_1 - \sigma M - \beta M = \beta M.$$

And, indeed, the 'transformation' of susceptibles into infectives occurs in mixed pairs at rate β (and only in mixed pairs), so the result was predictable.

Exercise 10.11 The verification is a matter of substitution of the expressions in (10.7) into the right-hand side of the system (10.3) and noting that this indeed yields zero if we ignore the βM terms. The logic is very simple. The pair formation process is completely independent of the (static, by assumption) S-I distinction, and so each category, according to this distinction, occupies the relevant fraction of singles or pairs. Indeed, $\overline{S}_1$ is just the susceptible fraction of singles and $\overline{I}_1$ is the infective fraction of singles. Pairs occur in three categories: {SS}, {SI}, and {II}. Consider, as an example, the {SI} category. We find $\overline{M} = 2\overline{P}\frac{S}{I}\frac{I}{N}$, since the first or the second member of the pair can be susceptible, provided the second or the first respectively is infective.

Exercise 10.12 Since $\overline{M} = 2\overline{P}\frac{S}{I}\frac{I}{N}$, we have that

$$C(N) = \frac{2\overline{P}}{N} = \frac{1 + 2\nu N - \sqrt{1 + 4\nu N}}{2\nu N}.$$

Multiplying the right-hand side by

$$\frac{1 + 2\nu N + \sqrt{1 + 4\nu N}}{1 + 2\nu N + \sqrt{1 + 4\nu N}},$$

we can check that the two expressions for $C(N)$ in (10.8) are equal to each other. It is easier to check the properties listed in the elaboration of Exercise 10.3 when using the second expression for $C(N)$.

Property i) is now self-evident. For property ii) we calculate the derivative

$$\frac{dC}{dN} = \frac{2\nu(1 + 2\nu N + \sqrt{1 + 4\nu N}) - 2\nu N \left(2\nu + 2\nu(1 + 4\nu N)^{-1/2}\right)}{(1 + 2\nu N + \sqrt{1 + 4\nu N})^2},$$

which, after some reordering of terms, gives

$$\frac{dC}{dN} = 2\nu \left((1 + \sqrt{1 + 4\nu N} - \frac{1}{\sqrt{1 + 4\nu N}} \right) \geq 0.$$

For property iii), note that for N much smaller than 1 we have that $1 + 2\nu N + \sqrt{1 + 4\nu N} \approx 1$ and therefore that $C(N) \approx 2\nu N$. Finally, for property iv) we calculate the limit

$$\lim_{N \to \infty} C(N) = \lim_{N \to \infty} \frac{2\nu}{\frac{1}{N} + 2\nu + \frac{\sqrt{1+4\nu N}}{N}} = \frac{2\nu}{2\nu} = 1.$$

Exercise 10.13 The model formulation of the exercise leads to the following system of differential equations:

$$\frac{dX_0}{dt} = \varepsilon_1 X_1 - \varepsilon_0 X_0,$$

$$\frac{dX_1}{dt} = -\varepsilon_1 X_1 + \varepsilon_0 X_0 - \rho X_1^2 + 2\sigma P,$$

$$\frac{dP}{dt} = \frac{1}{2}\rho X_1^2 - \sigma P.$$

Carrying out the same analysis as in Exercise 10.9, we express the steady state $\overline{P}$ as $\overline{P} = \frac{1}{2}\nu \overline{X}_1^2$. The consistency condition is now given by $\overline{X}_0 + \overline{X}_1 + 2\overline{P} = N$. The steady states $\overline{X}_0$ and $\overline{X}_1$ are obtained by putting the right-hand sides of the corresponding differential equations equal to zero. This leads to the equations

$$\overline{X}_0 = \frac{\varepsilon_1}{\varepsilon_0}\overline{X}_1$$

and

$$\nu \overline{X}_1^2 + \left(\frac{\varepsilon_1}{\varepsilon_0} + 1\right)\overline{X}_1 - N = 0.$$

The positive solution to this quadratic equation is

$$\overline{X}_1 = \frac{-\left(\frac{\varepsilon_1}{\varepsilon_0} + 1\right) + \sqrt{\left(\frac{\varepsilon_1}{\varepsilon_0} + 1\right)^2 + 4\nu N}}{2\nu},$$

and therefore

$$\overline{P} = \frac{1}{2}\left(N - \left(\frac{\varepsilon_1}{\varepsilon_0} + 1\right)\overline{X}_1\right)$$

$$= \frac{\left(\frac{\varepsilon_1}{\varepsilon_0} + 1\right)^2 - \left(\frac{\varepsilon_1}{\varepsilon_0} + 1\right)\sqrt{\left(\frac{\varepsilon_1}{\varepsilon_0} + 1\right)^2 + 4\nu N} + 2\nu N}{4\nu},$$

and finally, using $C(N) = 2P/N$,

$$C(N) = \frac{\left(\frac{\varepsilon_1}{\varepsilon_0} + 1\right)^2 - \left(\frac{\varepsilon_1}{\varepsilon_0} + 1\right)\sqrt{\left(\frac{\varepsilon_1}{\varepsilon_0} + 1\right)^2 + 4\nu N} + 2\nu N}{2\nu N}.$$

We see that by either taking $\varepsilon_1 = 0$ or the limit $\varepsilon_0 \to \infty$ (i.e. by letting the resting period be infinitesimally short), we recover the expression derived in Exercise 10.9.

Exercise 10.14 Let $\overline{X}$ be the singles available for contact. If we repeat the Holling argument from Exercise 10.4, with X instead of N (and ρ instead of a), we find

$$Z = \rho \overline{X} T_s = \rho \overline{X}(T - Z T_h).$$

With $Y := Z/T$, this leads to

$$Y = \rho\overline{X}(1 - YT_h) \Rightarrow Y = \frac{\rho\overline{X}}{1 + \rho T_h\overline{X}}.$$

Since T_h is the average duration of a contact, we can write $\rho T_h = \nu$ to connect to the notation of Exercise 10.9.

Now note that, in contrast to the predator-prey case, the singles are themselves time-limited and the available singles out of the population of size N are a fraction of N determined by the fraction of the time that individuals are not participating in a contact. The total time that an individual is expected to be in contact is ZT_h (i.e. the number of contacts per individual times their average duration). Therefore the time not spent in contacts is $T - ZT_h$ and the fraction of time not spent in contacts is $(T - ZT_h)/T$. Therefore

$$\overline{X} = N\left(\frac{T - ZT_h}{T}\right) = N(1 - YT_h).$$

So actually the relation for Y is not $Y = \rho\overline{X}(1 - YT_h)$ but

$$Y = \rho N(1 - YT_h)(1 - YT_h). \tag{12.7}$$

Now note that

$$\overline{P} = \frac{1}{2}NYT_h, \tag{12.8}$$

since $YT_h = \frac{ZT_h}{T}$ is the fraction of time spent in contact for a given individual, which is identical to the fraction of the population engaged in contact at any given time (Exercise 10.2), i.e. NYT_h gives the average number of individuals in the population that is engaged in contact at any particular time. The number of pairs is then half this number. If we substitute the expression (12.7) into (12.8), we obtain

$$\overline{P} = \frac{1}{2}N^2(1 - YT_h)^2\rho T_h = \frac{1}{2}\nu\overline{X}^2.$$

Together with the conservation condition $2\overline{P} + \overline{X} = N$, this gives the same equations from which we calculated $C(N)$ in Exercise 10.8.

Exercise 10.15 Start a clock at the moment that a pair consisting of a susceptible and an infective is created, and let t be the time according to this clock. To calculate the probability of transmission within this pair, we have to know the probability that transmission occurs before the pair reaches age t and take the weighted average of that over all t, where the weight is given by the probability that the pair lasts exactly t units of time. Given that the transmission rate is β during the period of existence of the pair, we see that $e^{-\beta t}$ is the probability that transmission has not occurred before time t, and therefore $1 - e^{-\beta t}$ is the probability that transmission has occurred by that time. Since pairs dissolve with exponential rate σ, the probability density function for pair duration is given by $\sigma e^{-\sigma t}$. So, the weighted average is

$$\int_0^\infty \left(1 - e^{-\beta t}\right)\sigma e^{-\sigma t}\,dt \;=\; \sigma\int_0^\infty \left(e^{-\sigma t} - e^{-(\sigma+\beta)t}\right)dt$$

$$= \sigma\left(\frac{1}{\sigma}\right) - \sigma\left(\frac{1}{\sigma+\beta}\right) = \frac{\beta}{\sigma+\beta}$$

$$= \frac{b}{1+b},$$

where we have defined $b := \beta/\sigma$. We conclude that one should, when taking the limit $\sigma \to \infty$, let the infection rate β go to infinity as well, but in such a way as to keep their ratio b constant. In that way, the pair duration becomes infinitesimally short, while at the same time the infection probability per pair remains constant.

Exercise 10.16 We have seen in the previous exercise that the right way to take the limit towards infinitesimally short-lived pairs is to let both σ and β go to infinity, while keeping their ratio b constant. We therefore rewrite the transmission term with $C(N)$ given by (10.8) as

$$\beta C(N)\frac{SI}{N} = \frac{2\beta\frac{\rho}{\sigma}N}{1+2\frac{\rho}{\sigma}N+\sqrt{1+4\frac{\rho}{\sigma}N}}\frac{SI}{N}$$

$$= \frac{2\rho bN}{1+2\frac{\rho}{\sigma}N+\sqrt{1+4\frac{\rho}{\sigma}N}}\frac{SI}{N}.$$

Now take the limit

$$\lim_{\sigma,\beta\to\infty}\frac{2\rho bN}{1+2\frac{\rho}{\sigma}N+\sqrt{1+4\frac{\rho}{\sigma}N}}\frac{SI}{N} = 2\rho bN\frac{SI}{N} = \widehat{\beta}SI,$$

where $\widehat{\beta} := 2\rho b$ is the new transmission rate constant.

Exercise 10.18 For the incidence dS_i/dt among i-individuals, we have the expression

$$\frac{dS_i}{dt} = -\left(\sum_{j=1}^{n}\pi_{ij}\phi_{ij}\frac{I_j}{N_j}\right)\frac{S_i}{N_i},$$

since, of the ϕ_{ij} contacts per unit of time, a fraction $\frac{I_j}{N_j}\frac{S_i}{N_i}$ is between susceptible individuals of type i and infective individuals of type j.

Exercise 10.19 Per low-tide period, the number of individuals that sunbathe at site k while being based at site j equals $\rho_{kj}N_j$. The probability that such an individual has contact with an individual based at site i is proportional to

$$\frac{\rho_{ki}N_i}{\sum_l \rho_{kl}N_l}.$$

Hence ϕ_{ij} is proportional to

$$\sum_k \frac{\rho_{ki}N_i\rho_{kj}N_j}{\sum_l \rho_{kl}N_l},$$

where the constant of proportionality measures the number of 'contacts' (e.g. lying within a distance of two metres from each other) per low tide period.

Exercise 10.20 The total number of contacts per unit of time between members of the aggregated groups l and r is found by adding contacts over all possible combinations between individuals of the subgroups out of which l and r are composed. This rate equals

$$\sum_{i \in s_l} \sum_{j \in s_r} c_{ij} N_i N_j.$$

Therefore the average contact rate is given by

$$\widehat{c}_{lr} = \frac{\sum_{i \in s_l} \sum_{j \in s_r} c_{ij} N_i N_j}{\widehat{N}_l \widehat{N}_r}.$$

If $c_{ij} = v_i v_j$, we obtain

$$
\begin{aligned}
\widehat{c}_{lr} &= \frac{\sum_{i \in s_l} \sum_{j \in s_r} v_i v_j N_i N_j}{\widehat{N}_l \widehat{N}_r} = \frac{\sum_{i \in s_l} v_i N_i \sum_{j \in s_r} v_j N_j}{\widehat{N}_l \widehat{N}_r} \\
&= \frac{\sum_{i \in s_l} v_i N_i}{\widehat{N}_l} \frac{\sum_{j \in s_r} v_j N_j}{\widehat{N}_r} =: \widehat{v}_l \widehat{v}_r.
\end{aligned}
$$

Exercise 10.21 i) Per unit time, the total number of contacts between type-i and type-j individuals equals $c_{ij} N_i N_j$. So, the total number of contacts of all type-i individuals equals, on the one hand, $\sum_j c_{ij} N_i N_j$, and, on the other hand, $c_i N_i$. Therefore $c_i = \sum_j c_{ij} N_j$. From the c_i contacts, only $c_{ii} N_i$ are within the own group, which amounts to a fraction

$$q_i = \frac{c_{ii} N_i}{c_i}.$$

ii) We have to show that $\widehat{R}_0 = \widehat{c} \leq R_0$, with

$$\widehat{c} = \frac{c_1 N_1 + c_2 N_2}{N_1 + N_2}$$

and

$$R_0 = \frac{c_1^2 N_1 + c_2^2 N_2}{c_1 N_1 + c_2 N_2}.$$

The following chain of equivalent inequalities leads from the one that we wish to prove to one that we know is true:

$$
\begin{aligned}
\frac{c_1 N_1 + c_2 N_2}{N_1 + N_2} &\leq \frac{c_1^2 N_1 + c_2^2 N_2}{c_1 N_1 + c_2 N_2}, \\
(c_1 N_1 + c_2 N_2)^2 &\leq (N_1 + N_2)(c_1^2 N_1 + c_2^2 N_2), \\
c_1^2 N_1^2 + 2 c_1 c_2 N_1 N_2 + c_2^2 N_2^2 &\leq c_1^2 N_1^2 + c_1^2 N_1 N_2 + c_2^2 N_1 N_2 + c_2^2 N_2^2, \\
c_1^2 + c_2^2 - 2 c_1 c_2 &\geq 0.
\end{aligned}
$$

iii) Now regard the situation without assuming proportionate mixing. When we wrote this chapter, we included this exercise in order to illustrate Adler's result (10.19) for a simple case that could be worked out explicitly. We expected it to be easy and therefore postponed writing down the solution to the last minute. But we readily admit that when we finally devoted some time to the problem, in the hectic week before handing in the manuscript, we were not able to come up with a proof! If you, dear reader, are successful, please let us know. Several special cases are easy (we expect).

Exercise 10.22 The next-generation matrix for the two-group case becomes

$$\begin{pmatrix} 10q & 30(1-q) \\ 10(1-q) & 30q \end{pmatrix},$$

and the dominant eigenvalue is given by

$$R_0 = 20q + 10\sqrt{4q^2 - 6q + 3},$$

whereas $\widehat{R}_0 = 10 \times 2 = 20$. Analysing the function

$$f(q) := 20q + 10\sqrt{4q^2 - 6q + 3} - 20$$

we find that f is zero at $p = \frac{1}{2}$ and that $f < 0$ for all $p < \frac{1}{2}$.

Exercise 10.23 Consider a newly infected k-individual. Its expected number of 'offspring' is $(k-1)\bar{q}$, since, in the initial phase, exactly one of its acquaintances, viz. the one by which it was infected itself, is not susceptible. Note that we assume independence between the type k and the amount of infectious material that is disseminated. Such an assumption deserves scrutiny when, for instance, k is the number of concurrent sexual partners and the agent is transmitted during sexual contact (the point being that there could be a correlation between sexual activity and skin injuries or ulcerations that could enhance transmission).

Each of the offspring has probability ν_m to be of type m. So the distribution of type at 'birth' is given by ν and is, notably, independent of the type of the 'mother'. To calculate R_0, we simply average the number of offspring with respect to ν.

Note that apparently we also assume independence between the type and the susceptibility along any particular connection. This assumption may be questionable in certain contexts for the reasons explained above. The overall susceptibility is, of course, proportional to k. This explains the occurrence of ν_k in the formula (10.17). The factor $k-1$ then incorporates the dependence of infectivity on type, and our crucial network assumption guarantees that the types have independent influence on the likelihood of a connection. Recalling Section 5.3.1, we understand that an explicit formula for R_0 had to be expected.

Exercise 10.24 We have

$$R_0 = \bar{q}\sum_{k=1}^{\infty} \frac{k^2 - k}{\bar{k}}\mu_k = \bar{q}\left(\frac{\text{variance}(k) + \bar{k}^2}{\bar{k}} - 1\right)$$

$$= \bar{q}\left(\bar{k} - 1 + \frac{\text{variance}(k)}{\bar{k}}\right).$$

In Exercise 5.10 we derived $R_0 = \text{mean} + \text{variance/mean}$. The slight difference is that in the present context, the first 'mean' is replaced by 'mean'-1 to reflect that, by becoming infected, the transmission opportunities diminish by one.

Exercise 10.25 A k-individual begets j offspring with probability

$$p_{jk} = \binom{k-1}{j} \bar{q}^j (1-\bar{q})^{k-1-j}, \qquad 0 \le j \le k-1.$$

By definition,

$$g_b(z) = \sum_{j=0}^{\infty} \left(\sum_{k=j+1}^{\infty} \nu_k p_{jk} \right) z^j.$$

Hence, interchanging the order of summation, we find

$$g_b(z) = \sum_{k=1}^{\infty} \nu_k \sum_{j=0}^{k-1} p_{jk} z^j = \sum_{k=1}^{\infty} \nu_k (1 - \bar{q} + \bar{q}z)^{k-1}.$$

Clearly

$$g_b'(1) = \sum_{k=1}^{\infty} (k-1)\bar{q}\nu_k = R_0.$$

Exercise 10.26 The probability of extinction can be related to π_b immediately after the first reproduction. Indeed, it equals π_b^j when the first individual begets j offspring. Hence it equals

$$\sum_{k=1}^{\infty} \mu_k \sum_{j=0}^{k} \bar{q}^j (1-\bar{q})^{k-j} \pi_b^j = \sum_{k=1}^{\infty} \mu_k (1 - \bar{q} + \bar{q}\pi_b)^k.$$

Exercise 10.27 On the one hand, we have

$$\begin{aligned}
1 - \eta &= \pi_b = g_b(\pi_b) = \sum_{k=1}^{\infty} \nu_k (1 - \eta\bar{q})^{k-1} \\
&= \sum_{k=1}^{\infty} \nu_k \left\{ 1 - (k-1)\eta\bar{q} + \frac{1}{2}(k-1)(k-2)\eta^2\bar{q}^2 + \cdots \right\} \\
&= 1 - R_0\eta + \frac{1}{2}\eta^2\bar{q}^2 \sum_{k=3}^{\infty} (k-1)(k-2)\nu_k + \cdots.
\end{aligned}$$

Rearranging this equality (i.e. eliminating the 1 on both sides and dividing by η) leads to

$$R_0 - 1 = \frac{1}{2}\eta^2\bar{q}^2 \sum_{k=3}^{\infty} (k-1)(k-2)\nu_k + \cdots.$$

On the other hand, we have

$$\begin{aligned}
s_{\infty} &= \sum_{k=1}^{\infty} \mu_k \left\{ 1 - k\eta\bar{q} + \frac{1}{2}(k)(k-1)\eta^2\bar{q}^2 + \cdots \right\} \\
&= 1 - \bar{k}\eta\bar{q} + \cdots,
\end{aligned}$$

which implies that
$$1 - s_\infty = \bar{k}\eta\bar{q} + \cdots.$$

Combining the two expressions, we can eliminate the first-order term in η and obtain (10.25), i.e.
$$1 - s_\infty = \frac{2\bar{k}}{\bar{q}\sum_{k=3}^\infty (k-1)(k-2)\nu_k}(R_0 - 1) + \cdots.$$

Exercise 10.28 The probability that a (k, ξ)-individual begets j offspring equals
$$\binom{k-1}{j}(1 - q(\xi))^{k-1-j}(q(\xi))^j.$$

Hence,
$$g_f(z) = \sum_{j=0}^\infty z^j \sum_{k=1}^\infty \nu_k \binom{k-1}{j} \int_\Omega (1 - q(\xi))^{k-1-j}(q(\xi))^j m(d\xi),$$

which, after changing the order of summation and integration to, from left to right, summation over k, integration over ξ and summation over j yields, by the binomial expansion formula,
$$g_f(z) = \sum_{k=1}^\infty \nu_k \int_\Omega (1 - q(\xi) + zq(\xi))^{k-1} m(d\xi).$$

Since g_f is the generating function, the probability that the line of descent from any infected individual is finite equals π_f, the root in the interval $(0, 1)$ of the equation $\pi = g_f(\pi)$. But when we speak of the probability of a minor outbreak, we have in mind that we start with infecting an arbitrary individual from 'outside'. This has two effects: the probability distribution of type k is given by $\{\mu_k\}$, not by ν_k, and a k-type individual can infect all of its k acquaintances, not just $k - 1$. Hence the probability of a minor outbreak equals

$$\sum_{k=1}^\infty \mu_k \sum_{j=0}^k \pi_f^j \binom{k}{j} \int_\Omega (1 - q(\xi))^{k-j}(q(\xi))^j m(d\xi) = \sum_{k=1}^\infty \mu_k \int_\Omega (1 - q(\xi) + \pi_f q(\xi))^k m(d\xi).$$

Exercise 10.29 By differentiation, we find
$$g_f'(z) = \sum_{k=1}^\infty (k-1)\nu_k \int_\Omega (1 - q(\xi) + zq(\xi))^{k-2} q(\xi)m(d\xi),$$

and hence $g_f'(1) = \sum_{k=1}^\infty (k-1)\nu_k\bar{q} = R_0$.

Exercise 10.30 From
$$\frac{d\mathcal{F}}{d\tau} = -\beta\mathcal{F}, \quad 1 \leq \tau \leq \xi,$$
$$\mathcal{F}(\tau, \xi) = 1, \quad 0 \leq \tau \leq 1,$$

we deduce that
$$\mathcal{F}(\tau,\xi) = e^{-\beta\min(\tau-1,\xi-1)},$$

and hence
$$q(\xi) = 1 - \mathcal{F}(\infty,\xi) = 1 - e^{-\beta(\xi-1)}.$$

Therefore
$$\bar{q} = \int_2^3 \left(1 - e^{-\beta(\xi-1)}\right) d\xi = 1 - \frac{1}{\beta}\left(e^{-\beta} - e^{-2\beta}\right).$$

As the expected length of the infectious period is $1\frac{1}{2}$, the reduction factor equals $\frac{2}{3}\frac{\bar{q}}{\beta}$.

Exercise 10.31

$$\begin{aligned}
\overline{\mathcal{F}}(\tau) &= \int_0^\infty e^{-\beta\min(\tau,\xi)}\alpha e^{-\alpha\xi}\,d\xi = \int_0^\tau e^{-\beta\xi}\alpha e^{-\alpha\xi}\,d\xi + \int_\tau^\infty e^{-\beta\xi}\alpha e^{-\alpha\xi}\,d\xi \\
&= \frac{\alpha}{\alpha+\beta}\left\{1 - e^{-(\beta+\alpha)\tau}\right\} + e^{-(\beta+\alpha)\tau}.
\end{aligned}$$

Exercise 10.32 The probability of transmission to any susceptible acquaintance, as a function of the time τ elapsed since the infected individual was itself infected, is given by
$$1 - \overline{\mathcal{F}}(\tau).$$

The infected individual has with probability ν_k exactly $k-1$ susceptible acquaintances. So the rate of producing 'offspring' is given by

$$\sum_{k=1}^\infty \nu_k(k-1)\frac{d}{d\tau}(1-\overline{\mathcal{F}}(\tau)),$$

and if we integrate $e^{-r\tau}$ over this quantity, the result should be one if r is to be the growth rate. This is exactly what equation (10.32) states.

Appendix A

Stochastic basis of the Kermack-McKendrick ODE model

In this appendix we aim to show two things. We show what precisely one assumes about a population if one imposes a mass-action contact structure, and we show how these ideas can be used to give a precise set of conditions that guarantee a good approximation of a stochastic epidemic model by the Kermack-McKendrick ODE model. This analysis is based on ideas of J.A.J. Metz.

The basic stochastic model we use as our point of departure has been encountered before in Exercise 1.33. We reiterate the derivation in a slightly more formal way. Regard a population of fixed size N without heterogeneity, and introduce the following notation:

$\uparrow$ symbolises a susceptible;

$\downarrow$ symbolises an infective;

$\odot$ symbolises a removed individual.

Define the p-state $U = (\# \uparrow, \# \downarrow, \# \odot)$, where by the symbol '$\#$' we denote the number of items in a certain set, so $\# \uparrow + \# \downarrow + \# \odot = N$.

Now define

$$P_{ijl}(t) := \Pr\{U = (i, j, l)\}.$$

There are two distinct events by which the system can enter state $U = (i, j, l)$, viz. infection and recovery. In symbols, these entail the following:

- In a population in p-state, $U = (i+1, j-1, l)$ at time t one susceptible may become infected. The probability of this event is $\lambda \triangle t$ (where λ may depend on the p-state).
- In a population in p-state $U = (i, j+1, l-1)$ at time t one infective is removed (i.e. dies or becomes immune). The probability of this event is $\theta \triangle t$ (where θ may depend on the p-state).

A third possibility is of course that the system is already in p-state $U = (i, j, l)$ at time t, and that there is no change in $\triangle t$.

Adding the probabilities of the various chains of events leading to the p-state $U = (i, j, l)$ at time $t + \triangle t$, we find

$$\begin{aligned}
P_{ijl}(t + \triangle t) &= \lambda_{i+1,j-1,l} P_{i+1,j-1,l}(t) \triangle t + \theta_{i,j+1,l-1} P_{i,j+1,l-1}(t) \triangle t \\
&\quad + P_{ijl}(t)(1 - \lambda_{ijl} \triangle t - \theta_{ijl} \triangle t).
\end{aligned}$$

If we rearrange this expression and let $\triangle t \to 0$ we get a differential equation for the P_{ijl}:

$$\frac{dP_{ijl}(t)}{dt} = \lambda_{i+1,j-1,l} P_{i+1,j-1,l}(t) + \theta'_{i,j+1,l-1} P_{i,j+1,l-1}(t)$$
$$-\lambda_{ijl} P_{ijl}(t) - \lambda'_{ijl} P_{ijl}(t). \tag{A.1}$$

Equation (A.1) represents the infection as a Markov process on $\{(i,j,l) \mid i,j,l \in \mathbb{N}, i+j+l = N\}$. Now note that, since N is constant, we can ignore removed individuals in the sense that $l = N - i - j$ for any p-state $U = (i,j,l)$. We suppress dependence on l in the notation. The task is now to specify conditions that guarantee that the parameters λ and θ in (A.1) are indeed well-defined Markovian transition probabilities per unit time. As we will argue below, what we need is precisely the mass-action assumption for λ. In the intermezzo below we show that it is natural to assume

$$\lambda_{i+1,j-1} = (i+1)(j-1)\beta(N), \tag{A.2}$$
$$\theta_{i,j+1} = (j+1)\alpha, \tag{A.3}$$

where $\beta(N)$ is the transmission rate constant and α is the probability per unit time to recover. With this choice, we obtain for (A.1)

$$\frac{dP_{ij}(t)}{dt} = (i+1)(j-1)\beta(N)P_{i+1,j-1}(t) + (j+1)\alpha P_{i,j+1}(t)$$
$$-ij\beta(N)P_{ij}(t) - j\alpha P_{ij}(t). \tag{A.4}$$

Readers wishing to proceed directly with the derivation of the Kermack-McKendrick ODE model from (A.4) can skip the intermezzo and return to it later.

Intermezzo on graphs and mass action We want to study the spread of an infection on a graph $G = (V, E)$, consisting of a set $V = \{1, \cdots, N\}$ of vertices (representing individuals), and a set E of edges connecting vertices (representing that the individuals involved can have contacts with each other). We assume G to be static (i.e. no vertices or edges disappear or are created). G is called complete if there is an edge between any two vertices.

We label the vertices of G with the label set $\{\uparrow, \downarrow, \odot\}$. A *labelling* of G is the assignment of an element of this set to each vertex, i.e. a map $V \to \{\uparrow, \downarrow, \odot\}$, which we shall denote by L. Unspecified elements of $\{\uparrow, \downarrow, \odot\}$ will be denoted by l. We write $q_x(t)$ for the state of the vertex x at time t, $q_x(t) \in \{\uparrow, \downarrow, \odot\}$. Let $\chi(\wp)$ be the indicator function of statement $\wp$, defined as

$$\chi(\wp) = \begin{cases} 1 & \text{if } \wp \text{ is true,} \\ 0 & \text{otherwise.} \end{cases}$$

In order to find expressions for λ and θ in the stochastic model, we first study the transition where one specific susceptible vertex becomes an infected vertex in a small time step $\triangle t$. Regard two vertices: x, susceptible (i.e. $q_x(t) = \uparrow$), and y, infective (i.e. $q_y(t) = \downarrow$), and assume they are connected by an edge.

Denote by μ_{xy} the probability per unit time that the state of vertex x changes to $q_x(t + \triangle t) =\downarrow$. The transition changes the labelling $\{L(1) = l_1, \cdots, L(x) =\uparrow , \cdots, L(N) = l_N\}$ of G into the labelling $\{L(1) = l_1, \cdots, L(x) =\downarrow, \cdots, L(N) = l_N\}$. The probability per unit time for a vertex x with $q_x(t) =\uparrow$ to become infected by any one of the infected vertices is

$$\mu_x = \sum_y \mu_{xy}\chi(q_y(t) =\uparrow). \tag{A.5}$$

In the same way, we get a transition $\{L(1) = l_1, \cdots, L(y) =\downarrow, \cdots, L(N) = l_N\} \rightarrow \{L(1) = l_1, \cdots, L(y) = \odot, \cdots, L(N) = l_N\}$, where α_y is the probability per unit time to recover. Now μ_x and α_y are both Markovian transition probabilities per unit time on the label set. As we can have such transitions for any $\uparrow$-vertex (respectively $\downarrow$-vertex), we have to sum (A.5) (respectively α_y) over all susceptible (respectively infected) vertices:

$$\mu = \sum_x \mu_x\chi(q_x(t) =\uparrow), \ \alpha = \sum_y \alpha_y\chi(q_y(t) =\downarrow). \tag{A.6}$$

Now $\mu + \alpha$ gives the probability per unit time that in a certain given labelling of our graph one vertex label is changed in $\triangle t$. This is a transition in a stochastic process on the space $\mathbf{X}$ given by

$$\mathbf{X} := \{\text{all labellings of } V \text{ by the label set}\{\uparrow, \downarrow, \odot\}\}.$$

The λ_{ij} and θ_{ij} we are looking for, however, belong to a different stochastic process on another space. For example, as introduced above, λ_{ij} is the probability per unit time that a labelling of G with $\#\uparrow= i$ and $\#\downarrow= j$ is transformed into a labelling with $\#\uparrow= i - 1$ and $\#\downarrow= j + 1$. So λ_{ij} should apply to a process on the space $\mathbf{Y}$ given by

$$\mathbf{Y} := \{(\text{labellings of } i \uparrow\text{- and } j \downarrow\text{-vertices}), i, j \geq 0, i + j \leq N\}.$$

The difference between the two spaces is that in $\mathbf{Y}$ the labellings are lumped together by their number of $\uparrow$ and $\downarrow$ labels, whereas in $\mathbf{X}$ they are not lumped. In other words, the elements of $\mathbf{X}$ are labellings of G, but the elements of $\mathbf{Y}$ are *sets* of labellings of G. For a given process on $\mathbf{X}$, the process on $\mathbf{Y}$ is in general not Markovian. This is because μ and α depend strongly on the precise labelling with $\#\uparrow= i$ and $\#\downarrow= j$ from which we start. For λ_{ij} to be a well-defined transition probability on $\mathbf{Y}$ one has to assume that all labellings with $\#\uparrow= i$ and $\#\downarrow= j$ can be considered to be equivalent and therefore can be lumped together. This is precisely the content of the *(strong) lumping condition* or condition of *exchangeability* (see next paragraph). If and only if this condition is satisfied are the λs and the θs well-defined transition probabilities of a Markov process on $\mathbf{Y}$; see Metz (1981).[1]

---

[1] J.A.J. Metz: *Mathematical representations of the dynamics of animal behaviour (an expository survey)*. PhD-thesis, University of Leiden, 1981.

The conditions entail the following. Let the spaces $\mathbf{X}$ and $\mathbf{Y}$ be as defined above. The elements of $\mathbf{X}$ are labellings of G and the elements of $\mathbf{Y}$ are sets of labellings of G. Let $f : \mathbf{X} \to \mathbf{Y}$ be a map that sends an element of $\mathbf{X}$ to the corresponding element of $\mathbf{Y}$. Let $x_1 \in \mathbf{X}$. Let $\pi_{x_1 \Rightarrow x_2}$ denote the probability per unit time that x_1 is transformed into $x_2 \in \mathbf{X}$. Finally let $\hat{\pi}_{y_1 y_2}$ denote the analogous probability for $y_1, y_2 \in \mathbf{Y}$. For example, let y_1 be the set of all labellings of G with $\#{\uparrow} = i$ and $\#{\downarrow} = j$, and y_2 the set of all labellings of G with $\#{\uparrow} = i-1$ and $\#{\downarrow} = j+1$. In other words, the condition then states that the probability per unit of time that an element of set y_1 is transformed into an element of set y_2 must be independent of the specific element in y_1 we start from and of the specific element in y_2 we go to. In formal language,

$$\forall_{y_1, y_2 \in \mathbf{Y}} \forall_{x_1 \in f^{-1}(y_1)} : \sum_{x_2 \in f^{-1}(y_2)} \pi_{x_1 \Rightarrow x_2} = \hat{\pi}_{y_1 y_2},$$

which is equivalent to

$$\forall_{y_1, y_2 \in \mathbf{Y}} \forall_{x_1, x_1^* \in f^{-1}(y_1)} : \sum_{x_2 \in f^{-1}(y_2)} \pi_{x_1 \Rightarrow x_2} = \sum_{x_2 \in f^{-1}(y_2)} \pi_{x_1^* \Rightarrow x_2}.$$

The lumping condition is fulfilled if and only if we make the following assumption about μ_{xy}:

$$\mu_{xy} = \beta(N) \text{ for all } x, y \in V \tag{A.7}$$

(see Metz (1981), loc. cit.). So our desire to have a Markov process on the right state space forces us to assume that the probability per unit time of infection is the same for all pairs of vertices. This implies that G has to be complete, i.e. every individual potentially has contact with any other individual in the population. The above condition on μ_{xy} and its consequence are what is usually referred to as the *mass-action assumption*.

If we substitute (A.7) into (A.5), and the result in (A.6), we find that

$$\lambda_{ij} = \beta(N) i j.$$

In the above, we have concentrated on the derivation of an expression for λ, but the same reasoning can be applied in the case of θ. This leads to

$$\theta_{ij} = \alpha j.$$

End of Intermezzo

In the deterministic approximation $N \to \infty$ to (A.4), which we want to justify, the observables vary in a continuum, while in (A.4) they are discrete. To reconcile this, we switch from a probability P to a probability *density* function $p(x)\, dx$. Define

$$w = w(t) = \frac{i(t)}{N}, \quad z = z(t) = \frac{j(t)}{N},$$

with $i(t)$ the number of vertices with label $\uparrow$ at time t and $j(t)$ the number of vertices with label $\downarrow$ at time t. We suppress the t dependence in the notation. If we switch to probability densities in (A.4), we get a partial differential equation for p:

$$\frac{\partial p(w,z)}{\partial t} = \beta(N)(wN+1)(zN+1)p\left(w+\frac{1}{N},z-\frac{1}{N}\right)$$

$$+\alpha(zN+1)p\left(w,z+\frac{1}{N}\right) - \beta(N)wzN^2p(w,z) - \alpha zNp(w,z).$$

Next, formally Taylor expand the right-hand side:

$$\frac{\partial p(w,z)}{\partial t} = \beta(N)(wN+1)(zN+1)\left\{p(w,z)+\frac{1}{N}\frac{\partial p(w,z)}{\partial w}-\frac{1}{N}\frac{\partial p(w,z)}{\partial z}\right.$$

$$\left.+\frac{1}{2N^2}\frac{\partial^2 p(w,z)}{\partial w^2}+\frac{1}{N^2}\frac{\partial^2 p(w,z)}{\partial w\partial z}+\frac{1}{2N^2}\frac{\partial^2 p(w,z)}{\partial z^2}+O\left(\frac{1}{N^3}\right)\right\}$$

$$+\alpha(zN+1)\left\{p(w,z)+\frac{1}{N}\frac{\partial p(w,z)}{\partial z}+\frac{1}{2N^2}\frac{\partial^2 p(w,z)}{\partial z^2}+O\left(\frac{1}{N^3}\right)\right\}$$

$$-\beta(N)wzN^2p(w,z) - \alpha zNp(w,z).$$

The above equation constitutes the 'full' stochastic model with mass action. We want to approximate this equation by the deterministic Kermack-McKendrick ODE equations, and we want, therefore, to take the limit $N \to \infty$. It turns out that unless we choose the parameter $\beta(N)$ to depend on N as $\beta(N) = \frac{\beta}{N}$, we lose in the limit either one or the other of the two processes that we wish to keep in the picture: individuals getting infected and individuals recovering. Furthermore, β has to be of the same order of magnitude as α: both $O(1)$ for $N \to \infty$. The mathematical argument for this choice of $\beta(N)$ is that only for this choice can the full stochastic model be 'transformed' into a first-order partial differential equation of the appropriate form.

With the proposed scaling, the Taylor expansion above becomes

$$\frac{\partial p(w,z)}{\partial t} = -\beta wp(w,z)+\beta zp(w,z)+\alpha p(w,z)-\frac{\beta}{N}p(w,z)$$

$$+\left(\beta wz-\frac{\beta}{N}w+\frac{\beta}{N}z-\frac{\beta}{N^2}\right)\frac{\partial p(w,z)}{\partial w}$$

$$+\left(-\beta wz+\frac{\beta}{N}w-\frac{\beta}{N}z+\frac{\beta}{N^2}+\alpha z+\frac{\alpha}{N}\right)\frac{\partial p(w,z)}{\partial z}+O\left(\frac{1}{N}\right).$$

Now let $N \to \infty$. The result is a first-order partial differential equation for p:

$$\frac{\partial p(w,z)}{\partial t} = -\beta wp(w,z)+\beta zp(w,z)+\beta wz\frac{\partial p(w,z)}{\partial w}-\beta wz\frac{\partial p(w,z)}{\partial z}$$

$$+\alpha p(w,z)-\alpha z\frac{\partial p(w,z)}{\partial z}$$

$$= -\frac{\partial(-\beta wzp(w,z))}{\partial w}-\frac{\partial(\beta wzp(w,z)-\alpha zp(w,z))}{\partial z}.$$

We finally find the Kermack-McKendrick ODE model as the equations for the characteristics (see Courant & Hilbert (1962)) that are used to solve the above first-order PDE:

$$\frac{dw}{dt} = -\beta wz, \quad \frac{dz}{dt} = \beta wz - \alpha z.$$

Notice that implicitly we have made an important assumption about $i = \# \uparrow$ and $j = \# \downarrow$ in taking the limit $N \to \infty$: we have to assume that $i(0)$ and $j(0)$ are of the same order of magnitude and go to infinity 'in the same way as N does'.

In the Kermack-McKendrick ODE model as derived above we interpret w and z as fractions. We can, however, just as well regard them as densities. To achieve this, we do not scale $\beta(N)$ with N but rather with A, the area occupied by the population. We then let both N and A tend to infinity while keeping the density $\frac{N}{A}$ fixed. In this case we still allow β to be dependent on the (fixed) density, so that, for example, less densely populated areas can have a lower transmission rate constant.

To summarise, we find that in order to be able to approximate the stochastic model (A.1) by the Kermack-McKendrick ODE model, we have to assume the following:

- The probability per unit time of meeting and infecting is the same for all pairs of individuals. As a consequence, each individual must, in principle, be able to come into contact with any other individual.
- We have to take the limit of the number of individuals going to ∞. Moreover, the numbers of susceptibles and infectives should go to ∞ in the same way.

Appendix B

Bibliographic skeleton

This bibliography lists mostly books. The aim is to provide the reader with pointers to a small selection of textbook sources from mathematics and population dynamics that contain much of the backbone of the material presented in the text. The books are chosen for direct relevance, and are biased by the present authors' preferences. The bibliography starts with a list of books devoted to various aspects of epidemic modelling.

Epidemic modelling

R.M. Anderson (ed.): *Population Dynamics of Infectious Diseases: Theory and Applications.* Chapman & Hall, London, 1982.

R.M. Anderson & R.M. May: *Population Biology of Infectious Diseases.* Springer-Verlag, Berlin, 1982.

R.M. Anderson & R.M. May: *Infectious Diseases of Humans: Dynamics and Control.* Oxford University Press, Oxford, 1991.

N.J.T. Bailey: *The Mathematical Theory of Infectious Diseases and its Applications.* Griffin, London, 1975.

N.J.T. Bailey: *The Biomathematics of Malaria.* Griffin, London, 1982.

M. Bartlett: *Stochastic Population Models in Ecology and Epidemiology.* Methuen, London, 1960.

N.G. Becker: *Analysis of Infectious Disease Data.* Chapman & Hall, London, 1989.

S. Busenberg & K. Cooke: *Vertically Transmitted Diseases: Models and Dynamics.* Springer-Verlag, Berlin, 1993.

V. Capasso: *Mathematical Structures of Epidemic Systems.* Springer-Verlag, Berlin, 1993.

A.D. Cliff & P. Haggett: *Atlas of Disease Distributions: Analytical Approaches to Epidemiological Data.* Blackwell, London, 1988.

D.J. Daley & J. Gani: *Epidemic Modelling: an Introduction.* Cambridge University Press, Cambridge, 1999.

J.C. Frauenthal: *Mathematical Modeling in Epidemiology.* Springer-Verlag, Berlin, 1980.

J.-P. Gabriel, C. Lefévre & P. Picard (eds.): *Stochastic Processes in Epidemic Theory.* Springer-Verlag, Berlin, 1990.

B.T. Grenfell & A.P. Dobson (eds.): *Ecology of Infectious Diseases in Natural Populations.* Cambridge University Press, Cambridge, 1995.

H.W. Hethcote & J.A. Yorke: *Gonorrhea Transmission Dynamics and Control.* Springer-Verlag, Berlin, 1984.

H.W. Hethcote & J.W. Van Ark: *Modelling HIV Transmission and AIDS in the United States.* Springer-Verlag, Berlin, 1992.

P.J. Hudson, A. Rizzoli, B.T. Grenfell, J.A.P. Heesterbeek & A.P. Dobson (eds.): *Ecology of Wildlife Diseases.* Oxford University Press, Oxford, to appear, 2000.

V. Isham & G. Medley: *Models for Infectious Human Diseases: Their Structure and Relation to Data.* Cambridge University Press, Cambridge, 1996.

J. Kranz (ed.): *Epidemics of Plant Diseases: Mathematical Analysis and Modelling.* Springer-Verlag, Berlin.

H.A. Lauwerier: *Mathematical Models of Epidemics.* Mathematisch Centrum, Amsterdam, 1981.

D. Mollison, G. Scalia-Tomba & J.A. Jacquez (eds.): *Spread of Epidemics: Stochastic Modeling and Data Analysis.* Special issue of *Math. Biosci.*, **107**, pp. 149-562, 1991.

D. Mollison (ed.): *Epidemic Models: Their Structure and Relation to Data.* Cambridge University Press, Cambridge, 1995.

I. Nåsell: *Hybrid Models of Tropical Infections.* Springer-Verlag, Berlin, 1985.

J. Radcliffe & L. Rass: *Spatial Deterministic Epidemics.* Mathematical Surveys and Monographs, American Mathematical Society, Providence, to appear, 2000.

R. Ross: *The Prevention of Malaria*, 2nd edition, Churchill, London, 1911.

M.E. Scott & G. Smith: *Parasitic and Infectious Diseases: Epidemiology and Ecology.* Academic Press, San Diego, 1994.

N. Shigesada & K. Kawasaki: *Biological Invasions: Theory and Practice.* Oxford University Press, Oxford, 1997.

J.E. Van der Plank: *Plant Diseases: Epidemics and Control.* Academic Press, New York, 1963.

Population dynamics

H. Caswell: *Matrix Population Models*. Sinauer, Massachusetts, 1989 (2nd Edition, 2000).

J.M. Cushing: *An Introduction to Structured Population Dynamics*. SIAM, Philadelphia, 1998.

L. Edelstein-Keshet: *Mathematical Models in Biology*. Birkhäuser (McGraw-Hill), New York, 1988.

M. Gilpin & I. Hanski (eds.): *Metapopulation Dynamics: Empirical and Theoretical Investigations*. Academic Press, London, 1991.

I. Hanski & M. Gilpin (eds.): *Metapopulation Biology: Ecology, Genetics and Evolution*. Academic Press, San Diego, 1997.

F. Hoppensteadt: *Mathematical Theories of Populations: Demographics, Genetics and Epidemics*. SIAM, Philadelphia, 1975.

S.A. Levin, T.G. Hallam & L.J. Gross (eds.): *Applied Mathematical Ecology*. Springer-Verlag, Berlin, 1989.

J.A.J. Metz & O. Diekmann: *The Dynamics of Physiologically Structured Populations*. Springer-Verlag, Berlin, 1986.

J. McGlade (ed.): *Advanced Theoretical Ecology: Principles and Applications*. Blackwell Science, London, 1999.

R.M. Nisbet & W.S.C. Gurney: *Modelling Fluctuating Populations*. Wiley, New York, 1982.

N. Shigesada & K. Kawasaki: *Biological Invasions: Theory and Practice*. Oxford University Press, Oxford, 1997.

S. Tuljapurkar & H. Caswell (eds.): *Structured Population Models in Marine, Freshwater and Terrestrial Systems*. Chapman & Hall, London, 1997.

Non-negative matrices and operators

A. Berman & R.J. Plemmons: *Nonnegative Matrices in the Mathematical Sciences*. Academic Press, New York, 1979.

M.A. Krasnosel'skij, Je.A. Lifshits, A.V. Sobolev: *Positive Linear Systems: The Method of Positive Operators*. Heldermann Verlag, Berlin, 1989.

H. Minc: *Nonnegative Matrices*. Wiley, New York, 1988.

E. Seneta: *Non-negative Matrices*. George Allen & Unwin, London, 1973.

Dynamical systems and bifurcations

J.K. Hale: *Ordinary Differential Equations*. Wiley, New York, 1969.

J.K. Hale & H. Koçak: *Dynamics and Bifurcations*. Springer-Verlag, New York, 1991.

M. Hirsch & S. Smale: *Differential Equations, Dynamical Systems and Linear Algebra*. Academic Press, New York, 1974.

Y.A. Kuznetsov: *Elements of Applied Bifurcation Theory*. 2nd edition, Springer-Verlag, New York, 1998.

Analysis

R. Courant & D. Hilbert: *Methods of Mathematical Physics*, Vol. II. Interscience, New York, 1962.

J. Dieudonné: *Foundations of Modern Analysis*. Academic Press, New York, 1969.

J.A. Jacquez: *Compartmental Analysis in Biology and Medicine*, 3rd edition. BioMedware, Ann Arbor, 1996.

A.N. Kolmogorov & S.V. Fomin: *Introductory Real Analysis*. Prentice-Hall, Englewood Cliffs, 1970/Dover, New York, 1975.

W. Rudin: *Real and Complex Analysis*, 2nd edition. McGraw-Hill, New York, 1974.

Stochastic processes

K.L. Chung: *Elementary Probability Theory with Stochastic Processes*. Springer-Verlag, Berlin, 1979.

N.S. Goel & N. Richter-Dyn: *Stochastic Models in Biology*. Academic Press, New York, 1974.

T.E. Harris: *The Theory of Branching Processes*. Springer-Verlag, Berlin, 1963.

P. Jagers: *Branching Processes with Biological Applications*. Wiley, London, 1975.

T.G. Kurtz: *Approximation of Population Processes*. SIAM, Philadelphia, 1981.

D. Ludwig: *Stochastic Population Theories*. Springer-Verlag, Berlin, 1974.

C.J. Mode: *Multitype Branching Processes, Theory and Applications*. Elsevier, New York, 1971.

H.M. Taylor & S. Karlin: *An Introduction to Stochastic Modelling*. Academic Press, Orlando, 1984.

Index